Interdisciplinary Approaches to the Oldowan

Vertebrate Paleobiology and Paleoanthropology Series

Edited by

Eric Delson
Vertebrate Paleontology, American Museum of Natural History,
New York, NY 10024, USA
delson@amnh.org

Ross D. E. MacPhee
Vertebrate Zoology, American Museum of Natural History,
New York, NY 10024, USA
macphee@amnh.org

Focal topics for volumes in the series will include systematic paleontology of all vertebrates (from agnathans to humans), phylogeny reconstruction, functional morphology, Paleolithic archaeology, taphonomy, geochronology, historical biogeography, and biostratigraphy. Other fields (e.g., paleoclimatology, paleoecology, ancient DNA, total organismal community structure) may be considered if the volume theme emphasizes paleobiology (or archaeology). Fields such as modeling of physical processes, genetic methodology, nonvertebrates or neontology are out of our scope.

Volumes in the series may either be monographic treatments (including unpublished but fully revised dissertations) or edited collections, especially those focusing on problem- oriented issues, with multidisciplinary coverage where possible.

Editorial Advisory Board

For other titles published in this series, go to: www.springer.com/series/6978

Interdisciplinary Approaches to the Oldowan

Edited by

Erella Hovers
The Institute of Archaeology
The Hebrew University of Jerusalem
Mt. Scopus
Jerusalem 91905, Israel

and

David R. Braun
Department of Archaeology
University of Cape Town
Rondebosch 7701, South Africa

Springer

Library of Congress Control Number: 2008937484

ISBN: 978-1-4020-9059-2 (HB)
ISBN: 978-1-4020-9060-8 (e-book)

Published by Springer,
P.O. Box 17, 3300 AA Dordrecht, The Netherlands.

www.springer.com

Cover illustration:
Refits from A. L. 894, Hadar, Ethiopia; photos: Erella Hovers.

Printed on acid-free paper

Acknowledgments

The symposium "Interdisciplinary approaches to the Oldowan" took place in March 2006 within the framework of the annual meeting of the Society of American Archaeology in San Juan, Puerto Rico. We organized this symposium with the aim of discussing questions pertaining to the Oldowan that we deemed interesting and pressing in view of the rapid developments in the field over the last decade. The twelve papers that were presented in the symposium tackled lithic technology, raw material studies, site formation and meat acquisition, paleo-environmental frameworks, and Oldowan occurrences within and outside the African continent. The answers of the participants to our initial questions were far from uniform, but all were illuminating, interesting and thought-provoking. Participants and audience alike urged us to publish the symposium as an edited volume. This volume is the outcome of that symposium and the enthusiasm of our colleagues, for which we thank them.

Unfortunately, some of the original speakers were unable to contribute to this project; nevertheless we wish to thank them for sharing with us their interest, results and knowledge. We thank the Wenner-Gren foundation for their financial support that enabled the participation of some presenters (notably students) in the symposium and thus contributed to the intellectual diversity of the meeting. We are grateful to Naama Goren-Inbar and Jack Harris, our discussants in the symposium, for selflessly relinquishing the floor for a general discussion that was one of the highlights of the meeting.

Our deepest thanks go to the contributors to the volume. Twenty-eight researchers of various disciplines joined forces to present the twelve contributions in this volume. They have always responded kindly (occasionally even in a timely fashion...) to our instructions, comments, requests for clarification and suggestions for revisions, and stuck it thorough the period of preparing this volume. We applaud twenty-two reviewers for their efforts and thank them all for their high standards in reviewing the papers and for their detailed comments on the manuscripts, which were of immense help in improving the quality of the volume. As a large number of the reviewers wished to remain anonymous, we decided to refrain from naming any names.

Special thanks are due to Eric Delson for suggesting this series as a venue for publishing the volume, for his diligent reading of the manuscripts and his attention to details, and for his help in all editorial and administrative aspects. Tamara Welschot and Judith Terpos at Springer dealt with the logistic and editorial sides of the project, and we thank them for taking us through the process.

Finally, on a personal note, DRB would like to thank his very supportive family and especially Kathryn Underwood, who always provided guidance, assistance, and sanity throughout the process of compiling this volume. EH wishes to thank her family and friends for their (sometimes hard-tried) patience, understanding and support.

Erella Hovers, Jerusalem
David Braun, Cape Town
July 2008

List of Contributors

Deborah Barsky
Centre Européen de Recherches Préhistoriques de Tautavel
Ave. Léon Jean-Grégory
66720, Tautavel, France
dbarsky@hotmail.fr

Laura C. Bishop
Research Centre in Evoltuionary Anthropology and Palaeoecology
School of Natural Sciences and Psychology
Liverpool John Moores University
Byrom Street, Liverpool, L3 3AF
United Kingdom
L.C.Bishop@ljmu.ac.uk

David R. Braun
Department of Archaeology
University of Cape Town
Rondebosch 7701, South Africa
david.braun@uct.ac.za

Adam Brumm
McDonald Institute for Archaeological Research
Downing Street
University of Cambridge
Cambridge CB2 3ER UK
arb74@cam.ac.uk

Eudald Carbonell
Institut Català de Paleoecologia Humana i Evolució Social
Àrea de Prehistòria
Universitat Rovira i Virgili
Plaça Imperial Tarraco 1
43005 Tarragona, Spain
eudald.carbonell@urv.net

Vincenzo Celiberti
Centre Européen de Recherches Préhistoriques de Tautavel
Ave. Léon Jean-Grégory
66720, Tautavel, France
vceliberti@hotmail.com

Parth R. Chauhan
Stone Age Institute
Indiana University
Bloomington, IN 47407-5097, USA
pchauhan@indiana.edu

Carmine Collina
CEPAM UMR 6130, CNRS
Université Nice -Sophia Antipolis
250 rue Albert Einstein, bât. 1
06560 Valbonne, France
and
Dipartimento di Scienze Storiche Archeologiche e Antropologiche dell'Antichità
Università di Roma "La Sapienza"
Via Palestro, 63
00185 Roma, Italy
barnum_it@yahoo.fr

Ignacio de la Torre
Institute of Archaeology
University College London
31-34 Gordon Square
London, WC1H-0PY, UK
i.torre@ucl.ac.uk

Peter W. Ditchfield
Research Laboratory for Archaeology
6 Keble Road
Oxford, OX1 3QT UK
peter.ditchfield@rlaha.ox.ac.uk

Manuel Domínguez-Rodrigo
Department of Prehistory
Complutense University
28040 Madrid, Spain
m.dominguez.rodrigo@gmail.com

Joseph V. Ferraro
Department of Anthropology
Baylor University
One Bear Place
Waco, TX 76798, USA
joseph_ferraro@baylor.edu

Rosalia Gallotti
Dipartimento di Scienze Storiche, Archeologiche e Antropologiche dell'Antichità
Università di Roma "La Sapienza"
Via Palestro, 63
00185 Roma, Italy
rosaliagallotti@yahoo.it

Denis Geraads
UPR 2147 CNRS
44 rue de l'Amiral Mouchez
75014 Paris, France
dgeraads@ivry.cnrs.fr

Talia Goldman-Neuman
The Institute of Archaeology
The Hebrew University of Jerusalem
Mt. Scopus
Jerusalem 91905, Israel
tali.neuman@mail.huji.ac.il

Sonia Harmand
Laboratoire de Préhistoire et Technologie
Maison de l'Archéologie et de l'Ethnologie
Boîte 3
21 allée de l'Université
92023 Nanterre Cedex France
sonia.harmand@mae.u-paris10.fr

Fritz Hertel
Department of Biology
California State University Northridge
18111 Nordhoff Street
Northridge, CA 91330-8303, USA
fritz.hertel@csun.edu

Erella Hovers
The Institute of Archaeology
The Hebrew University of Jerusalem
Mt. Scopus
Jerusalem 91905, Israel
hovers@mscc.huji.ac.il

Guy Kieffer
22 rue Paul Gaubin
63450 Le Crest, France
and UMR 6042 CNRS

John D. Kingston
Department of Anthropology
Emory University
1557 Pierce Drive
Atlanta, GA 30322, USA
jkingst@emory.edu

François-Xavier Le Bourdonnec
Institut de Recherche sur les Archéomatériaux
UMR 5060 CNRS-Université Bordeaux 3
Maison de l'Archéologie
Esplanade des Antilles
33607 Pessac, France
francois.lebourdonnec@etu.u-bordeaux3.fr

Mark W. Moore
Archaeology and Palaeoanthropology
University of New England
Armidale, New South Wales, 2351, Australia
mmoore2@une.edu.au

Rafael Mora
CEPAP-UAB
Facultad de Letras
Universidad Autónoma de Barcelona
08193, Bellaterra, Barcelona, Spain
rafael.mora@uab.es

Marcello Piperno
Dipartimento di Scienze Storiche, Archeologiche e Antropologiche dell'Antichità
Università di Roma "La Sapienza"
Via Palestro, 63
00185 Roma, Italy
m.piperno@tin.it

Thomas W. Plummer
Department of Anthropology
Queens College, CUNY and NYCEP
65-30 Kissena Blvd
Flushing, NY 11367, USA
thomas.plummer@qc.cuny.edu

Gerard Poupeau
Institut de Recherche sur les Archéomatériaux
UMR 5060 CNRS-Université Bordeaux 3
Maison de l'Archéologie
Esplanade des Antilles
33607 Pessac, France
g.poupeau@u-bordeaux3.fr

Jean-Paul Raynal
CNRS, UMR 5199 PACEA, IPGQ
Université de Bordeaux 1, Bâtiment B18
avenue des Facultés
33405 Talence, France
jpraynal@wanadoo.fr

Robert Sala
Institut Català de Paleoecologia Humana i Evolució Social
Àrea de Prehistòria
Universitat Rovira i Virgili
Plaça Imperial Tarraco 1
43005 Tarragona, Spain
robert.sala@urv.cat

Contents

1. Introduction: Current Issues in Oldowan Research

David R. Braun*
Department of Archaeology
University of Cape Town
Rondebosch 7701
South Africa
david.braun@uct.ac.za

Erella Hovers*
Institute of Archaeology
The Hebrew University of Jerusalem
Mt. Scopus Jerusalem 91905
Israel
hovers@mscc.huji.ac.il

Keywords: lithic technology, cultural evolution, decision-making, raw material, ecological adaptation

Abstract

The Oldowan industry represents the oldest known manifestation of material culture. The expression of these earliest tools is marked by its diversity. Here we review various approaches to this material. In particular we focus on how the history of research on these industries has changed dramatically from the earliest investigations in East Africa through the last thirty years and finally we describe the approaches found in the volume. This volume is distinct in its incorporation of many industries that would not classically be considered Oldowan, and subsequently provides a unique perspective on early stone tool manufacture and use. In particular the conceptual framework around the evolution of different technical skills provides a major theoretical step in constructing models of early hominin behavioral evolution. Although this chapter reviews the material found in each of the chapters of this volume it does not reiterate the information found in these chapters. Rather we provide a conceptual perspective for the various themes that link the contributions together. We identify the challenges for future Oldowan research. We see the synthesis of different schools of thought in the study of Oldowan stone artifacts as one of the most important contributions that future work can make for the study of the earliest technologies.

The discipline of paleoanthropology identifies the causes, origins and beginnings of behavioral and biological phenomena associated with human evolution. Using the theoretical framework of evolution, paleoanthropology tries to explain processes of biological and cultural change in the hominin lineage. Where biological anthropologists talk about gradual change, stasis or punctuated equilibrium, Paleolithic archaeologists discuss transitions and revolutions in attempts to identify both the essence and the temporal framework of distinct behavioral processes, sometimes referred to as "Human Revolutions", that made humans a unique culture-dependent species.

To some, this revolution is as late as the Neolithic, with its growing social and political complexities and the shift from hunting and gathering to a food production economy. Others focus on the period between 60 to 30 Ka ago, when cultural and social innovations allegedly enabled *Homo sapiens*' colonization of practically every ecological niche in the Old World. This shift is coincident with our species becoming the sole surviving representative of the genus *Homo*. In fact, however, the term "Human Revolution" was coined originally in reference to the moment when stone tools were first made (Montagu 1965:15), which constitutes the first step in a long

---

* Addresses for correspondence

E. Hovers and D.R. Braun (eds.), *Interdisciplinary Approaches to the Oldowan*, DOI: 10.1007/978-1-4020-9059-2,

process of economic intensification and heralds the onset of hominin material culture.

Montagu identified the "Human Revolution" at this particular point in human evolution not the least because he wished to make a political and social statement about the ancient unity of modern humanity against the background of global events of the first half of the twentieth century (Gamble 2007). Unrelated to non-scientific considerations, few paleoanthropologists today will contest the notion that the appearance of stone tool-making was a turning point in human evolution. The Oldowan is sometimes viewed as a threshold phenomenon in terms of the changes that it brought in subsequent prehistory (Isaac 1986). Yet after a major surge of interest in the 1970s, Oldowan studies seem to have elicited little excitement and lukewarm responses from the larger paleoanthropological community until relatively recently. While findings of skeletal remains of hominins contemporaneous with Oldowan assemblages have always excited the scientific and lay public alike, this has rarely been the case with the material culture that constitutes Montagu's "Human Revolution". We identify the dynamics of research as a main cause for the ebb and flow of interest in Oldowan studies over the last three decades. These dynamics influence the level of interest in Oldowan studies in several ways.

A truism shared by all historical sciences, paleoanthropology included, is that time is an active player in shaping the record. The further back in time we go, the more difficult it becomes to provide meaningful middle range accounts of what happened in the past. By that dictum the Oldowan is doomed to be the least understood of all the material culture manifestations encountered in prehistory. It can be approached by many new and exciting research techniques (for example, isotopic studies of diets; Sponheimer et al. 2006, 2007) but, inevitably, less so than other fields of paleoanthropological research. Blatantly put, Oldowan stratigraphic sequences, fauna and stone tools have a hard time standing their ground in a discipline that is becoming blasé about Neandertal DNA.

On another level, it is the history of Oldowan research that may have stalled interest in this field. Perhaps more than in any other branch of paleoanthropological inquiry, post-early 1980s Oldowan scholars are standing on the shoulders of giants but also in their long-cast shadow. The Leakeys' work in Olduvai Gorge provided the chronological, stratigraphic and material culture framework on which rested the definition of the Oldowan as an archaeological entity with bounded space (eastern Africa) and time (Early Pleistocene, ca. 1.8-1.58 Ma) dimensions. Thus the Oldowan fulfilled the criteria of what was defined as an archaeological culture (Clarke 1968). Capitalizing on the chronological markers dispersed throughout the Olduvai sequence and the changing paleo-landscape deduced from geological studies, Mary Leakey interpreted the spatial associations of artifacts and faunal remains in Olduvai Beds I and II as windows on an emergent human way of life. At the same time she also created the typological guidelines for the characteristics that define the lithic assemblages of the then earliest material culture.

Working in broadly contemporaneous deposits in Koobi Fora, Glynn Isaac and his team expanded the scope of research to include, in addition to collection of archaeological and geological data, formal model building. Enlisting actualistic research, experimental archaeology, ethnographic analogs and primatology (to name but a few disciplines) this group of workers aimed to explain the dynamics of the formation of the archaeological record in deep time. They concerned themselves with epistemological issues, namely questions of archaeological visibility, how it is affected by the vagaries of time, and how its changing quality interferes with paleoanthropologists' reading of the past. Taking such issues into consideration, Isaac and colleagues constructed testable hypotheses about the ecological and economic conditions that may have triggered and influenced early stone tool-making, and attempted to reconstruct the social behaviors of early hominins. Interestingly, the identified behaviors of meat acquisition by hunting, food sharing and central place occupation, which we often tend to associate with "humanness", may be part and parcel of primate behavior prior to the speciation of the hominin lineage rather than a defining characteristic of the Oldowan (Cameron 1993; Cameron and Groves 2004). The same could not be said of sharp-edged stone tool-making, which remained a tale tell sign, decisively separating humans from other primates.

Between these two major projects, the groundwork was laid out for research on the Oldowan in the course of the following twenty years. As a rule, Oldowan researchers conducted their work under the umbrella of evolutionary theory and/or its cultural counterparts and derivatives (the latter discussed in Ziman 2000; Jablonka and Lamb 2004; Richerson and Boyd 2005; Whiten 2005). However, this hardly led to the crystallization of comprehensive models that addressed questions of Oldowan origins, technology, economic planning and hominin cognition as parts of a whole. As can be seen in the research on the origins of anatomically modern humans, for example, such theory building is not a trivial matter. In the latter case, the emergence of two clear, competing theories (i.e. regional continuity vs. late single-species models) prodded researchers to construct hypotheses and come up with the means to test them, propelling the study of modern human origins and behavior. In contrast, Oldowan studies from the late 1970s through the 1990s were conducted mostly in relation to the *specifics* of questions put forth by the Olduvai and Koobi Fora teams and in their wake, at the expense of developing *new* theoretical frameworks and dynamic models (but see, for example, Potts 1991, 1994). Much of the prolific literature on bone modification, for example, fixated on analytical debates. Similarly, followers of the pioneer typological and technological frameworks too often did so without considering the fact that such frameworks were constructed in attempts to understand local or regional phenomena and were not meant to be overarching explanatory scenarios.

The prolonged "stasis" in Oldowan work has winnowed interest in this time period to a few specialists. This has recently changed as, over the last decade and half, the Oldowan has again become a dynamic field of research. The process is driven by new fieldwork that has yielded exciting results as well as analytical developments that offer better insight into previously unthinkable details and sometimes completely modified previously established concepts in the field.

The change of highest visibility to professionals and the lay public alike is undoubtedly the broadening of the Oldowan time span. On current archaeological evidence the Oldowan goes back to ca. 2.6 Ma, with a number of occurrences that are dated 2.6-1.9 Ma, i.e. prior to the time of the traditional key sites for this technocomplex. Lithic classes, site categories and bone modification patterns observed in Olduvai and Koobi Fora no longer constitute the most fundamental expressions of behaviors of stone tool-making hominins. Minimally, these occurrences need to be re-thought in a new context, asking whether there are differences through time in all or some of the archaeological aspects. Furthermore, if differences can be recognized, they should be described formally. This way the variation expressed in the Oldowan can be investigated and hypotheses can be developed about the behavioral implications of this variation.

A clear pattern that has recently begun to emerge from the burgeoning Oldowan database is the considerable variability among assemblages and sites. In the early days of the study of Oldowan assemblage, the differences between the contemporaneous KBS industries of Koobi Fora and the Oldowan of Olduvai Gorge represented the known range of variability. The recently augmented number of sites, many of which cluster in the time frame of 2.3-2.0 Ma, increases the diversity encountered in specific patterns of raw material selection and acquisition, technological practices, and the organization of lithic production and use across the landscape (e.g. de la Torre and Mora 2005; Hovers et al. 2008; Stout et al. 2008; Braun et al. 2008a,b). In the past, researchers focused on the evolution of the Oldowan into later industries. Now it is possible, given the combination of an extended duration and cultural variability, to ask whether there are discernible developmental trajectories throughout the long time span of the Oldowan or whether it represents merely "stasis" (Semaw 2000). Moreover with a constantly enhanced understanding of paleo-environments associated with Oldowan occupations, Oldowan lithic assemblages can now be analyzed as a system (or possibly systems) of hominin adaptations (e.g. Plummer 2004). The high levels of variability, beginning with the earliest Oldowan occurrences, are not a trivial matter. The very early lithic assemblages in the growing Oldowan database already followed at least rudimentary principles of geometric and spatial planning of lithic reduction. For many, this combination is an indication that the Oldowan does not represent the earliest stages of hominin material culture or even an incipient stage of stone tool production and use.

When taken together with the tool *use* behaviors documented for several genera of non-hominin primates as well as evidence for tool-*making* among some species (Whiten et al. 1999; Biro et al. 2004; de A. Moura and Lee 2004; Fragaszyet al. 2004), this led to the notion that tool-making is entrenched in the evolutionary past of hominins. This insight resulted in suggestions that tool-making should have occurred among hominins prior to the known Oldowan (Panger et al. 2002). The model raises the fundamental issue of the tempo and mode of evolution of stone tool-making. Supporters of "older than the Oldowan" models thus endorse scenarios of gradual evolution from a common tool-making ancestor of humans and chimps, arguing for a "missing phase" in the archaeological record. According to this model stone tools made during this early stage may have been very simple technologically. It is possible that these earliest stone artifacts would not be recognizable as the product of hominin activity. Much of the earliest toolkits may have been made of non-lithic, organic materials that do not preserve. The gradual model thus poses some demanding analytical and methodological challenges for students of Oldowan material culture.

The notion of homologous tool-making traits of hominin and chimp tool production is not unanimous, however (see Sayers and Lovejoy 2008 for a critical discussion of chimpanzee referential models). To some, the emergence of the Oldowan can best be described as a punctuated event due to its relatively sudden emergence. For paleoanthropologists, stone tool-making is an early defining trait of an emergent human way of life (hence a "Human Revolution"). However, there is the possibility that the development of stone tool use is an exaptation. According to this scenario, cognitive and biomechanical abilities that are necessary for stone tool-making evolved due to positive selection for traits unrelated to tool-making. Subsequent application of these traits provided a predisposition to the production of stone artifacts. In this rendition of the emergence of lithic technology, stone tool-making in human and non-human primates are at most analogous phenomena. The 2.6 Ma assemblages may indeed represent the earliest stages of lithic production.

This scenario requires rigorous hypothesis building and testing with regard to the advantages of stone tools in Late Pliocene eastern Africa, to explain both the initial drive to *use* such items as well as identify the reasons for the rapid spread of this innovation in varied ecological contexts. Such hypotheses inevitably have to touch on the issue of the taxonomic identification of the tool-makers, since biological and potentially cognitive differences in the abilities of various species would have played a major role as "pre-adaptations". In addition, demographic modeling is called for to understand processes of information transmission among the (most likely) small groups surviving on the landscape.

A third development in Oldowan studies has taken place outside the African continent. The first out-of-Africa dispersals were thought to be of Early Pleistocene Age and associated with advantages conferred by Acheulian technology. Yet new evidence indicates unequivocally that the earliest out-of-Africa dispersal of hominins took place during the Plio-Pleistocene. The earliest archaeological occurrences in Dmanisi are dated ca. 1.78 Ma (Ferring et al. 2008), whereas some Chinese sites go back to only slightly younger dates (1.66 Ma; Zhu et al. 2004). These occurrences clearly predate the first known Acheulian assemblages ca. 1.6-1.5 Ma (Asfaw et al. 1992; Texier et al. 2006; Semaw et al. 2008). Accordingly, the lithic assemblages in the earliest sites off the African continent suggest that the first sorties were made by hominins whose taxonomic identity is not fully understood (de Lumley et al. 2005) and who were equipped solely with Oldowan core and flake technology ("Mode 1" as defined by Clark 1970; see Schick et al. 1991; Shen and Qi 2004; de Lumley et al. 2005).

A geographical boundary that represents different behavioral repertoires is expressed in the concept of the Movius Line, east of which Acheulian technology did not seem to have advanced (Movius 1948, 1969; Corvinus 2004). This boundary has been explained as the result of raw material availability (Pope 1989), ecological differences between the Mode 2 Acheulian world to the west and the Mode 1 world to the east, failure of information transmission, or cultural differences between dispersing groups (as reviewed in Schick 1994). Many of these explanations cannot be applied to the Eurasian record that is unfolding at sites such as Dmanisi (de Lumley et al. 2005) and Atapuerca (Carbonell et al. 2008). Two implications of the new evidence are pertinent to our discussion of the Oldowan. On the one hand, it is possible that the Oldowan was a more complex system of technological adaptation than is suggested by its seemingly simple production sequences and basic morphological themes. A contrasting implication is that while the Oldowan was indeed a simple, possibly functionally-limited technological system, the earliest out-of-Africa dispersal(s) were not mediated by technological competence.

These developments provided much of the rationale for this volume, as well as the symposium in the 2006 SAA meetings in Puerto Rico, from which the book stems. When organizing the symposium we attempted to include research projects that study the nature of the Oldowan as a lithic production system with a strong underlying ecological signature, as well as those that emphasize its systematics. Some of the original speakers focused on the evolutionary dynamics of the Oldowan in relation to chimpanzee behavior and archaeology, whereas others discussed the "globalization" of the Oldowan. Unfortunately, not all the speakers could write for the present volume, which explains the regrettable absence of papers discussing the Chinese early record. On the other hand, researchers who had been unable to take part in the oral presentations agreed to contribute papers to the book. Although not all the aspects are equally covered by the papers, we feel that the volume as it stands fulfills its original aim. This collection of papers is best perceived as an "interim report" on the state of affairs rather than as an authoritative text book. Our ultimate goal is to provide the reader with the range of current perspectives on the challenges in the field of Oldowan studies, and to present the theoretical frameworks and analytical tools applied in response to such challenges. Many contributions to the volume review the strengths and potentials of the various approaches as guidelines for future work.

The majority of authors did not necessarily aim to address all of the broad themes that we touched upon above. Reading the papers as a single corpus emphasizes, however, that in some important ways these issues are inseparable from one another. Some common conclusions as well as a number of problems emerge from this compilation. Hence we decided not to divide the contributions into separate, stand-alone sections. Instead, we review below what seem to us the main points that this collection of papers emphasizes.

All the contributors to this volume agree that the study of Oldowan behavior is vital to understanding early hominin evolution, but not all would agree on a definition of the Oldowan. One of the insights to be gained from this volume is that such disagreements are not a matter of different emphases on the specifics of a definition. Rather, they stem from a more fundamental problem of diverse, sometimes incompatible, criteria of definition.

The original, traditional definition of the Oldowan pertains exclusively to material culture, more specifically to a lithic industry with defined morphological and (to a lesser degree) technological characteristics, bounded in both time and space. Some workers have chosen to define this industry on the basis of the mental aptitudes of its makers (Perles 1993). Yet previous descriptions of the Oldowan tool-makers as having poor associations between the technical gestures of tool-making and the practical results, appear to be changing. Recent research has shown the recurrent use of technical rules and a distinctive technological competence (de la Torre 2004; Delagnes and Roche 2005; Roche 2005), countering the idea of the Early Stone Age as representative of an opportunistic method of tool use and manufacture (Harmand 2004, 2005).

Some contributors of this volume struggle to come up with an adequate new definition of the Oldowan. Such attempts mostly revolve around assemblage characteristics. Barsky looks at the dominance of small-size debitage in most sites that have been described as Oldowan. In her analysis, this feature links the classic Oldowan sites from Olduvai and Koobi Fora to other Early Stone Age East African localities as well as those described by Chauhan (2009) in Djibouti and younger assemblages from Europe (Carbonell et al. 2009). Other defining features usually involve some modification of the original core forms described by Leakey (1971), which include polyhedrons, discoids, spheroids, heavy and light-duty scrapers (sometimes referred to as *nucléus racloirs* or *rostro carénés*). Not all of these tool forms can be found at every Oldowan sites, but most localities include the majority of these forms. Other core forms are delimited geographically. For example, Barsky (2009) considers spheroids as part of an African "tradition", although experimental research has suggested that they are a product of intensive reduction on particular types of raw materials (Sahnouni et al. 1997).

Strict typological definitions of the Oldowan would necessarily include industries that traditionally would not be considered part of this technocomplex. Moore and Brumm argue that the multiplatform cores, radial cores, and bifacial centripetal cores found in association with the *Homo floresiensis* specimen at Liang Bua are virtually indistinguishable from Oldowan polyhedrons, choppers and discoids respectively. Further, they believe the emphasis on retouched forms bears a resemblance to the retouched flakes at Olduvai (Moore and Brumm 2009). Although de la Torre and Mora have questioned the importance, and in some cases the validity, of retouched flakes at Olduvai (de la Torre and Mora 2005), their appearance at sites at Melka Kunture (Piperno et al. 2009) as well as some of the localities in Djibouti (Chauhan 2009) suggests that they are important part of many Oldowan assemblages.

The resemblance of the Late Pleistocene Liang Bua industry to the Olduvai collection calls into question the previously definitive temporal limits of the Oldowan. The "Last Appearance Datum" of the Oldowan seems to differ geographically, but it also depends on one's criteria for defining the technocomplex. The oldest Oldowan in Africa is well known ca. 2.6 Ma at Gona in Ethiopia (Semaw et al. 1997, 2003), and its end in Africa is broadly defined as the appearance of Mode 2 industries. Previously this boundary was strictly defined as the appearance of Acheulian technology even if according to this criterion it was not synchronous throughout Africa. For instance, in the Middle Awash this shift may have occurred only during the Middle Pleistocene (Clark et al. 1994). Some researchers now emphasize the elaboration of the technological process – the appearance of systematization, predetermination and the expression of symmetry – rather than the morphological shift as the real distinction of this next major step in technological development (Barsky 2009). Using this definition, the end of the Oldowan is at Konso in Ethiopia (Asfaw et al. 1992) and Kokiselei 5 in West Turkana (Texier et al. 2006). For Eurasia this chronology appears to be very different. The Oldowan may begin as far back as 2.0 Ma at sites like Renzidong (Dong 2006) or Riwat (Gaillard 2006) and ends as late as 1 Ma at sites like Isampur (Petraglia et al. 1999), yet earlier sites do show affinities for predetermination and systematization and contain Acheulian tools, for example 'Ubeidiya (Bar-Yosef and Goren-Inbar 1993; Barsky 2009).

Against this global background, the technological (as opposed to strictly typological) approach leads some of the contributors to this volume to reconsider the status of the Oldowan as an archaeological taxon. These authors reject the idea of a clear delineation between the Mode 1 and Mode 2 technologies (Barsky 2009; Carbonell et al. 2009; Piperno et al. 2009). Instead they prefer to recognize changes through the Early Stone Age/Lower Paleolithic as a progression of technological developments, leading toward higher levels of technological complexity within the Oldowan. The variability expressed in the earliest known stone tools leads Carbonell et al. (2009) to hypothesize the existence of a precursor ("Mode 0") to the Gona assemblages, which would be marked by lack of variability. Their analysis preempts the notion of a "pre-Oldowan" between 2.6-2.0 Ma. This model of gradual, directional change is open to critique on theoretical and analytical grounds, and may not be acceptable to all researchers. It does have the important distinction of offering a way to test "older than the Oldowan" scenarios from an analysis of the lithics themselves rather than indirectly through ethological studies or analogical reasoning. As a consequence, it lays the ground for formulating concrete expectations about the hypothetical earliest lithics, something that was lacking in previous discussions.

Virtually all the contributors to the volume write under the assumption that Oldowan lithics play a role in the ecological adaptations of early hominins, although not all of them address this issue explicitly. The model presented by Carbonell et al. (2009) elucidates distinct technological changes but it is not necessarily an "evolutionary" model. Since we cannot track the heritability of these changes, it does not follow strict Darwinian explanations of evolution. On the other hand, de la Torre and Mora (2009) emphasize that cognitive approaches to tool use allow the investigation of individual behaviors that are the focus of natural selection.

For us to fully understand the evolutionary consequences of hominin behavior it will be necessary to understand the selection pressures that are driving these changes. If we can understand the aspects of technological change which impact on the genetic fitness of the manufacturers of certain technological forms it may be possible to link evolutionary mechanisms with technological change (Bamforth and Bleed 1997; Ferraro et al. 2006).

Carbonell et al. (2009) argue that initial tool manufacture was based on flake production, but the selective pressures behind this sequence are not yet fully understood. This issue is taken up by Plummer et al. (2009), who regard tool use as an expansion of the foods available to hominins, conferring an adaptive advantage to any hominin species. Tool use provided a buffer from resource variation that was a result of environmental instability in the Plio-Pleistocene (de la Torre and Mora 2009).

At one time the sharp edges on Oldowan tools were thought to be part of an adaptive complex associated with the first appearance of carnivory in hominins (Isaac and Crader 1981), a notion seriously challenged by Domínguez-Rodrigo's (2009) extensive review of the zooarchaeological and taphonomic signatures in Pliocene and Plio-Pleistocene archaeological sites. Considering the lack of unambiguous evidence for meat consumption in the course of the first 600 kyr of stone tool production, microwear evidence for plant use needs to be reconsidered (Beyries 1993; Plummer 2004). Purposeful tool-making may possibly be less a means of economic *intensification* (i.e. obtaining more meat) than a *risk reduction* strategy (namely, a way to ensure the supply of vegetal food staples).

The mystery behind the use of the earliest stone tools further complicates the question of the makers of the industries. This issue is tackled by the various authors of this volume either implicitly or very cursorily, and with good reason. Considering the number of potential authors of the earliest tool forms it is difficult to assign one hominin species as the maker of the Oldowan. Experimental work in the neuroscience and biomechanics of tool-making (Marzke 1997, 1998; papers in Roux and Bril 2005; Stout and Chaminade 2007) provides us with rudimentary ideas about the ways modern humans approach stone knapping and how their brains handle the multi-tasking involved in the process, but for obvious reasons offers less help in investigating the brains of ancient hominins belonging to other species and genera. Some physical anthropologists have studied the functional hand anatomy of potential authors of the Oldowan in relation to stone knapping abilities (Susman 1991, 1998; Marzke 1997, 1998; Tocheri et al. 2008), yet differentiation between the various hominin taxa remains a difficult task due to sample size limitations. Among the several authors in this volume who raise the issue of hominin taxa and their relation to stone tool-making, Harmand (2009) is the most explicit. Her comparison of selectivity in the contemporaneous assemblages of Lokalalei 2C and 1 suggests differential ability to implement core reduction strategies that cannot be explained by raw material variability. She carefully infers that possibly more than a single hominin species was engaged in tool-making in West Turkana.

Whether or not archaeologists can use tools to investigate phylogenetic relationships is a different matter, and of considerable debate (Moore and Brumm 2009), despite the fact that several researchers have advocated this approach (Wood 1997; Foley and Lahr 2003). The association of Oldowan-like tools with the small-brained hominin (Morwood et al. 2004, 2005) at Liang Bua cautions the link between tool technology and species.

A second major theme throughout the volume is the distinction between gradual changes and long-term variability. Many of the contributors regard the analysis of tool technology as a unique window into the evolution of behavior in the Plio-Pleistocene. Among the contributors to this volume, supporters of gradual processes of change are more vociferous and explicit. The issue of nondirectional variability is not addressed specifically by the papers in this volume, yet it underlines many of the discussions of raw material selectivity and lithic technology. Thus Barsky (2009) argues that the Oldowan record can be read as a series of evolutionary steps, and Carbonell et al. (2009) emphasize the fact that environmental factors triggered the "latent evolutionary potential" of tool use, rejecting the concept of stasis (Semaw 2000) in the Oldowan archaeological record. Carbonell et al. (2009) and Barsky (2009) lay out an evolutionary trajectory beginning with a homogenous phase, which is similar to chimpanzee tool use and defined by the absence of long operational chains of production (although others have described extensive variability in chimpanzee tool use (McGrew 1992; Carvalho et al. 2008). Carbonell et al. (2009) argue that chimpanzees select tool forms based on their inherent properties as opposed to the potential abilities, such that chimps select nut cracking anvils based on their observed existing properties, whereas hominins recognize that they can turn a cobble into something useful (e.g. a chopper) by actively changing its original properties. The second stage in this system is described as "variability" and is supposedly built upon improvements from the homogenous nature of the earliest tool use. The assemblages at Fejej fall into this category because they lack truly "configured" tools (Barsky 2009). This expands to a phase of "diversity", defined by an increase in multidirectional flaking and exemplified by assemblages such as Olduvai, Ain Hanech, and Kokiselei 5 (Carbonell et al. 2009). Finally "multiplicity" is reached with the onset of a large-flake economy and standardization of tool forms. This phase is similar to descriptions of Mode 2 industries. Interestingly this three-stage model tracks an increase in technological complexity on a temporal axis as well as geographic axis, once hominins making "variable" Oldowan assemblages dispersed out of Africa.

Proponents of both directional evolutionary scenarios and of variability scenarios rely on dynamic approaches (as opposed to strictly typological ones) to the study of lithics. For advocates of directional evolutionary scenarios, technology as a dynamic process provides the prime mover that sets the process of cumulative change in motion and determines its direction from less to more complex lithic phenomena. For the proponents of nondirectional variability the dynamic approach to lithics allows the study of behavioral responses to local constraints. It can be argued that the incorporation of operational sequence approaches to the analysis of Oldowan assemblages is what enabled the development of these two alternative points of view.

De la Torre and Mora (2009), Harmand (2009), Braun et al. (2009), Carbonell et al. (2009) and Barsky (2009) focus explicitly on the study of technology that, until recently, has not been applied to the earliest industries. They rely on the *chaîne opératoire* approach, which integrates all aspects of the study of stone artifacts, linking the spatial and temporal contexts of selection and acquisition of raw material and the elaboration, utilization and discard of artifacts. A main difference from previous typological approaches is that it views tools as the result of a process of tool manufacture. Furthermore, it provides some of the best insights into the strategies of tool manufacture because it involves processes of problem-solving and decision-making by the individual stone knapper (Perlès 1993). De la Torre and Mora (2009) suggest that earlier processual methods of studying the Oldowan stagnated because they were reductionist and constrained the comprehension of knapping methods by blindly applying isolated attributes to varied assemblages (Kimura 1997, 1999, 2002; Ludwig and Harris 1998; Ludwig 1999). Indeed the application of attributes in a vacuum of contextual information has proved to be an unsatisfactory approach to the past (Braun et al. 2005). Following the ethnological school as represented by Lemonnier (1990), de la Torre and Mora (2009) see merit in the close focus of the *chaîne opératoire* approach on the individual context while at the same time viewing any artifact as the representation of a socially transmitted mental image. This requires stone tools to be representative of mental templates, a point of considerable debate in the context of the earliest stone tool industries.

Mental templates are themselves a modified typological classification. Even though *chaîne opératoire* integrates the concept of tool manufacture it requires artifacts to fall into categories (e.g. unidirectional, bidirectional) and therefore necessarily limits the amount of variation observable in the archaeological record. Like all typologies, *chaîne opératoire*, is a data reduction technique and therefore provides a set of categories that are more manageable for investigation. At the same time this approach disregards the variation between and within technical systems.

Viewing stone artifacts as the product of a series of manufacturing steps has many advantages. When artifacts are understood as a product of a dynamic manufacturing process it is possible to investigate the techniques and rules associated with that process. Many contributors have outlined certain technological strategies which are representative of a dynamic system. Barsky (2009) and Carbonell et al. (2009) have shown an elaboration of unidirectional flaking into discoidal or unifacial centripetal flaking in many Oldowan knapping systems. This also allows an investigation of cognitive systems such as the preparation of cores to produce specific debitage products. For example, the production of "configured" tools appears to be the precursor to Mode 2 technology and represents a conceptual division of the core into striking platform and flake removal surface (Carbonell et al. 2009; but see Delagnes and Roche 2005 for potential evidence of this conceptual distinction at Lokalalei 2C).

The *chaîne opératoire* approach to lithic production provides an important investigation into knapping strategies throughout the Oldowan archaeological record. Carbonell et al. (2009) focus on the diversity of different reduction sequences seen at different stages in the Oldowan. Beginning with the Gona assemblage, lithic production consists of the removal of unstandardized flakes through the use of a largely unidirectional and recurrent flaking strategy, dominated exclusively by unifacial orthogonal and unifacial discoidal technical systems, with rare core rotation. Carbonell et al. (2009) consider the Gona debitage sequences to be long but not exhaustive (although see Toth et al. 2006 for alternative views) At Lokalalei 2C, higher quality raw material with abundant natural platforms was used, allowing long debitage sequences without the need to engage in platform preparation (Harmand 2009; see also Delagnes and Roche 2005; Roche 2005). The end result was a heterogeneous debitage with unidirectional recurrent removals on opposed or orthogonal platforms (Harmand 2009; Carbonell et al. 2009). Raw material selection played a major role also in the reduction strategies at Hadar (Goldman-Neuman and Hovers 2009). Stones of particular shapes and raw material quality were selected as the major first step in the reduction sequence. Piperno and colleagues (2009) also describe the integration of raw material selection as part of "flaking" and "shaping" technical systems at Gombore and Garba IV, introducing an analytical distinction between these different technological systems. The lithic assemblages at Dmanisi, Fejej, and Garba IV all show a great diversity of different knapping strategies. Barsky (2009), Carbonell et al. (2009) and Piperno et al. (2009) note the lack of true multifacial flaking at many of the early Plio-Pleistocene sites. At some sites (Garba IV, Fejej, Lokalalei 1, and some of the Orce sites) the technological behavior was not standardized and produced irregular debitage. Similar techniques were used to produce the tools at Mata Menge and Liang Bua, but the use of flakes as the blanks for cores resulted in different tool forms (Moore and Brumm 2009).

The set of papers relating to raw material selection and acquisition as part of the technological system also bears on the competing views on directional changes and nondirectional variability. Harmand's (2009) description of the technical rules governing raw material selection and acquisition in two contemporaneous sites – Lokalalei 1 and Lokalalei 2C – underlines the possibility that lithic variability is not necessarily the result of evolutionary changes through time in hominins' technical perceptions of stone flaking. It could be linked to specific biological differences between the tool-makers, as hinted by Harmand (2009). For example, in Lokalalei 1 the inadequate choice of raw material is accompanied by high frequencies of flakes derived from knapping accidents (e.g. Siret or hinged flakes) and early core exhaustion and discard (Harmand 2009; Kibunjia 1994). Such differences in knapping skills between the two closely located, contemporaneous sites could also reflect the use of the two localities by individuals more (Lokalalei 2C) or less (Lokalalei 1) experienced in stone knapping. Ekshtain (2006) suggested that there may be a correlation between the proximity to raw materials and the liberty of novices to experiment with selection and knapping of stone, whereas Hovers (in press) emphasizes that the main difference between novices and experts is not in that the latter do not make mistakes, but in that they are adept in rectifying them sufficiently to continue core reduction. Piperno et al. (2009) and Braun et al. (2009) demonstrate how even slight variation in availability of high quality raw material drives technological variation. Yet in demonstrating the degree of flexibility of early hominins treatment of raw materials (discussed also by Goldman-Neuman and Hovers 2009), the influence of raw material availability does not take the form of determinism, emphasizing instead that Oldowan-making hominins were engaged in active problem-solving.

The importance of raw material availability, quality, selection and transport is indeed another major focus of many of the contributions to this volume. There is no doubt that the nature of raw material resources around Oldowan sites plays a major role in shaping the strategies of reduction. To fully integrate raw material variability into a study of Oldowan technology requires that we describe raw materials in a way that is relevant to technological questions. Goldman-Neuman and Hovers note that a strict lithological classification is insufficient because of the variability within a single rock lithology. Harmand dealt with this variation by describing several different aspects of raw materials including color, grain size, texture, ground mass, and the homogeneity of the rock matrix, a practice implemented in the study of other Oldowan assemblages (e.g. Goldman 2004; Stout et al. 2005). Hominins were likely cognizant of these features because many sites record the use of high quality raw materials such as phonolite, trachyte, rhyolite and basalt, the highest quality material being the obsidian from Balchit Massif that was identified at Melka Kunture. These high quality materials were selected for artifact manufacture in sites as old as the Plio-Pleistocene Yemeni sites (Chauhan 2009) to the Pleistocene Mata Menge and Liang Bua localities (Moore and Brumm 2009). Geochemical characterization of sources allows the identification of specific sources of raw material (Braun et al. 2008a,b).

But the availability of raw material resources is difficult to model. Goldman-Neuman and Hovers (2009) describe the use of conceptual "source area" as a unit of analysis in their raw material study. Like Harmand (2009), they quantify the availability of different types of stone at nearby conglomerates (Goldman 2004; Stout et al. 2005). This is important because at many Oldowan localities the distribution of raw material mimics the local availability of certain types of stone (Merrick and Merrick 1976; Schick 1987; Kimura 2002). Different raw materials at Melka Kunture and Gran Dolina are often subject to different reduction strategies based on the availability of certain materials. This exemplifies the need to analytically separate different types of stone.

Based on refit sets (Delagnes and Roche 2005), Harmand (2009) calculates the number of cobbles transported into the Lokalalei sites as anywhere between 90 and 110 cobbles of known sizes. When refitting is not feasible or cannot be carried out, it is sometimes possible to determine size of transported items by the size and shape of the detached pieces, as in the cases of Garba IV and Liang Bua (Piperno et al. 2009; Moore and Brumm 2009).

The patterns of raw material transport are directly relevant to the discussion of the evolutionary origins of the repertoire of Oldowan stone tool-making behaviors. Although most Oldowan sites focus on local sources of raw material, transport of resources is a consistent behavioral feature of the Early Stone Age hominins (Potts 1991). Oldowan hominins clearly did not always just settle for what was locally available (Blumenschine et al. 2003; Plummer 2004). The spatial fragmentation of lithic reduction sequences at Gona (Semaw et al. 2003; Toth et al. 2006) attests to at least minimal transport at the very earliest localities. Most Oldowan sites show that costs of provisioning a site with raw material represent transport of less than 1 kilometer, a pattern that sometimes has been equated to the behaviors of chimpanzees with regards to raw material transport. Late Pliocene tool-makers transport raw material over distances that are on par with those of chimpanzees who never move lithic objects more than 17% of their home range (estimated as a function of body size), but this pattern changes slightly after 2.0 Ma, when raw materials are transported over distance greater than 100% of the estimated home range (Goldman-Neuman and Hovers 2009; see Marwick 2003). But even the earliest tool-makers differ from chimpanzees in that the latter transport lithics as finite tools (e.g. hammerstones). In contrast, Oldowan knappers transport stone from sources of raw material or from other sites based on its future potential as sources of sharp edges (Toth et al. 2006; Braun et al. 2008a,b; Hovers et al. 2008).

One final aspect of raw materials that is reviewed by many contributors to this volume is the concept of raw material selection, bearing directly on the issue of cognitive abilities as they are expressed in planning depth, decision-making parameters and active problem-solving. Understanding of the degree of selectivity among hominins is presently beyond our understanding, but the contributions here make it clear that Oldowan knappers recognized the differences in the physical properties of different types of stone and were able to act upon this knowledge. Identified selection patterns usually show non-random discrepancies in the appearance of certain raw material characteristics between archaeological and geological sources (Goldman-Neuman and Hovers 2009). In practice this is expressed as differences between conglomerates sampled in sediments that are contemporaneous with archaeological sites and the archaeological assemblage (Harmand 2009; and see Stout et al. 2005; Braun et al. 2008a,b). Unfortunately, as discussed in this book by Goldman-Neuman and Hovers and Harmand, with Oldowan raw materials being transported over short distances at some sites, selectivity is less readily demonstrable or quantified. Even more difficult is comparing levels of selectivity between archaeological sites. Because artifacts are quantified usually by number of pieces and conglomerates are often quantified by the volume of rock, it is sometimes difficult to determine the comparability of these two databases (Blumenschine et al. 2007; Goldman-Neuman and Hovers 2009). Often the exact reason for hominin selection of certain raw materials is not clear. Papers in this volume argue almost unanimously that selectivity is a compromise between considerations of suitability for flaking as well as their ability to resist wear – two qualities that do not necessarily overlap – but also of the ease of recognizing the appropriate raw material. At sites like Dmanisi, Garba IV, and Lokalalei 2C homogenous raw materials (fine-grained tuff, obsidian, and phonolite, respectively) were selected for artifact manufacture despite the abundance and accessibility of other rock types (Barsky 2009; Harmand 2009; Piperno et al. 2009). In the case of Lokalalei 2C internal foliation of phonolites made the fracture mechanics of this material more predictable and hominins focused tool production on this material. Other raw materials are often selected for different features. At Olduvai and Lokalalei, basalt cobbles were selected for their ability to be used as pounding or percussion tools (Harmand 2009; see de la Torre and Mora 2005). Piperno et al. show that in Gombore I, the vast majority of artifacts are "percussion materials" and are made from a group of several different raw materials other than obsidian, while at both Melka Kunture sites discussed in their paper obsidian was selected to produce retouched flakes as a specific component of the toolkit. The use of different technical systems for different rock types is evident in the late Liang Bua assemblages as well as in those from early sites such as Dmanisi, Barranco Leon and Fuenta Nueva. Barsky suggests the selection of quartz materials at Fejej may be a culturally transmitted phenomenon because of the appearance of this type of material in the nearby Omo assemblages (Merrick 1976; de la Torre 2004; de Lumley and Beyene 2004).

But not all selection involved understanding the lithological properties of rocks. Goldman-Neuman and Hovers (2009) emphasize the fact that lithologically, many raw materials at Hadar were similar, and the distinctions that hominins made for selection involved the shape of the raw material. The understanding of the correct angles needed for stone manufacture is a prerequisite for artifact manufacture and something that is evident in the Gona materials (Semaw et al. 1997, 2003; Semaw 2000; Stout et al. 2005). For Carbonell et al. (2009) this behavior is related to the emergence of the phase of "variability" in their evolutionary model, as cobbles of certain morphologies are selected for the realization of particular technical systems. The hominins who produced the artifact collection at Dmanisi also selected cobbles from a nearby alluvial deposit that had natural angles for artifact production. But this does not seem to be the case at Lokalalei 1 where rock types were selected that did not have the proper platforms for the production of extensive debitage sequences. This heightens the contrast with Lokalalei 2C where hominins carefully selected flaking surfaces that allowed long and productive debitage sequences.

The contributions to this volume obviously show that research on the earliest technologies is beginning to enter a new arena where we are now learning much about early hominin behavior. The approach that started with Isaac (1971, 1972, 1976, 1978a,b,1986; Blumenschine 1991a) and continued with his students (Harris and Bishop 1976; Harris and Gowlett 1980; Toth 1982, 1985a,b, 1987, 1991; Harris 1983, 1986; Harris and Capaldo 1993; Schick and Toth 1993; Harris et al. 2002) is now taking a new turn by integrating a technological approach to lithics with principles of ecological and economic theories (Braun and Harris 2001, 2002; Hovers et al. 2002; Stout 2002; de la Torre et al. 2003; de la Torre 2004; Harmand 2004, 2005; Braun et al. 2005, 2008a, b; de la Torre and Mora 2005; Stout et al. 2005; Hovers 2007). The contribution of de la Torre and Mora (2009) provides a review of the changing attitudes to the study of the Oldowan, illuminating the strengths of each approach. These authors clearly favor the *chaîne opératoire* method of studying the Oldowan. Yet, as we have very little information on the social milieu of Oldowan groups, Hovers (2007) has recently suggested that in the context of the early industries the term "reduction sequence" may be more apt for this type of study. Researchers are currently struggling to find the balance between the two main ways of studying Oldowan artifacts. An exclusively technological approach focuses so closely on the individual context that the broader evolutionary and ecological context disappears from the discussion of Oldowan behavior. At the same time, it is important that evolutionary implications be made in relation to the principles of adaptation that drive Darwinian evolution. The integration of technical systems with the broader frame of activities carried out on a landscape is needed (Plummer et al. 2009). But in order to do that we need to understand what those activities mean.

The study of Oldowan stone artifacts needs further experimentation in order to place a real understanding on what artifact morphology means cognitively and practically. The best evidence to support this research trajectory comes from the history of research on Oldowan zooarchaeology. Without the experimental work of Blumenschine (1986a,b, 1987, 1988a,b,c, 1989, 1991b, 1995), Bunn (1981, 1986; Bunn and Blumenschine 1987; Bunn et al. 1988, 1991; Bunn and Ezzo 1993) and Domínguez-Rodrigo (1997, 1999, 2001, 2003; Domínguez-Rodrigo and Piqueras 2003), understanding of hominin diets would be based almost purely on conjecture about the effect of hominins and carnivores on large mammal bones. Indeed Domínguez-Rodrigo's contribution to this volume shows that twenty-five years after Bunn's initial experiments, deciphering the pattern of marks on specific bone portions still requires further experimentation. Toth and Schick's experiments (Toth 1982, 1985a,b; Schick 1987; Toth and Schick 2006) were a major step for this type of work, but further work is needed before we can integrate the type of intricate detail that Delagnes and Roche (2005) have provided for Oldowan technical systems with the ecological and cultural constraints that defined Oldowan tool manufacture throughout prehistory.

The papers in this book do not speak in one voice. Were we to single out one take-home message, it would be that the Oldowan is not a monolithic behavioral phenomenon, and that it exhibits behavioral plasticity of the type that will become the hallmark of later material culture signatures. Variation is the substance on which natural selection operates, and the Oldowan seems to have a lot to offer in this respect. It is from the integration of ecological possibilities and constraints, cognitive and anatomical abilities and their behavioral outcomes that discussion of the Oldowan can move forward. Then we can begin to tackle questions about the mechanisms that caused an individual's discovery of purposeful stone knapping to become an immensely successful innovation that spread and survived among small hominin groups in Africa and beyond.

References

Asfaw, B., Beyene, Y., Suwa, G., Walter. R.C., White, T.D., Wolde-Gabriel. G., Yemane, T., 1992. The earliest Acheulean from Konso-Gardula. Nature 360, 732–735.

Bamforth, D.B., Bleed. P., 1997. Technology, flaked stone technology, and risk. In: Barton, C.M., Clark, G.A., (Eds.), Rediscovering Darwin : Evolutionary Theory and Archeological Explanation, 7th edn. American Anthropological Association, Washington, pp. 109–139.

Bar-Yosef, O., Goren-Inbar, N., 1993. The Lithic Assemblages of Ubeidiya: A Lower Paleolithic Site in the Jordan Valley. Monographs of the Institute of Archaeology, 34. Jerusalem, Institute of Archaeology, The Hebrew University of Jerusalem.

Barsky, D., 2009. An overview of some African and Eurasian Oldowan sites: evaluation of hominin cognitive levels, technological advancement and adaptive skills. In: Hovers, E., Braun, D.R. (Eds.), Interdisciplinary Approaches to the Oldowan. Springer, Dordrecht, pp. 39–47.

Beyries, S.,1993. Are we able to determine the function of the earliest Palaeolithic tools? In: Berthelet, A., Chavaillon, J. (Eds.), Use of Tools by Human and Non-human Primates. Clarendon Press, Oxford, pp. 225–236.

Biro, D., Inoue-Nakamura, N., Tonooka, R., Yamakoshi, G., Sousa, C., Matsuzawa, T., 2004. Cultural innovation and transmission of tool use in wild chimpanzees: evidence from field experiments. Animal Cognition 6, 213–223.

Blumenschine, R.J., 1986a. Carcass consumption sequences and the archaeological distinction of scavenging and hunting. Journal of Human Evolution 15, 639–659.

Blumenschine, R.J., 1986b. Early Hominid Scavenging Opportunities: Implications of Carcass Availability in the Serengeti and Ngorongoro Ecosystems. BAR International Series 283. Oxford

Blumenschine, R.J., 1987. Characteristics of an early hominid scavenging niche. Current Anthropology 28, 383–394.

Blumenschine, R.J., 1988a. An experimental model of the timing of hominid and carnivore influence on archaeological bone assemblages. Journal of Archaeological Science 15, 483–502.

Blumenschine, R.J., 1988b. Reinstating an early hominid scavenging niche: a reply to Potts. Current Anthropology 29, 483–486.

Blumenschine, R.J., 1988c. A taphonomic trio for archaeology: mechanics, energetics and ecology. Society for American Archaeology 53rd Annual Meeting. Society for American Archaeology, Phoenix.

Blumenschine, R.J., 1989. A landscape taphonomic model of the scale of prehistoric scavenging opportunities. Journal of Human Evolution 18, 345–371.

Blumenschine, R.J., 1991a. Breakfast at Olorgesailie: the natural history approach to Early Stone Age archaeology. Journal of Human Evolution 21, 307–327.

Blumenschine, R.J., 1991b. Hominid carnivory, foraging strategies, and the socio-economic function of early archaeological sites. Philosophical Transactions of the Royal Society of London B 334, 211–221.

Blumenschine, R.J., 1995. Percussion marks, tooth marks, and experimental determinations of the timing of hominid and carnivore access to long bones at FLK Zinjanthropus, Olduvai Gorge, Tanzania. Journal of Human Evolution 29, 21–51.

Blumenschine, R.J., Masao, F.T., Tactikos, J., Ebert, J., 2007. Effects of distance from stone source on landscape-scale variation in Oldowan artifact assemblages in the Paleo-Olduvai Basin, Tanzania. Journal of Archaeological Science 35, 76–86.

Blumenschine, R.J.,, Peters, C.R., Masao, F.T., Clarke, R.J., Deino, A.L., Hay, R.L., Swisher, C.C., Stanistreet, I.G., Ashley, G.M., McHenry, L.J., Sikes, N.E., van der Merwe, N.J., Tactikos, J.C., Cushing, A.E., Deocampo, D.M., Njau, J.K., Ebert, J.I., 2003. Late Pliocene *Homo* and hominid land use from western Olduvai Gorge, Tanzania. Science 299, 1217–1221.

Braun, D.R., Harris, J.W.K., 2001. Oldowan technology: new insights from digital image analysis of Koobi Fora lithic assemblages. Journal of Human Evolution 40, A4.

Braun, D.R., Harris, J.W.K., 2003. Technological developments in the Oldowan of Koobi Fora: innovative techniques of artifact analysis. In: Moreno, J.M., Torcal, R.M., Sainz, I.T. (Eds.), Oldowan: Rather More than Smashing Stones. University of Barcelona Press, Barcelona, pp. 117–144.

Braun, D.R., Harris, J.W.K., Maina, D., 2008a. Oldowan raw material procurement and use: evidence from the Koobi Fora Formation. Archaeometry. doi: 10.1111/j.1475-4754.2008.00393.

Braun, D.R., Plummer, T.W., Ditchfield, P.W., Bishop, L.C., Ferraro, J.V., 2009. Oldowan technology and raw material variability at Kanjera South. In: Hovers, E., Braun, D.R. (Eds.), Multidisciplinary Approaches to the Oldowan. Springer, Dordrecht, pp. 99–110.

Braun, D.R. Plummer, T., Ditchfield, P., Ferraro, J.V., Maina, D., Bishop, L.C., Potts, R., 2008b. Oldowan behavior and raw material transport: perspectives from the Kanjera Formation. Journal of Archaeological Science 35, 2329–2345.

Braun, D.R., Tactikos, J.C., Ferraro, J.V., Harris, J.W.K., 2005. Flake recovery rates and inferences of Oldowan hominin behavior: a response to Kimura 1999, 2002. Journal of Human Evolution 48, 525–531.

Bunn, H.T., 1981. Archaeological evidence for meat-eating by Plio-Pleistocene hominids from Koobi Fora and Olduvai Gorge. Nature 291, 574–577.

Bunn, H.T., 1986. Patterns of skeletal representation and hominid subsistence activities at Olduvai Gorge, Tanzania, and Koobi Fora, Kenya. Journal of Human Evolution 15, 673–690.

Bunn, H.T., 2007. Meat made us human. In: Ungar, P.S. (Ed.), Evolution of the Human Diet. The Known, the Unknown and the Unknownable. Oxford University Press, Oxford, pp. 191–211.

Bunn, H.T., Bartram, L.E., Kroll, E.M., 1988. Variability in bone assemblage formation from Hadza hunting, scavenging, and carcass processing. Journal of Anthropological Archaeology 7, 412–457.

Bunn, H.T., Blumenschine, R.J., 1987. On "Theoretical framework and tests" of early hominid meat and marrow acquisition: a reply to Shipman. American Anthropologist 89, 444–448.

Bunn, H.T., Ezzo, J.A., 1993. Hunting and scavenging by Plio-Pleistocene hominids: nutritional constraints, archaeological patterns, and behavioral implications. Journal of Archaeological Science 20, 365–398.

Bunn, H.T., Kroll, E.M., Bartram, L.E., 1991. Bone distribution on a modern East African landscape and its archaeological implications. In: Clark, J.D. (Ed.), Cultural beginnings: Approaches to understanding early hominid life-ways in the African savanna, Mainz, Germany, pp. 33–54.

Cameron, D.W., 1993. The Pliocene hominid and protochimpanzee behavioral morphotypes. Journal of Anthropological Archaeology 12, 386–414.

Cameron, D.W., Groves, C.P., 2004. Bones, Stones and Molecules: Out of Africa and Human Evolution. Elsevier Academic Press, Amsterdam.

Carbonell, E., Bermudez de Castro, J.M., Pares, J.M., Perez-Gonzalez, A., Cuenca-Bescos, G., Olle, A., Mosquera, M., Huguet, R., van der Made, J., Rosas, A., Sala, R., Vallverdu, J., Garcia, N., Granger, D.E., Martinon-Torres, M., Rodriguez, X.P., Stock, G.M., Verges, J.M., Allue, E., Burjachs, F., Caceres, I., Canals, A., Benito, A., Diez, C., Lozano, M., Mateos, A., Navazo, M., Rodriguez ,J., Rosell J., Arsuaga J.L., 2008. The first hominin of Europe. Nature 452, 465–469.

Carbonell, E., Sala, R., Barsky, D., Celiberti, V., 2009. From homogeneity to multiplicity : a new approach to the study of archaic stone tools. In : Hovers, E., Braun, D.R. (Eds.), Interdisciplinary Approaches to the Oldowan. Springer, Dordrecht, pp. 25–37.

Carvalho, S., Cunha, E., Sousa, C.U., Matsuzawa, T., 2008. Chaines opératoires and resource-exploitation strategies in chimpanzee *(Pan troglodytes)* nut cracking. Journal of Human Evolution 55, 148–163.

Chauhan, P.R., 2009. Early *Homo* occupations near the "Gate of Tears": examining the paleoanthropological records of Djibouti and Yemen. In: Hovers, E., Braun, D.R. (Eds.), Interdisciplinary Approaches to the Oldowan. Springer, Dordrecht, pp. 49–59.

Clark, G., 1970. Aspects of Prehistory. University of California Press, Berkeley.

Clark, J.D., de Heinzelin, J., Schick, K.D., Hart, W.K., White, T.D., WoldeGabriel, G., Walter, R.C., Suwa, G., Asfaw, B., Vrba, E., Selassie, Y., 1994. African *Homo erectus*: old radiometric ages and young Oldowan assemblages in the Middle Awash Valley, Ethiopia. Science 264, 1907–1910.

Clarke, D.L., 1968. Analytical Archaeology. Methuen, London.

Corvinus, G,. 2004. *Homo erectus* in East and Southeast Asia, and the questions of the age of the species and its association with stone artifacts, with special attention to handaxe-like tools. Quaternary International 117, 141–151.

de A. Moura, A.C., Lee, P.C., 2004. Capuchin stone tool use in Caatinga dry forest. Science 306, 1909.

de la Torre, I., 2004. Omo revisited: evaluating the technological skills of Pliocene hominids. Current Anthropology, 45, 439–465.

de la Torre, I., Mora, R., 2005a. Unmodified lithic material at Olduvai Bed I: manuports or ecofacts? Journal of Archaeological Science 32, 273–285.

de la Torre, I., Mora, R., 2005b. Technological Strategies in the Lower Pleistocene at Olduvai Beds I and II. ERAUL, Liège.

de la Torre, I., Mora, R., Domínguez-Rodrigo, M., de Luque, L., Alcala, L., 2003. The Oldowan industry of Peninj and its bearing on the reconstruction of the technological skills of Lower Pleistocene hominids. Journal of Human Evolution 44, 203–224.

de la Torre, I., Mora, R., 2009. Remarks on the Current Theoretical and Methodological Approaches to the Study of Technological Strategies of Early Humans in Eastern Africa. In: Hovers, E., Braun, D.R. (Eds.), Interdisciplinary Approaches to the Oldowan. Springer, Dordrecht, pp. 15–24.

de Lumley, H., Beyene., Y. (Eds.), 2004. Les sites préhistoriques de la région de Fejej, sud-Omo, Éthiopie, dans leur contexte stratigraphique et paléontlogique. ADPF association, Paris.

de Lumley, H., Nioradze, M., Barsky, D., Cauche, D., Celiberti, V., Nioradze, G., Notter, O., Zvania, D., Lordkipanidze, D,. 2005. Les industries lithiques preoldowayennes du debut du Pleistocene inferieur du site de Dmanissi en Georgie. L'Anthropologie 109, 1–182.

Delagnes, A., Roche, H., 2005. Late Pliocene hominid knapping skills: the case of Lokalalei 2C, West Turkana, Kenya. Journal of Human Evolution 48, 435–472.

Domínguez-Rodrigo, M., 1997. Meat-eating by early hominids at the FLK 22 Zinjanthropus site, Olduvai Gorge (Tanzania): an experimental approach using cut-mark data. Journal of Human Evoltuion 33, 669–690.

Domínguez-Rodrigo, M., 1999. Distinguishing between apples and oranges: the application of modern cut-mark studies to the Plio-Pleistocene (a reply to Monahan). Journal of Human Evolution 37, 793–800.

Domínguez-Rodrigo, M., 2001. A study of carnivore competition in riparian and open habitats of modern savannas and its implications for hominid behavioral modelling. Journal of Human Evolution 40, 77–98.

Domínguez-Rodrigo, M., 2003. On cut marks and statistical inferences: methodological comments on Lupo & O'Connell (2002). Journal of Archaeological Science 30, 381–386.

Domínguez-Rodrigo, M., 2009. Are all Oldowan sites palimpsests? If so, what can they tell us about hominid carnivory? In: Hovers, E., Braun, D.R. (Eds.), Interdisciplinary Approaches to the Oldowan. Springer, Dordrecht, pp. 129–147.

Domínguez-Rodrigo, M., Piqueras, A., 2003. The use of tooth pits to identify carnivore taxa in tooth-marked archaeofaunas and their relevance to reconstruct hominid carcass processing behaviours. Journal of Archaeological Science 30, 1385–1391.

Dong, W., 2006. Recent discoveries of hominid fossils in southern China. L'Anthropologie 110, 241–257.

Ekshtain, R., 2006. A Comparative Study of Knapping Accidents in the Levantine Mousterian. Unpublished M.A. thesis, The Hebrew University of Jerusalem.

Ferraro, J.V., Braun, D.R., Tactikos, J.C., 2006. Contextualizing Oldowan lithic strategies: a biological perspective. Society of American Archaeology Meetings, San Juan

Ferring, R., Lordkipanidze, D., Berna, F., Ohms, O. 2008. Geology and formation processes at Dmanisi in the Georgian Caucasus. Abstracts of the 73rd Annual Meeting of the Society of American Archaeology, Vancouver, British Columbia, 195.

Foley, R., Lahr, M.M., 2003. On stony grounds: lithic technology, human evolution, and the emergence of culture. *Evolutionary Anthropology* 12, 109–122.

Fragaszy D., Isar P., Visalberghi E., Ottoni E. B., de Oliveira, M.G., 2004. Capuchin monkeys (*Cebus libidinosus*) use anvils and stone pounding tools. American Journal of Primatology 64, 359–366.

Gaillard, C., 2006. Les premiers peuplements d'Asie du Sud: vestiges culturels. Comptes Rendus Palevol 5, 359–369.

Gamble, C., 2007. Origins and Revolutions. Human Identity in Earliest Prehistory. Cambridge University Press, Cambridge.

Goldman, T., 2004. Raw Material Selection Patterns in A.L., 894, Hadar, Ethiopia. Unpublished M. A. thesis, The Hebrew University.

Goldman-Neuman, T., Hovers, E., 2009. Methodological issues in the study of Oldowan raw material selectivity: insights from A. L. 894 (Hadar, Ethiopia). In: Hovers, E., Braun, D.R. (Eds.), Interdisciplinary Approaches to the Oldowan. Springer, Dordrecht, pp. 71–84.

Harmand, S., 2004. Raw materials and economic behavior of Late Pliocene Hominids: the case of Lokalalei 2C and Lokalalei 1 sites, West Turkana (Kenya). Paleoanthropology Society, Montreal.

Harmand, S., 2005. Matières premières lithiques et comportements techno-économiques des hominiés Plio-Pléistocènes du Turkana occidental, Kenya. Ph. D. dissertation, Université de Paris X - Nanterre, Paris, p 317.

Harnamd, S., 2009. Variability in raw material selectivity at the Late Pliocene sites of Lokalalei, West Turkana, Kenya. In: Hovers, E., Braun, D. R. (Eds.), Interdisciplinary Approaches to the Oldowan. Springer, Dordrecht, pp. 85–97.

Harris, J.W.K., 1983. Cultural beginnings: Plio-Pleistocene archaeological occurrences from the Afar, Ethiopia. African Archaeological Review 1, 3–31.

Harris, J.W.K., 1986. Oldowan archaeological findings in the Afar Rift. L'Anthropologie 90, 3–21.

Harris, J.W.K., Bishop, W.W., 1976. Sites and assemblages from the early Pleistocene beds of Karari and Chesowanja, Kenya. Proceedings of the 9th International Congress of Prehistoric and Protohistoric Sciences, Université de Nice. CNRS, Paris, pp. 70–117.

Harris, J.W.K., Braun, D.R., McCoy, J.T., Pobiner, B.L., Rogers, M.J., 2002. New directions and preliminary results from a landscape approach to the study of archaeological traces for the behavior of Plio-Pleistocene hominids at Koobi Fora. Journal of Human Evolution 42, A15.

Harris, J.W.K., Capaldo, S.D., 1993. The earliest stone tools. In: Berthelet A, Chavaillon J (Eds.), The Use of Tools by Human and Non-human Primates. Clarendon Press, Oxford, pp. 196–220.

Harris, J.W.K., Gowlett, J.A.J., 1980. Evidence of early stone industries at Chesowanja, Kenya. In: Leakey, R.E.F., Ogot, B.A. (Eds.), Proceedings of the 8th Panafrican Congress of Prehistory and Quaternary Studies, Nairobi, pp. 208–212.

Hovers, E., 2007. Review of "Technological strategies in the Lower Pleistocene at Olduvai Beds I and II" (Ignacio de la Torre and Rafael Mora). PaleoAnthropology 2007, 53–57.

Hovers, E., in press. Learning from mistakes: flaking accidents and knapping skills in the assemblage of A.L. 894 (Hadar, Ethiopia). In: Schick, K. and Toth, N. (Eds.), The Cutting Edge: New Approaches to the Archaeology of Human Origins. Stone Age Institute Press Publications Series 3. Stone Age Institute, Gosport.

Hovers, E., Campisano, C.J., Feibel C.S., 2008. Patterns of land use by Late Pliocene hominins in Hadar, Ethiopia. PaleoAnthropology, A14.

Hovers, E., Schollmeyer, K., Goldman, T., Eck, G., Reed, K., Johanson, D., Kimbel, W., 2002. Late Pliocene archaeological sites in Hadar, Ethiopia. Journal of Human Evolution 42, A17-A17.

Isaac, B. (Ed.), 1989. The Archaeology of Human Origins: Papers by Glynn Isaac. Cambridge University Press, Cambridge.

Isaac, G.L., 1971. The diet of early man: aspects of archaeological evidence from Lower and Middle Plestocene sites in Africa. World Archaeology 2, 278–299.

Isaac, G.L., 1972. Some experiments in quantitative methods for characterizing assemblages of Acheulian artifacts. VIième Congrès Panafricain de Préhistoire et de l'Etude du Quaternaire, 1967. Imprimeries Réunies de Chambéry, Chambéry.

Isaac, G.L., 1976. The activities of early African hominids: a review of archaeological evidence from the time span two and a half to one million years ago. In: Isaac, G.L., McCown, E. (Eds.), Human Origins: Louis Leakey and the East African Evidence. W. A. Benjamin, Menlo Park, California, pp. 483–514.

Isaac, G.L., 1978a. The archaeological evidence for the activities of early African hominids. Early hominids of Africa. St Martin's Press, New York, pp. 219–254.

Isaac, G.L., 1978b. Food sharing and human evolution: archaeological evidence from the Plio-Pleistocene of East Africa. Journal of Anthropological Research 34, 311–325.

Isaac, G.L., 1986. Foundation stones: early artefacts as indicators of activities and abilities. In: Bailey, G.N., Callow, P. (Eds.), Stone Age Prehistory. Cambridge University Press, Cambridge, pp. 221–241.

Isaac, G.L., Crader, D.C., 1981. To what extent were early hominids carnivorous? an archaeological perspective. In: Harding, S.O., Teleki, G. (Eds.), Omnivorous Primates. Columbia University Press, New York, pp. 37–103.

Jablonka, E., Lamb, M.J., 2005. Evolution in Four Dimensions. Genetic, Epigenetic, Behavioral, and Symbolic Variation in the History of Life. MIT Press, Cambridge MA.

Kerr R.A., 2008. A time war over the period we live in. Science 319, 402–403.

Kibunjia, M., 1994. Pliocene archaeological occurrences in the Lake Turkana basin. Journal of Human Evolution 27, 159–171.

Kimura, Y., 1997. The MNK chert factory site: the chert-using strategy by early hominids at Olduvai Gorge, Tanzania. African Study Monographs 18, 1–28.

Kimura, Y., 1999. Tool-using strategies by early hominids at Bed II, Olduvai Gorge, Tanzania. Journal of Human Evolution 37, 807–831.

Kimura, Y., 2002. Examining time trends in the Oldowan technology at Beds I and II, Olduvai Gorge. Journal of Human Evolution 43, 291–321.

Leakey, M.D., 1971. Olduvai Gorge. Cambridge University Press, Cambridge.

Lemonnier, P., 1990. Topsy-turvy techniques. Remarks on the social representation of techniques. Cambridge Archaeological Review 9, 27–37.

Ludwig, B.V., 1999. A technological reassessment of East African Plio-Pleistocene lithic artifact assemblages. Ph. D. dissertation, Rutgers University, New Brunswick.

Ludwig, B.V., Harris, J.W.K., 1998. Towards a technological reassessment of East African Plio-Pleistocene lithic assemblages. In: Petraglia, M., Paddaya, K. (Eds.), The Rise and Diversity of the Lower Paleolithic. Academic Press, New York, pp. 84–106.

Marzke, M.W., 1997. Precision grips, hand morphology, and tools. American Journal of Physical Anthropology 102, 91–110.

Marzke, M.W., Toth, N., Schick, K.D., Reece, S., Steinberg, B., Hunt, K., Lindscheid, R.L., An, K.-N., 1998. EMG study of hand muscle recruitment during hard hammer percussion manufacture of Oldowan tools. American Journal of Physical Anthropology 105, 315–322.

McGrew, W.C., 1992. Chimpanzee Material Culture. Implications for Human Evolution. Cambridge University Press, Cambridge.

Merrick, H.V., 1976. Recent archaeological research in the Plio-Pleistocene deposits of the lower Omo Valley, southwestern Ethiopia. Perspectives on Human Evolution 3, 461–481.

Merrick, H.V., Merrick, J.P.S., 1976. Archeological occurrences of Earlier Pleistocene age from the Shungura formation. In: Coppens, Y., Howell, F.C., Isaac, G.L., Leakey R.E.F. (Eds.), Earliest Man and Environments in the Lake Rudolf Basin . University of Chicago Press, Chicago, pp. 574–584.

Montagu, A., 1965. The Human Revolution. The World Publishing Company, Cleveland and New York.

Moore, M.W., Brumm, A., 2009. *Homo floresiensis* and the African Oldowan. In: Hovers, E., Braun, D.R. (Eds.), Interdisciplinary Approaches to the Oldowan. Springer, Dordrecht, pp. 61–69.

Morwood, M.J., Brown, P., Jatmiko, Sutikna, T., Wahyu Saptomo, E., Westaway, K.E., Awe Due, R., Roberts, R.G., Maeda, T., Wasisto, S., Djubiantono, T., 2005. Further evidence for small-bodied hominins from the Late Pleistocene of Flores, Indonesia. Nature 437, 1012–1017.

Morwood, M.J., Soejono, R.P., Roberts, R.G., Sutikna, T., Turney, C.S.M., Westaway, K.E., Rink, W.J., Zhao, J.-X., van den Bergh, G.D., Due, R.A., Hobbs, D.R., Moore, M.W., Bird, M.I., Fifield, L.K., 2004. Archaeology and age of a new hominin from Flores in eastern Indonesia. Nature 431, 1087–1091.

Movius, H., 1948. The Lower Paleolithic cultures of southern and eastern Asia. Transactions of the American Philosophical Society 38, 329–420.

Movius, H., 1969. Lower Paleolithic archaeology in southeastern Asia and the Far East. Reprinted In: Howells, W.W. (Ed.), Early Man in the Far East. Studies in Physical Anthropology No. 1. Humanities Press, New York, pp.17–82.

Panger M.A., Brooks A., Richmond B.G., Wood B., 2002. Older than the Oldowan? rethinking the emergence of hominin tool use. Evolutionary Anthropology 11, 235–245.

Perlès, C., 1993. Ecological determinism, group strategies, and individual decisions in the conception of prehistoric stone assemblages. In: Berthelet, A., Chavaillon, J. (Eds.), Use of Tools by Human and Non-human Primates. Clarendon Press, Oxford, pp. 267–277.

Petraglia, M., LaPorta., P., Paddayya, K., 1999. The first Acheulian quarry in India: stone tool manufacture, biface morphology, and behaviors. Journal of Anthropological Research 55, 39–70.

Piperno, M., Collina, C., Galloti, R., Raynal, J.-P., Kieffer, G., Bourdonnec, F.-X., Poupeau, G., Geraads, D., 2009. Obsidian exploitation and utilization during the Oldowan at Melka Kunture (Ethiopia). In: Hovers, E., Braun, D.R. (Eds.), Interdisciplinary Approaches to the Oldowan. Springer, Dordrecht, pp. 111–128.

Plummer, T., 2004. Flaked stones and old bones: Biological and cultural evolution at the dawn of technology. Yearbook of Physical Anthropology 47, 118–164.

Plummer, T.W., Bisop, L.C., Ditchfield, P.W., Ferraro, J.V., Kingstone, J.D., Hertel, F., Braun, D.R., 2009. The environmental context of Oldowan hominin activities at Kanjera South, Kenya. In: Hovers, E., Braun, D.R. (Eds.), Interdisciplinary Approaches to the Oldowan. Springer, Dordrecht, pp. 149–160.

Pope, G.G., 1989. Bamboo and human culture. Natural History (October 1989), 48–57.

Potts, R., 1991. Why the Oldowan? Plio-Pleistocene toolmaking and the transport of resources. Journal of Anthropological Research 47,153–176.

Potts, R., 1994. Variables versus models of early Pleistocene hominid land use. Journal of Human Evolution 27, 2–24.

Richerson, P.J., Boyd, R., 2005. Not by Genes Alone. How Culture Transformed Human Evolution. Chicago University Press, Chicago.

Roche, H., 2005. From simple flaking to shaping: stone-knapping evolution among ealry hominins. In: Roux, V., Bril, B. (Eds.), Stone Knapping: The Necessary Conditions for a Uniquely Hominin Behavior. McDonald Institute for Archaeological Research, Cambridge, pp. 35–53.

Roux, V., Bril, B. (Eds.), 2005. Stone Knapping. The Necessary Conditions for a Uniquely Hominin Behavior. McDonald Institute for Archaeological Research, Cambridge.

Sahnouni, M., Schick, K.D., Toth, N., 1997. An experimental investigation into the nature of faceted limestone ''spheroids'' in the Early Palaeolithic. Journal of Archaeological Science 24, 701–713.

Sayers, K., Lovejoy C.O., 2008. The chimpanzee has no clothes: a critical examination of *Pan troglodytes* in models of human evolution. Current Anthropology 49, 87–114.

Schick, K.D., 1987. Modeling the formation of early stone age artifact concentrations. Journal of Human Evolution 16, 789–807.

Schick, K.D., 1994. The Movius Line reconsidered. Prespectives on the earlier Paleolithic of eastern Asia. In: Corrucini, R.S., Ciochon, R.L. (Eds.), Integrative Paths to the Past. Paleoanthropological Advances in Honor of F. Clark Howell, pp. 569–596. London: Prentice Hall.

Schick, K.D., Toth, N., 1993. Making Silent Stones Speak: Human Evolution and the Dawn of Technology. Simon and Schuster, New York.

Schick, K.D., Toth, N., Wei, Q., Clark, J.D., Etler, D., 1991. Archaeological perspectives in the Nihewan Basin, China. Journal of Human Evolution 21, 13–26.

Semaw, S., 2000. The world's oldest stone artefacts from Gona, Ethiopia: their implications for understanding stone technology and patterns of human evolution between 2.6-1.5 million years ago. Journal of Archaeological Science 27, 1197–1214.

Semaw, S., Renne, P., Harris, J.W.K., Feibel, C.S., Bernor, R.L., Fesseha, N., Mowbray, K.M., 1997. 2.5-million-year-old stone tools from Gona, Ethiopia. Nature 385, 333–336.

Semaw, S., Rogers, M.J., Quade, J., Renne, P.R., Butler, R.F., Domínguez-Rodrigo, M., Stout, D., Hart, W.S., Pickering, T., Simpson, S.W., 2003. 2.6-million-year-old stone tools and associated bones from OGS-6 and OGS-7, Gona, Afar, Ethiopia. Journal of Human Evolution 45, 169–177.

Semaw, S., Rogers, M.J., Stout, D., Quade, J., Levine, N., Renee, P.R., Butler, R., Kidane, T., Simpson, S.W. 2008. The Oldowan -Acheulian transition: new insights from Gona, Ethiopia. Paleoanthropology 2007, A28.

Shen, C., Qi, W., 2004. Lithic technological variability of the Middle Pleistocene in the eastern Nihewan Basin, northern China. Asian Perspectives, 43, 281–301.

Sponheimer, M., Lee-Thorp, J., de Ruiter, D., 2007. Icarus, isotopes, and australopith diets. In: Ungar, P.S. (Ed.), Evolution of the Human Diet. Oxford University Press, Oxford, pp. 132–149.

Sponheimer, M., Passey, B.H., de Ruiter, D.J., Guatelli-Steinberg, D., Cerling, T.E., Lee-Thorp, J.A., 2006. Isotopic evidence for dietary variability in the early hominin *Paranthropus robustus.* Science 314, 980–982.

Stout, D., 2002. Skill and cognition in stone tool production - an ethnographic case study from Irian Jaya. Current Anthropology 43, 693–722.

Stout, D., Chaminade, T., 2007. The evolutionary neuroscience of tool making. Neuropsychologia 45, 1091–1100.

Stout, D., Quade, J., Semaw, S., Rogers, M.J., Levin, N.E., 2005. Raw material selectivity of the earliest stone toolmakers at Gona, Afar, Ethiopia. Journal of Human Evolution 48, 365–380.

Stout, D., Toth, N., Schick, K.D., Stout, J., Hutchins, G., 2000. Stone tool-making and brain activation: positron emission tomography (PET) studies. Journal of Archaeological Science 27, 1215–1223.

Stout, D., Semaw, S., Rogers, M.J., Cauche, D., 2008. Technological variation in the earliest Oldowan (2.6 Ma) from Gona, Afar, Ethiopia. PaleoAnthropology 2007, A31.

Susman, R.L., 1991. Who made Oldowan tools? fossil evidence for tool behavior in Plio-Pleistocene hominids. Journal of Anthropological Research 47, 132–151.

Susman, R.L., 1998. Hand function and tool behavior in early hominids. Journal of Human Evolution 35, 28–46.

Texier, J.-P., Roche, H., Harmand, S., 2006. Kokiselei 5, Formation de Nachukui, West Turkana (Kenya): un temoignange de la variabilité ou de l'évolution des comportements techniques au Pleistocene ancien? VIVth International Congress for the Union for Prehsitoric and Protohistoric Sciences. British Archaeological Reports International Series, Liège, pp. 11–22.

Tocheri, M.W., Orr, C.M., Jacofsky, M. C., Marzke, M.W., 2008. The evolutionary history of the hominin hand since the last common ancestor of *Pan* and *Homo*. Journal of Anatomy 212, 544–562.

Toth, N., 1982. The Stone Age Technology of Early Hominids at Koobi Fora, Kenya: an experimental approach. Ph. D. dissertation, University of California, Berkeley.

Toth, N., 1985a. Archaeological evidence for preferential right-handedness in the Lower and Middle Pleistocene, and its possible implications. Journal of Human Evolution 14, 607–614.

Toth, N., 1985b. Oldowan reassessed: a close look at early stone artifacts. Journal of Archaeological Science 12, 101–120.

Toth, N., 1987. Behavioral inferences from early stone artifact assemblages: an experimental model. Journal of Human Evolution 16, 763–787.

Toth, N., 1991. Importance of experimental replicative and functional studies in Palaeolithic archaeology. In: Clark. J.D. (Ed.), Cultural Beginnings: Approaches to Understanding Early Hominid Life-ways in the African Savanna. Rudolf Habelt Verlag, Bonn, pp. 109–124.

Toth, N., Schick, K.D. (Eds.), 2006. The Oldowan: Case Studies into the Earliest Stone Age. Stone Age Institute Press, Bloomington

Toth, N., Schick, K.D., Semaw, S. 2006. A comparative study of the stone tool-making skills of *Pan, Australopithecus* and *Homo sapiens*. In: Toth, N., Schick K.D. (Eds.), The Oldowan: Case Studies into the Earliest Stone Age. Stone Age Institute Press, Bloomington, pp. 155–222.

Whiten, A., 2005. The second inheritance system of chimpanzees and humans. Nature, 437, 52–57.

Whiten A., Goodall J., McGrew W.C., Nishida T., Reynolds V., Sugiyama Y., Tutin C.E.G., Wrangham R., Boesch C., 1999. Cultures in chimpanzees. Nature 399, 682–685.

Wood, B., 1997. The oldest whodunnit in the world. Nature 385, 292–293.

Zhu, R.X., Potts, R., Xie, F., Hoffman, K.A., Deng, C.L., Shi, C.D., Pan, Y.X., Wang, H.Q., Shi, R.P., Wang, Y.C., Shi, G.H., Wu, N.Q., 2004. New evidence on the earliest human presence at high northern latitudes in northeast Asia. Nature, 431, 559–562.

Ziman, J. (Ed.), 2000. Technological Innovation as an Evolutionary Process. Cambridge University Press, Cambridge.

2. Remarks on the Current Theoretical and Methodological Approaches to the Study of Early Technological Strategies in Eastern Africa

Ignacio de la Torre*
Institute of Archaeology University College London
31-34 Gordon Square
London, WC1H-0PY
United Kingdom
i.torre@ucl.ac.uk

Rafael Mora
CEPAP-UAB
Facultad de Letras
Universidad Autónoma de Barcelona
08193 Bellaterra, Barcelona Spain
rafael.mora@uab.es

Keywords: East Africa, Early technologies, Theoretical approaches, Oldowan, Acheulian

Abstract

This paper explores the theoretical and methodological backgrounds that sustain the current knowledge of early East African technologies. The typological approach dominated lithic studies along the 1960s and 1970s and was later on replaced by processual tendencies, whose paradigms still prevail nowadays. Nonetheless, the present scene in Plio-Pleistocene archaeology is not monolithic, and the theoretical-methodological background of the academic school to which each researcher belongs, has influenced our understanding and interpretations of the technological abilities by early humans. In this article similarities and differences between schools of thought are discussed, and the collections from Olduvai are used as a case study for reflecting on the variety of theoretical and methodological approaches and their relevance for reconstructing early African technologies.

2.1 Introduction

Lithic technological organization consists of the selection and integration of strategies employed for manufacturing, using, transporting and discarding the instruments as well as materials used in their fabrication and maintenance (Nelson 1991:57). Therefore, the concept of "technological strategies" as used in the title of this paper refers to the processes of problem-solving and decision-making by tool-makers. Such behavioral processes are responses to conditions created by the interaction between human societies and the environment in which they live. In this point of view analyses of predictability, distribution, periodicity, productivity and variability of resources are crucial for understanding the cultural adaptations of human groups (Nelson 1991). Thus this approach to technological organization focuses

* Address for correspondence: i.torre@ucl.ac.uk

E. Hovers and D.R. Braun (eds.), *Interdisciplinary Approaches to the Oldowan*, DOI: 10.1007/978-1-4020-9059-2,

on spatial and temporal aspects of artifacts manufacture, use and abandonment within a particular behavioral and/or cultural system (Kelly 1988). At the same time, the organization of technology is conceived as a cultural subsystem interacting dynamically with other subsystems; hence organizational efficiency fluctuates according to limitations imposed by other variables (Ricklis and Cox 1993).

A different perspective on the study of lithic technology emphasizes the fundamental role of social aptitudes rather than ecological constraints. This approach minimizes the environmental determinism, since human societies use their settings variably according to diverse traditions and technologies (Geneste 1991). In this spirit Lemonnier (1990) underlines the relevance of studying the social aspects of technology, and not only its economical organization. According to this author any artifact is a representation that emerges from mental images and thoughts, which are socially (and not necessarily consciously) transmitted and therefore shared by a group. Thus, representations are set off through the technical action typical of each society (Lemonnier 1990).

This cursory overview of some general concepts shows the wide range of perspectives and approaches that can nowadays be applied to the analysis of technical strategies by Paleolithic societies. However, our aim here is neither to present a worldwide review of approaches to lithic analysis, nor analyze systematically their relationships with different theoretical tendencies. Such topics are far beyond the scope of this paper, which is actually focused on a single period and region, the Early Stone Age in eastern Africa, and on the alternative ways by which those lithics have been analyzed during recent decades. Hence, our objective is to show that such variability of approaches has a remarkable influence on the knowledge of what we denominate the technological strategies of early humans in eastern Africa. Borrowing the categorization by Karlin and Julien (1994), all examples that will be presented in the following sections could be synthesized into two approaches that in principle are complementary; first, the study of technical production, focused on the analysis of methods applied to knapping. And second, the understanding of cultural, ecological and economical constraints that such technical production depends on.

2.2 The Typological Approach to the Oldowan and Acheulian from East Africa

As was the case in the rest of Old World prehistory at the time, lithic studies of the Plio-Pleistocene in East Africa between the 1950s and the 1970s were strongly influenced by the development of Bordes' (1961) typology (e.g., Kleindienst 1962; Clark et al. 1966; Leakey 1967). At that time it was necessary to homogenize nomenclatures, as well as to define the characteristics of early industries (VVAA 1967) in order to create a common vocabulary applicable to the various archaeological sites known in those days.

This typological approach gave rise to two works that still stand as some of the most relevant ones on the early African industries; i.e. Leakey's (1971) report on Olduvai and Isaac's (1977) on Olorgesailie. The volume on Beds I and II at Olduvai (Leakey 1971) was — and still is — of paramount importance. Because Olduvai contained sites across the entire stratigraphic column, it was possible to place assemblages on a diachronic sequence and to understand cultural changes following a temporal trend. Moreover, in the 1970s Olduvai was one of few sites in Africa with reliable radiometric dates, providing a temporal dimension for the early archaeological and paleontological record. Influenced by Bordes' typologies, Leakey (1971) identified at Olduvai a wide range of morphotypes, and proposed a unilinear evolution of artifacts related to the emergence of successive cultures.

As suggested by several of his articles, by the mid-1970s Isaac had already changed his conception of the study of lithic industries. Yet his report on the stone tools from Olorgesailie (Isaac 1977) — based on his thesis from the 1960s — was still a typological analysis. In spite of this, in the Olorgesailie report Isaac (1977) underlined the difficulty of applying a typological classification to bifaces in the East African Acheulian, avoided unilinear interpretations based on historical-cultural grounds, and insisted on stochastic processes as an explanation of the observed variability.

The typological approach developed in the seminal works on Olorgesailie (Isaac, 1977) and Olduvai (Leakey, 1971) also guided the analysis of assemblages elsewhere in eastern Africa, such as those excavated at Koobi Fora (Harris 1978), Gadeb (Kurashina 1978; Clark 1979), Melka Kunture (Chavaillon and Chavaillon 1976, 1980; Chavaillon et al. 1979), Kalambo Falls (Clark and Kleindienst, 1974) and others. In studies of these assemblages, as well as in those that reinterpreted the evidence from Olduvai (i.e. Bower 1977; Davis 1980; Stiles 1980), the analysis was focused on the metrics and typologies of artifacts, whose frequencies were eventually compared with types recorded in other sites. Even though from the 1980s onward tendencies in the study of lithics would change their perspective, some recent works still focused on metrical (i.e. Crompton and Gowlett 1993; Gowlett and Crompton 1994; Roe 1994) and typological (Leakey 1994; Clark 2001) analyses, aiming to identify patterns of variability across early African assemblages. However, in the last decades a new analytical framework, based on a processual worldview, has superseded the typological perspective. In the next section we discuss the foundations of this approach.

2.3 The Paleoecological Approach to African Early Technologies

The typological approach permitted a formal characterization of the eastern African Oldowan and Acheulian industries with certain independence from the classifications originally developed in South Africa. However, this typological perspective, useful for building a cultural sequence, did not provide information about site function, human organizational strategies, site formation processes, etc. Aware of such limitations, and through the empirical support provided by his own excavations in Koobi Fora, Isaac began in the 1970s a research program devoted to the construction of frames of reference for understanding human activities in the Plio-Pleistocene. Clearly analyses of lithic assemblages had to play a relevant role in those endeavors, but such studies could not stem from the traditional cultural historical approach of the previous decades. Isaac insisted on the necessity of studying artifacts within the contexts they were recovered in, and on the importance of including stone tools in the general frame of the overall activities carried out by hominins.

> "It is clear that stone artifacts do provide a rich record of cultural transmission patterns ('culture-history') and that their characteristics are also related to economy and ecology. However, Archaeology is not well served by unrealistic attempts to squeeze too much blood from stones alone. We need to concentrate our efforts on situations where the stones are only a part of a diverse record of mutually related traces of behavior and adaptation".
>
> (Isaac 1977b:11)

With this declaration of intent, the processual approach to the study of lithic industries burst into East Africa; lithics became yet another manifestation of environmental adaptations by hominin societies, and therefore could not be understood outside of the paleoecological system in which they were made and used.

Isaac aimed to develop a new descriptive system alternative to the list of types proposed by Leakey (1971), by then used by most researchers in their classifications of eastern African early stone assemblages. As suggested explicitly in some of his last articles (Isaac 1984, 1986), Isaac intended to reduce the variability comprised in Leakey's typology, which in his opinion was an artificial construct rather than the shape of a real tendency. Moreover, Leakey (1971) included subjective functional connotations in the definition of each morphotype, something that Isaac (1984, 1986) was trying to avoid. Starting from the technological genesis of each artifact, Isaac lumped all stone tools in three main categories: flaked, detached and pounded pieces. The philosophy of this classification system was similar to the Analytic Logic System developed at that time by Carbonell et al. (1983). Like the Analytic Logic System, Isaac's (1984, 1986) categorization had the advantage of suppressing subjective functional meanings in the naming of stone tools, and linking the denomination of objects with the stage of the production process from which they derived. However, both systems also share some flaws. In his eagerness of reductionism, Isaac (1984, 1986) simplified too much the variability of stone tools. He thus constrained the comprehension of knapping methods, making too generic a comparison between lithic assemblages.

Be that as it may, Isaac's perspective had changed with respect to his own early works; he was not interested in comparisons between groups of objects anymore, but in integrating artifacts into the general strategies of environmental adaptation.

Experimental studies played a crucial role in this new outlook. In the field of lithic technology it was Toth (1982) who employed this research strategy in his thesis. The philosophy of Toth's work fit well in the newly emerging paleoecological framework. He focused on the study of the different stages of artifact reduction and their implications for the comprehension of landscape management by early humans. His conclusions (Toth 1982, 1985) were significant in that sense; Oldowan knapping methods were simple, but strategies of procurement, transport and management of raw material were rather complex. Besides, whereas Leakey's (1971) typological approach had considered choppers and other core pieces as the desired artifacts, Toth (1982, 1985) insisted on the opposite. He regarded heavy-duty objects as the cores from which the intended artifacts, i.e. flakes, were detached. Toth thus described a paradox already mentioned by Isaac: flakes had been categorized as waste material while they were actually the intended products (Isaac 1984:50).

Isaac's (1984, 1986) and Toth's (1982, 1985) contributions have had paramount influence on the current status of the Oldowan technology. Thus, in the last decades nearly all authors within the American academy (e.g., Potts 1988, 1991; Kimura 1997, 1999, 2002; Isaac et al. 1997; Semaw 1997, 2000; Ludwig 1999; Noll 2000) have adopted Isaac's (1986) lithic categories and/or the flake type classification advocated by Toth (1982) as a model. Moreover, most of these scholars have worked along the principles of Isaac's paleoecological approach, in which the main aim is to understand the place of raw material management within the general frame of landscape activities by humans.

Some differences are glaring, though. While Isaac's earlier historical-cultural comparisons among assemblages were based on typological assumptions, they were made from first hand experience with the collections (Isaac 1967, 1977; Isaac et al. 1976). On the other hand, his later processual approach to the study of stone tools (Isaac 1977b, 1984, 1986) stemmed from particular objectives, related to the relationships between artifacts along the different types of sites and the broad use of landscape by early humans (Isaac 1978, 1981, 1984; Isaac and Marshall 1981). Toth (1982, 1985) grounded his research on a holistic experimental approach, which provided a thorough empirical understanding of stone-making processes. Such primary research of actual collections and/or experimental background is rarely exhibited in some of the recent publications, which have limited the technological study just to the division of assemblages according to Isaac's (1986) categories and to the formal classification of flakes following Toth's (1982) types and eluded more far-reaching conclusions. Isaac's and Toth's contributions have sometimes been applied in a rather mechanic and rigid way, devoid of the theoretical and methodological approaches that had guided these two authors. For example, Toth (1982) developed his methodology for Koobi

Fora, a region with idiosyncratic characteristics, and wherein cortex percentages and their implications for the circulation of artifacts in the landscape were circumscribed to the features displayed by the local river cobbles. However, in recent years many authors — ourselves included — have applied Toth's types uncritically in areas with very different characteristics and raw materials (Kimura 1999, 2002; Ludwig 1999; Noll 2000; de la Torre et al. 2003).

To sum up, one could claim that nowadays there is a certain stagnation of processual methodologies of lithic analysis in eastern African early prehistory. After the initial analytical innovations (Toth 1982; Isaac 1986) it seems that the paleoecological approach to lithic analysis is in need of renovation and of broadening its interests. In recent years, only works with an explicit inter-site application (Potts et al. 1999; Stout et al. 2005) have provided relevant information to the understanding of hominin behavior via the study of lithic industries. The paleoecological perspective, nowadays predominant in eastern African lithic studies, has not given much attention to the methods of stone production and their implicit technical strategies, and from that point of view it shows a deficit in the comprehension of cultural adaptations by early tool-makers.

2.4 The *Chaîne Opératoire* Perspective and the Identification of Technical Aptitudes in the East African Early Stone Age

The processual approach has focused on what Karlin and Julien (1994) called the spatial and economic implications of technical production. The French school, in contrast, has centered on the study of lithic reduction methods from the perspective of the reconstruction of *chaînes opératoires*. In East Africa this approach has been mainly applied by Roche and colleagues in Acheulian and Oldowan assemblages (Delagnes and Roche 2005; Roche and Texier 1991, 1996; Roche et al. 1999; Texier 1995; Texier and Roche 1995). The study of *chaînes opératoires* that Roche and associates have applied to some eastern African assemblages has its roots in the philosophy of technique advocated by Leroi-Gourhan (1964, 1971), which has been conceptually developed in the study of Lower and Middle Paleolithic lithics by Pelegrin (1985, 1990); Geneste (1985, 1991); Pelegrin et al. (1988); Boëda (1991; Boëda et al. 1990), and others.

From this viewpoint, the analysis of knapping systems is understood as the spatial and temporal concatenation of contexts involved in the acquisition, elaboration, utilization and discarding of artifacts. The *chaîne opératoire* perspective is considered as a suitable tool for studying technological adaptations by early humans, and for investigating their degree of cognitive complexity. This techno-psychological approach, as Boëda (1991) named it, assumes that the study of the cognitive setting is an essential aspect embodied in the organization of human activities. The analysis of the techno-psychological background determines the characterization of technical traditions -understanding lithic knapping as a knowledge transmitted within a social collective- and, therefore, obeys patterns of information sanctioned by such collective (Boëda 1991).

This approach assumes that technical activities have three levels of interaction: first among elements of a concrete technical activity, second between technical activities practiced by a particular society and an assemblage (hence making up a technical system), and third between such technical system and other components of a social organization (Pelegrin et al. 1988). The *chaîne opératoire* is conceived as a dynamic, non-linear succession of gestures, i.e. a consecutive series of choices of particular alternatives from a range of parameters related to the geometric characteristics of pieces (Pelegrin 1985). Thus, the main concern is the understanding of technical processes (know -how or *savoir faire*) and the specific knowledge embodied in the analyzed artifacts (Pelegrin 1990, 1991).

In spite of the appeal of this theoretical contribution to the understanding of knapping techniques, its practical application to the Plio-Pleistocene record in Africa has been limited (i.e. Pelegrin 1993; Roche and Texier 1996; de la Torre 2004; de la Torre and Mora 2004; Delagnes and Roche 2005; Roche 2005). Pelegrin (1993) suggested that the ideational know-how (i.e. the knowledge arising from intelligence and memory) was rather rudimentary in the Oldowan, being characterized by the blurred nature of mental images, and by the unequivocal relationships between technical gestures and immediate practical results. There would be a standardization of gestures, with a high variability of forms but not too many technical concepts (Pelegrin 1990).

The description of Lokalalei 2C (Roche et al. 1999; Delagnes and Roche 2005) certainly casts doubts on this point of view, indicating as it does that in at least some Late Pliocene assemblages simple technical formulae were applied via suitable motor skills (Pelegrin and Roche 2000). Lokalalei 2C knappers followed recurrent technical rules and displayed a distinctive technological competence (Delagnes and Roche 2005). Therefore, there would be a relatively developed ideational *savoir faire*, in which the ability of evoking connections of modulated gestures and anticipation of actions is included (Pelegrin and Roche 2000).

This perspective, which highlights a cognitive approach to lithic production methods and aims to reconstruct the *chaînes opératoires* of any given assemblage, underpinned studies of the Plio-Pleistocene from Peninj (de la Torre and Mora 2004) and Nyabusosi (Texier 1995), the Oldowan from West Turkana (Roche et al. 1999; Delagnes and Roche 2005) and Omo (de la Torre 2004), the Acheulian from Isenya (Roche and Texier 1991, 1996; Texier and Roche 1995), and others. In these studies the objective is the understanding of knapping techniques and the reconstruction of reduction sequences. This approach, which borrows concepts from the ethnography (Karlin et al. 1991), is useful for describing the technical knowledge embodied in any knapping scheme, and is therefore suitable for evaluating cognitive skills among early tool-makers. However, most (if not all) of the aforementioned studies are focused on the study of knapping methods but do not deal with the "knowledge of the natural environment" *sensu* Karlin et al. (1991), e.g., the cultural, spatial and economical implications of such technical production (Karlin and Julien 1994). Hence this approach, which we have named here "the study of technical aptitudes," has its own shortcomings; it focuses on a monographic analysis of knapping skills, but lacks a wider perspective integrating such technical methods in the broad range of human activities across the landscape.

Of course not all the cognitive approaches applied to the eastern African early stone tools have been undertaken within the *chaîne opératoire* tradition, and other scholars have attempted to follow alternative cognitive approaches (i.e. Wynn 1979, 1981, 1989, 1993, 2002; Stout 2002, 2005). Wynn's approach is actually distant from the theoretical and methodological frameworks of both the processual and the *chaîne opératoire* tendencies. Using topological geometry and several concepts developed in Piaget's genetic epistemology, Wynn (1981; Wynn and McGrew 1989) asserts that Oldowan tools are distinctive of a preoperational organization comparable to that of chimpanzees, whereas early Acheulian bifaces indicate a more complex topological capacity (Wynn 1989). On the other hand, late Acheulian handaxes would show a spatial organization typical of modern human cognitive skills (Wynn 1979).

Despite its undisputable interest for evaluating cognitive aptitudes, Wynn's approach has also been criticized. As pointed out by Isaac (1986), Wynn focuses on the geometric analysis of flaked pieces, assuming that choppers, heavy-duty tools, etc, are intentionally-made forms. Davidson (2002) also criticized the a priori assumption of artifacts as preconceived morphotypes, whereas Graves (1990) viewed Wynn's approach as purely typological. The main flaw of Wynn's approach derives precisely from his formal study of artifacts, since he focuses on the analysis of a few morphologies without investigating the technological origin of pieces and the relationships among different categories.

2.5 Olduvai, A Case Study on the Different Approaches to the Analysis of Lithic Assemblages

Because of the relevance of the archaeological record unearthed in Olduvai Beds I and II in the 1960s, lithic assemblages from these sites have been restudied over and over again by different scholars with diverse backgrounds and a variety of research interests. Olduvai thus constitutes the perfect case for exemplifying the diversity of theoretical and methodological approaches mentioned in the sections above, and for reviewing the evolution of historiography in recent times.

The publication of the Beds I and II report (Leakey 1971) constituted a detailed, ordered and understandable study that placed lithics along a stratigraphic sequence and proposed a typological evolution of stone tools. Combined with its radiometric chronological control, at the time almost unique in eastern Africa, the sequence of typological development based on the record of Olduvai Beds I and II became a key reference for studying the evolution of early lithic artifacts everywhere. Therefore, subsequent research during the 1970s (i.e. Clark and Kleindienst 1974; Harris and Isaac 1976; Isaac 1976, 1977; Bower 1977; Harris 1978; Kurashina 1978; Chavaillon et al. 1979), but also during the 1980s (Davis 1980; Roche 1980; Stiles 1980; Toth 1982; Gowlett 1986; Potts 1988; Willoughby 1987) and in recent years (Schick and Toth 1993; Jones 1994; Isaac et al. 1997; Kyara 1999; Ludwig 1999; Kimura 1999, 2002; de la Torre and Mora 2004, 2005) has turned to Leakey's typology for supporting or challenging her classification.

Following the then-dominant typological paradigm, researchers who reviewed Olduvai assemblages in the 1970s were not interested in the analysis of entire collections but of particular artifacts such as the choppers (Bower 1977; Dies and Dies 1980; Roche 1980). In some works of the 1980s this typological viewpoint was still present, and particular objects such as polyhedrons, subspheroids, spheroids and others, were analyzed along the Olduvai sequence (e.g., Wynn 1981; Willoughby 1987; Sahnouni 1991). While including a wide range of analytical variables, these studies followed Leakey's original classification, and their conclusions were typologically-oriented.

The paleoecological perspective that Isaac and his associates had been developing for the Koobi Fora research was applied to the Olduvai remains from the 1980s. However, discussions stemming from this approach centered mainly on the faunal materials rather than on stone tools. Notable exceptions are the studies by Kroll and Isaac (1984), who focused on the spatial analysis of the FLK Zinj site through the study of stone refitting.

To the best of our knowledge, the first complete review of lithic collections from entire sites — as opposed to specific types and isolated objects — was the study carried out by Potts (1988). Potts followed Leakey's (1971) original classification but modified some of her type counts for Bed I assemblages. In so doing he adhered to Leakey's typological categories and therefore did not create a technological reconstruction of lithic-related activities at each site. Nonetheless, Potts' study relied on an integrative perception of assemblages, aiming to understanding site formation processes and the paleoecological contexts. He analyzed raw material procurement at each site and reconstructed the input and output dynamics of artifacts across Olduvai Bed I. Even though it was not a technological analysis, Pott's (1988) study benefited from a comprehensive approach that constituted a successful attempt to understand stone tool management by Olduvai Bed I hominins.

Such integrative perspective is lacking in some recent works. Kyara's (1999) study of lithic raw materials in Bed II follows Leakey's original categories. Adopting his conclusions by default requires acceptance of Leakey's morphotypes. Moreover, Kyara focused on a diachronic comparison of typological categories along the sequence, rather than on the synchronic reconstruction of raw material dynamics at each locality. Therefore, we can only gain insights into the changing frequencies of stone tools whose validity as morphotypes can be challenged.

This problem of "diachronic fixation" is evident in other modern reviews. Ludwig's (1999) case is particularly illustrative, as his revision of nearly all Beds I and Bed II sites is devoted to a formal comparison of decontextualized lithic attributes (amount of cortex, number of scars or metric attributes) throughout the sequence. Similarly, Kimura (1997, 1999, 2002) states explicitly her interest in diachronic variations along Beds I and II, and focuses on comparisons of isolated attributes and quantitative tendencies, with little attention to the knapping methods and technical behaviors which would have generated the observed patterns. She is therefore unable to provide descriptions of the technological strategies characterizing each site, which can be conceived as a necessary first step in making inter-site comparisons afterwards.

In recent years, the required synchronic viewpoint at Olduvai has followed two perspectives that should become complementary; theoretical models developed by OLAPP (Blumenschine and Peters 1998) and their application to Oldowan lithics (Tactikos 2005) are grounded on the paleoecological paradigm. On the other hand, our review of some Beds I and II assemblages (de la Torre and Mora 2005) aims to complement the typical processual quantitative approach applied to the Olduvai record with a detailed report of the knapping systems and the *chaînes opératoires*. This re-study has thus related to site formation process, the technical skills embodied in different knapping methods, the relevance of percussion activities, and the validity of the artifact types defined by Leakey (de la Torre and Mora 2005a, b; Mora and de la Torre 2005). This latter issue is particularly relevant with regard to some of the technical and behavioral interpretations of the Olduvai sequence; both second-hand studies - either typological (Stiles 1979, 1980; Davis 1980; Callow 1994; Roe 1994) or processual (Brantingham 1998; McNabb 1998) — and some of the first hand revisions — which add new variables to already preexisting types (Ludwig 1999; Kimura 1999, 2002) — have based on Leakey's (1971) original classification. However, such classification can be entirely challenged, as we have argued elsewhere (de la Torre and Mora 2005).

2.6 Conclusions: Decoding Early Technological Strategies

We have reviewed the diversity of approaches to tackling the study of early technologies in East Africa. While not directly equivalent, the paleoecological approach and the one based on the *chaînes opératoires* could respectively be included in what Perlès (1993) called determinist and individualist perspectives for explaining variability among lithic assemblages. The processual approach presumes that the nature of any lithic industry depends on ecological variables and is conditioned by factors beyond the decisions made by individuals or social groups. In contrast, the "individual's variability" perspective (Perlès 1993) constitutes a monographic study of individual lithic assemblages. In this approach each technical sequence is viewed as the result of successive decisions by the tool-maker through a critical reflection, based on his experience and the technical tradition the knapper socially belongs to (Perlès 1993).

Young and Bonnichsen's (1984) general division of approaches to the study of lithic industries is also helpful for categorizing theoretical and methodological tendencies in eastern Africa. Like Perlès (1993), they suggest a dual classification of types of lithic studies. The "normative approach" (Young and Bonnichsen, 1984) views the group as the basic unit of analysis. Individual phenomena are considered as erratic compared to the group, which contains predictable patterns and structures. The appropriate analytical tools are thus statistical studies and a basically quantitative perspective. Most of recent works (Kimura 1997, 1999, 2002; Semaw 1997, 2000; Ludwig 1999; Noll 2000) fall within this research paradigm. For Young and Bonnichsen (1984), the cognitive approach starts from the opposite view, and assumes that the basic unit of analysis is the individual rather than the group. According to this perspective, individual variation is real and should be taken into account because it provides the internal cultural dynamic that fuels technical change. Therefore, this approach emphasizes the reconstruction of particular decisions made by the tool-maker through the study of every single artifact. The studies of lithic assemblages in Lokalalei (Roche et al. 1999; Delagnes and Roche 2005), Isenya (Roche and Texier 1991, 1996; Texier and Roche 1995) and Nyabusosi (Texier 1995), in which narrative descriptions of knapping methods predominate over quantitative discussion, are clearly aligned with this approach.

However, a sharp dichotomy between the "ecological" and "cognitive" approaches as outlined by Perlès (1993) or Young and Bonnichsen (1984) could become too simplistic. For example, beyond his interests in the ecological adaptations by humans, Isaac (1978, 1982, 1984) was deeply engaged in the reconstruction of social parameters as mechanisms of cultural change. His approach clearly cannot be branded as deterministic. Similarly, the concept of *chaîne opératoire* is not entirely holistic. Edmonds (1990) pointed out that this approach refers to several conceptions of practical knowledge status, which are restricted to a methodological level, since the emphasis is on detailed descriptions. Certainly, such descriptions do not focus on the artifacts by themselves — which would then be considered a typological approach — but on the ensemble of decisions and technical actions involved in their making, rendering it a technological approach. Thus, the *chaîne opératoire* concept enables a referential framework for describing the logic of particular sequences of actions in material, temporal and spatial terms (Edmonds 1990) and is crucial for understanding human cognitive abilities. Edmonds (1990) accurately notices, however, that through the *chaîne opératoire* perspective it is possible to understand how the production was organized, but not the social context where it was generated. He suggests that this problem hampers the understanding of the material and historical contexts in which a lithic assemblage is embodied and narrows theoretical proposals that seek to reconstruct the relationships between human action, social practice and social structure (Edmonds 1990). In addition to these shortcomings, we should include the lack of attention in the perspective of the *chaîne opératoire* to human interaction with the environment, which is crucial for understanding hominin adaptations in the Plio-Pleistocene.

We can therefore summarize virtues and flaws of the two main tendencies that replaced the typological approach in the study of early African industries. The paleoecological approach has the advantage of its holistic perspective, in which research is focused on the landscape and the human adaptations to ecological settings. Its main flaw is the scarce attention — if any — to the particular methods of interaction between the knapper and the raw material block being worked, which hampers such pursued holistic perspective. The *chaîne opératoire* approach, as has been applied in East Africa so far, enables a detailed documentation of reduction methods and thus the evaluation of the technical competence of tool-makers. However, this approach at the moment does not devote attention to the dynamics of interaction between hominin technology and the paleoenvironment.

Is it possible to synthesize both approaches in the archaeology of African Plio-Pleistocene? In theory, this should be perfectly feasible; the thorough technological analysis of industries of each site has been combined with inter-site studies and detailed descriptions of landscapes in other regions and for other time periods by authors from different schools of thought (e.g., Henry 1995; Marks 1977; Geneste 1985), giving rise to splendid reconstructions of hominin technological adaptations in different temporal and geographical ranges. In eastern Africa, where preservation of the archaeological and paleoecological Plio-Pleistocene record is better than in many other regions of the Old World, such an integrative approach would be therefore particularly viable.

Ideally, future works over eastern African early industries should combine the strongest points of each approach. The comprehension of knapping methods and technical know-how is an important objective; stone knapping requires a conceptual scheme, that is, a cognitive frame that includes an abstract plan of integrated actions, which can be modified depending on the variable circumstances along the different reduction sequences (Perlès 1992). In order to accomplish this, it is important to develop analytical descriptions of lithics, aiming to place them into the particular stage of the *chaîne opératoire* they come from. It is fundamental to analyze every lithic element in relation to the rest of objects that make up an assemblage. Any site should be studied individually, since every assemblage is subjected to specific and punctual circumstances. Only after the comprehension of each particular collection via analyzing the relationships between categories that form such assemblage, it is feasible to elaborate conclusions that then can be extrapolated and compared with observations carried out in other sites.

Altogether with this intrasite orientation, it is fundamental that a wider perspective be maintained in order to place any lithic assemblage into the contextual and paleoecological frame it belongs to. To sum up, our goal should be the combination of a systematic study of exploitation methods and the reconstruction of lithic *chaînes opératoires*, with a more general view that integrates such technical systems in the broader frame of activities carried out by hominins across the landscape. We assume that the manufacture of any lithic artifact is the result of a series of technical, economic, social and symbolic options, which can be grouped within the term "strategies" (Perlès 1992). It is towards the study of such technological strategies that we should direct in the study of early stone tools.

Acknowledgments. We thank the editors for inviting us to participate in this volume. The editorial help by E. Hovers has been invaluable, and we thank her, D. Braun and four reviewers for their comments on previous drafts of this paper.

References

Blumenschine, R.J., Peters, C.R., 1998. Archaeological predictions for hominid land use in the paleo-Olduvai Basin, Tanzania, during lowermost Bed II times. Journal of Human Evolution 34, 565–607.

Boëda, E., 1991. Approche de la variabilité des systèmes de production lithique des industries du Paléolithique inférieur et moyen: chronique d´une variabilité attendue. Techniques et Culture 17–18, 37–79.

Boëda, E., Geneste, J.M., Meignen, L., 1990. Identification de chaînes opératoires lithiques du Paléolithique ancien et moyen. Paléo 2, 43–80.

Bordes, F., 1961. Typologie du Paléolithique Ancien et Moyen. Institut de Préhistoire de l´Université de Bordeaux, Bordeaux.

Bower, J.R.F., 1977. Attributes of Oldowan and Lower Acheulean tools: "tradition" and design in the Early Lower Palaeolithic. The South African Archaeological Bulletin 32, 113–126.

Brantingham, P.J., 1998. Mobility, Competition, and Plio-Pleistocene Hominid Foraging Groups. Journal of Archaeological Method and Theory 5, 57–98.

Callow, P., 1994. The Olduvai bifaces: technology and raw materials. In: Leakey, M.D., Roe, D.A. (Eds.), Olduvai Gorge. Volume 5. Excavations in Beds III, IV and the Masek Beds, 1968–1971. Cambridge University Press, Cambridge, pp. 235–253.

Carbonell, E., Guilbaud, M., Mora, R., 1983. Utilización de la lógica analítica para el estudio de tecno-complejos a cantos trabajados. Cahier Noir 1, 1–64.

Clark, J.D., 1979. An analysis of Earlier Stone Age bifaces from Gadeb (Locality 8E), Northern Bale Highlands, Ethiopia. South African Archaeological Bulletin 34, 93–109.

Clark, J D. (Ed.), 2001. Kalambo Falls Prehistoric Site, III: The Earlier Cultures: Middle and Earlier Stone Age. Cambridge University Press, Cambridge.

Clark, J.D., Cole, G.H., Isaac, G.L., Kleindienst, M.R., 1966. Precision and definition in African Archaeology. South African Archaeological Bulletin XXI, 114–121.

Clark, J.D., Kleindienst, M.R., 1974. The Stone Age cultural sequence: terminology, typology, and raw material. In: Clark, J.D. (Ed.), Kalambo Falls Prehistoric Site, Vol 2. Cambridge University Press, Cambridge, pp. 71–106.

Crompton, R.H., Gowlett, J.A.J., 1993. Allometry and multidimensional form in Acheulean bifaces from Kilombe, Kenya. Journal of Human Evolution 25, 175–199.

Chavaillon, J., Chavaillon, N., 1976. Le Paléolithique ancien en Ethiopie. Caractères techniques de l´Oldowayen de Gomboré I a Melka-Konturé. In: Clark, J.D., Isaac, G.L. (Eds.), Colloque V: Les plus anciennes industries en Afrique. UISPP IX Congrés, Nice, pp. 43–69.

Chavaillon, J., Chavaillon, N., 1980. Evolution de l´Acheuléen a Melka-Kunturé (Ethiopie). Anthropologie XVIII, 153–159.

Chavaillon, J., Chavaillon, N., Hours, F., Piperno, M., 1979. From the Oldowan to the Middle Stone Age at Melka-Kunturé (Ethiopia). Understanding Cultural Changes. Quaternaria XXI, 87–114.

Davidson, I., 2002. The Finished Artefact Fallacy: Acheulean Handaxes and Language Origins. In: Wray, A. (Ed.), Transitions to Language. Oxford University Press, Oxford, pp. 180–203.

Davis, D.D., 1980. Further Consideration of the Developed Oldowan at Olduvai Gorge. Current Anthropology 21, 840–843.

Delagnes, A., Roche, H., 2005. Late Pliocene hominid knapping skills: The case of Lokalalei 2C, West Turkana, Kenya. Journal of Human Evolution 48, 435–472.

Dies, K F., Dies, M.B., 1980. Researches on the palaeotechnology of artefacts (Developed Oldowan B and Acheulean) with mathematical statistics. In: Leakey, R. E. F., Ogot, B. A. (Eds.), Proceedings of the VIIIth Panafrican Congress of Prehistory and Quaternary Studies, Nairobi 1977. The International Louis Leakey Memorial Institute for African Prehistory, Nairobi, pp. 41–42.

de la Torre, I. , 2004. Omo Revisited. Evaluating the rechnological skills of Pliocene hominids. Current Anthropology 45, 439–465.

de la Torre, I., Mora, R., 2004. El Olduvayense de la Sección Tipo de Peninj (Lago Natron, Tanzania). CEPAP, vol. 1, Barcelona.

de la Torre, I., Mora, R., 2005a. Technological Strategies in the Lower Pleistocene at Olduvai Beds I & II. ERAUL 112, Liege.

de la Torre, I., Mora, R., 2005b. Unmodified lithic material at Olduvai Bed I: manuports or ecofacts? Journal of Archaeological Science 32, 273–285.

de la Torre, I., Mora, R., Domínguez-Rodrigo, M., Luque, L., Alcalá, L., 2003. The Oldowan industry of Peninj and its bearing on the reconstruction of the technological skills of Lower Pleistocene hominids. Journal of Human Evolution 44, 203–224.

Edmonds, M., 1990. Description, understanding and the chaine operatoire. Archaeological Review from Cambridge 9, 55–70.

Geneste, J.-M., 1985. Analyse lithique d´industries mousteriennes du Périgord: une approche technologique du comportement des groupes humains au Paléolitique Moyen. Unpublished Ph.D., Université de Bordeaux I, Bordeaux.

Geneste, J.-M., 1991. L´approvisionnement en matières premières dans les systemes de production lithique: la dimension spatiale de la technologie. In: Mora, R., Terradas, X., Parpal, A., Plana, C. (Eds.), Tecnología y cadenas operativas líticas. Treballs d´Arqueologia, 1, Universidad Autónoma de Barcelona, Barcelona, pp. 1–36.

Gowlett, J.A.J., 1986. Culture and conceptualisation: the Oldowan-Acheulian gradient. In: Bailey, G.N., Callow, P. (Eds.), Stone Age Prehistory: studies in memory of Charles McBurney. Cambridge University Press, Cambridge, pp. 243–260.

Gowlett, J.A.J., Crompton, R.H., 1994. Kariandusi: Acheulean morphology and the question of allometry. The African Archaeological Review 12, 3–42.

Graves, P., 1990. Sermons in stones: an exploration of Thomas Wynn´s The Evolution of Spatial Competence. Archaeological Review from Cambridge 9, 104–125.

Harris, J.W.K., 1978. The Karari industry: Its Place in the East African prehistory. Unpublished Ph.D. thesis, Berkeley, University of California.

Harris, J.W.K., Isaac, G.L., 1976. The Karari industry: Early Pleistocene archaeological evidence from the terrain east of Lake Turkana, Kenya. Nature 262, 102–107.

Henry, D.O. (Ed.), 1995. Prehistoric Cultural Ecology and Evolution. Insights from the Southern Jordan. Plenum Press, New York.

Isaac, G.L., 1967. The stratigraphy of the Peninj Group- Early Middle Pleistocene Formations West of Lake Natron, Tanzania. In: Bishop, W. W., Clark, J. D. (Eds.), Background to Evolution in Africa. University of Chicago Press, Chicago, pp. 229–257.

Isaac, G.L., 1976. Plio-Pleistocene artifact assemblages from East Rudolf, Kenya. In: Coppens, Y., Howell, F.C., Isaac, G.L., Leakey, R E.F. (Eds.), Early Man and Environments in the Lake Rudolf Basin: Stratigraphy, Paleoecology and Evolution. University of Chicago Press, Chicago, pp. 552–564.

Isaac, G.L., 1977a. Olorgesailie. Archeological Studies of a Middle Pleistocene Lake Basin in Kenya. University of Chicago Press, Chicago.

Isaac, G.L., 1977b. Squeezing blood from stones. In: Wright, R.V.S. (Ed.), Stone Tools as Cultural Markers: Change, Evolution, and Complexity. Humanities Press, New Jersey, pp. 5–12.

Isaac, G.L., 1978. The food-sharing behavior of protohuman hominids. Scientific American 238, 90–108.

Isaac, G.L., 1981. Emergence of human behavior patterns. Archaeological tests of alternative models of early hominid behavior: excavation and experiments. Philosophical Transactions of the Royal Society, London 292, 177–188.

Isaac, G.L., 1982. The earliest archaeological traces. In: Clark, J.D. (Ed.), Cambridge History of Africa. Volume 1. From the Earliest Times to c. 500 BC. Cambridge University Press, Cambridge, pp. 157–247.

Isaac, G.L., 1984. The archaeology of human origins: studies of the Lower Pleistocene in East Africa 1971–1981. In: Wendorf, F., Close, A. E. (Eds.), Advances in World Archaeology. Academic Press, Orlando, pp. 1–87.

Isaac, G.L., 1986. Foundation stones: early artifacts as indicators of activities and abilities. In: Bailey, G.N., Callow, P. (Eds.), Stone Age Prehistory: Studies in Memory of Charles McBurney. Cambridge University Press, Cambridge, pp. 221–241.

Isaac, G.L., Crader, D.C., 1981. To what extent were early hominids carnivorous? An archaeological perspective. In: Harding, R.S.O., Teleki, G. (Eds.), Omnivorous Primates: Gathering and Hunting in Human Evolution. Columbia University Press, New York, pp. 37–103.

Isaac, G.L., Harris, J.W.K., Crader, D., 1976. Archaeological evidence from the Koobi Fora Formation. In: Coppens, Y., Howell, F. C., Isaac, G.L., Leakey, R.E.F. (Eds.), Earliest Man and Environments in the Lake Rudolf Basin. University of Chicago Press, Chicago, pp. 533–551.

Isaac, G.L., Harris, J.W.K., Kroll, E.M., 1997. The Stone artefact assemblages: a comparative study. In: Isaac, G.L. (Ed.), Koobi Fora Research Project. Volume 5: Plio-Pleistocene Archaeology. Oxford University Press, Oxford, pp. 262–362.

Jones, P.R., 1994. Results of experimental work in relation to the stone industries of Olduvai Gorge. In: Leakey, M.D., Roe, D.A. (Eds.), Olduvai Gorge. Volume 5. Excavations in Beds III, IV and the Masek Beds, 1968–1971. Cambridge University Press, Cambridge, pp. 254–298.

Karlin, C., Bodu, P., Pelegrin, J., 1991. Processus techniques et chaînes opératoires; comment les préhistoriens s´approprient un concept élaboré par les ethnologues. In: Balfet, H. (Ed.), Observer l Áction Technique. pp. 101–117.

Karlin, C., Julien, M., 1994. Prehistoric technology: a cognitive science? In: Renfrew, C., Zubrow, E.B.W. (Eds.), The Ancient Mind. Elements of Cognitive Archaeology. Cambridge University Press, Cambridge, pp. 152–164.

Kelly, R.L., 1988. The three sides of a biface. American Antiquity 53, 717–734.

Kimura, Y., 1997. Studies on the Tool-Using and Ranging Patterns of Early Hominids at Olduvai Gorge, Tanzania. University of Wisconsin, Unpublished Ph.D. Madison.

Kimura, Y., 1999. Tool-using strategies by early hominids at Bed II, Olduvai Gorge, Tanzania. Journal of Human Evolution 37, 807–831.

Kimura, Y., 2002. Examining time trends in the Oldowan technology at Beds I and II, Olduvai Gorge. Journal of Human Evolution 43, 291–321.

Kleindienst, M.R., 1962. Component of the East African Acheulian assemblage: an analytic approach. In: Mortelmans, G., Nenquin, J. (Eds.), Actes du IV Congrès Panafricain de Préhistoire et de l´Etude du Quaternaire, Leopoldville, 1959. Belgie Annalen, Musée Royal de l´Afrique Centrale, Tervuren, pp. 81–108.

Kroll, E.M., Isaac, G.L., 1984. Configurations of artifacts and bones at early Pleistocene sites in East Africa. In: Hietala, H.J. (Ed.), Intrasite Spatial Analysis in Archaeology. Cambridge University Press, Cambridge, pp. 4–31.

Kurashina, H., 1978. An Examination of Prehistoric Lithic Technology in East-Central Ethiopia. Unpublished Ph.D., University of California, Berkeley.

Kyara, O.A., 1999. Lithic Raw Materials and Their Implications on Assemblage Variation and Hominid Behavior During Bed II, Olduvai Gorge, Tanzania. Unpublished Ph.D. University of Rutgers, New Brunswick.

Leakey, M.D. 1967. Preliminary survey of the cultural material from Beds I and II, Olduvai Gorge, Tanzania. In Bishop, W.W., Clark, J.D. (Eds.), Background to Evolution in Africa. University of Chicago Press, Chicago, pp. 417–446.

Leakey, M.D., 1971. Olduvai Gorge. Vol 3. Excavations in Beds I and II, 1960–1963. Cambridge University Press, Cambridge.

Leakey, M.D., Roe, D.A., 1994. Olduvai Gorge. Volume 5. Excavations in Beds III, IV and the Masek Beds, 1968–1971. Cambridge University Press, Cambridge.

Lemonnier, P., 1990. Topsy-turvy techniques. Remarks on the social representation of techniques. Archaeological Review from Cambridge 9, 27–37.

Leroi-Gourhan, A., 1964. Le geste et la parole. I- technique et language. II- La mémoire et les rythmes. Albin Michel, Paris.

Leroi-Gourhan, A., 1971. L´homme et la matière. Albin Michel, Paris.

Ludwig, B.V., 1999. A Technological Reassessment of East African Plio-Pleistocene Lithic Artifact Assemblages. Unpublished Ph.D., University of Rutgers, New Brunswick.

Marks, A.E., (ed.), 1977. Prehistory and Paleoenvironments in the Central Negev, Israel. Southern Methodist University, Dallas.

McNabb, J., 1998. On the move. Theory, time averaging and resource transport at Olduvai Gorge. In: Ashton, N., Healy, F., Pettit P. (Eds.), Stone Age Archaeology. Essays in Honour of John Wymer. Oxbow Monograph 102, Oxford, pp. 15–22.

Mora, R., Torre, I. de la, 2005. Percussion tools in Olduvai Beds I and II (Tanzania): Implications for early human activities. Journal of Anthropological Archaeology 24, 179–192.

Nelson, M.C., 1991. The Study of Technological Organization. In: Schiffer, M. B. (Ed.), Archaeological Method and Theory, Vol. 3n°1. University of Arizona Press, Tucson, pp. 57–100.

Noll, M.P., 2000. Components of Acheulean lithic assemblage variability at Olorgesailie, Kenya. Unpublished Ph.D., University of Illinois at Urbana-Champaign,

Pelegrin, J., 1985. Réflexions sur le comportement technique. In: Otte, M. (Ed.), La Signification Culturelle des Industries Lithiques. Actes du Colloque de Liège du 3 an 7 octobre 1984. BAR International Series 239, Oxford, pp. 72–91.

Pelegrin, J., 1990. Prehistoric lithic technology: some aspects of research. Archaeological Review from Cambridge 9, 116–125

Pelegrin, J., 1991. Les savoir-faires: une très longue histoire. Terrain 16, 106–113.

Pelegrin, J., 1993. A framework for analysing prehistoric stone tool manufacture and a tentative application to some early stone industries. In: Berthelet, A., Chavaillon, J. (Eds.), The Use of Tools by Human and Non-Human Primates. Clarendon Press, Oxford, pp. 302–317.

Pelegrin, J., Karlin, K., Bodu, P., 1988. "Chaînes opératoires": un outil pour le préhistorien. Technologie Préhistorique. Notes et Monographies Techniques 25, 55–62.

Pelegrin, J., Roche, H., 2000. Evolution et cognition: le temoignage des pierres taillées. In: Journée Scientifique de l´Association pour

la Recherche Cognitive, Ecole Nationale Supérieure des Télécommunications, Paris- 8 décembre 2000. Paris, electronic resource.
Perlès, C., 1992. In search of lithic strategies. A cognitive approach to prehistoric chipped stone assemblages. In: Gardin, J. C., Peebles, C. S. (Eds.), Representations in Archaeology. Indiana University Press, Bloomington, pp. 223–247.
Perlès, C., 1993. Ecological determinism, group strategies, and individual decisions in the conception of prehistoric stone assemblages. In: Berthelet, A., Chavaillon, J. (Eds.), The Use of Tools by Human and Non-humans Primates. Clarendon Press, Oxford, pp. 267–277.
Potts, R., 1988. Early Hominid Activities at Olduvai. Aldine and Gruyter, New York.
Potts, R., 1991. Why the Oldowan? Plio-Pleistocene toolmaking and the transport of resources. Journal of Anthropological Research 47, 153–176.
Potts, R., Behrensmeyer, A.K., Ditchfield, P., 1999. Paleolandscape variation and Early Pleistocene hominid activities: Members 1 and 7, Olorgesailie Formation, Kenya. Journal of Human Evolution 37, 747–788.
Ricklis, R.A., Cox, K.A., 1993. Examining lithic technological organization as a dynamic cultural subsystem: the advantages of an explicitly spatial approach. American Antiquity 58, 444–461.
Roche, H., 1980. Prémiers outils taillés d´Afrique. Societé d´ethnographie, Paris.
Roche, H., 2005. From simple flaking to shaping: stone-knapping evolution among early hominins. In: Roux, V., Bril, B. (Eds.), Stone knapping. The necessary Conditions for a Uniquely Hominin Behaviour. McDonald Institute Monographs, Cambridge, pp. 35–48.
Roche, H., Delagnes, A., Brugal, J.-P., Feibel, C., Kibunjia, M., Mourre, V., Texier, P.-J., 1999. Early hominid stone tool production and technical skill 2.34 Myr ago in West Turkana, Kenya. Nature 399, 57–60.
Roche, H., Texier, P.-J., 1991. La notion de complexité dans un ensemble lithique. Application aux séries acheuléenes d´Isenya (Kenya). In: 25 ans d´études technologiques en Préhistoire. APDCA, Juan-les-Pins, pp. 99–108.
Roche, H., Texier, P.-J., 1996. Evaluation of technical competence of *Homo erectus* in East Africa during the Middle Pleistocene. In: Bower, J.R.F., Sartorno, S. (Eds.), Human Evolution in its Ecological Context. Royal Netherlands Academy of Arts and Sciences, Leiden, pp. 153–167.
Roe, D.A., 1994. A metrical analysis of selected sets of handaxes and cleavers from Olduvai Gorge. In: Leakey, M.D., Roe, D.A. (Eds.), Olduvai Gorge. Volume 5. Excavations in Beds III, IV and the Masek Beds, 1968–1971. Cambridge University Press, Cambridge, pp. 146–234.
Sahnouni, M., 1991. Étude comparative des galets taillés polyédriques, subsphériques et sphériques des gisements d´Ain Hanech (Algérie Orientale) et d´Olduvai (Tanzanie). L´Anthropologie 97, 51–68.
Schick, K.D., Toth, N., 1994. Early Stone Age technology in Africa: a review and case study into the nature and function of spheroids and subspheroids. In: Corruchini, R.S., Ciochon, R.L. (Eds.), Integrative Paths to the Past. Paleoanthropological Advances in Honor of F. Clark Howell. Prentice Hall, New Jersey, pp. 429–449.
Semaw, S., 1997. Late Pliocene Archaeology of the Gona River Deposits, Afar, Ethiopia. Unpublished Ph.D., Rutgers University, New Jersey.
Semaw, S., 2000. The world´s oldest stone artefacts from Gona, Ethiopia: their implications for understanding stone technology and patterns of human evolution between 2.6-1.5 million years ago. Journal of Archaeological Science 27, 1197–1214.
Stiles, D., 1979. Early Acheulean and Developed Oldowan. Current Anthropology 20, 126–129.
Stiles, D., 1980. Industrial taxonomy in the Early Stone Age of Africa. Anthropologie XVIII, 189–207.
Stout, D., 2002. Skill and cognition in stone tool production: an ethnographic casestudy from Irian Jaya. Current Anthropology 43, 693–722.
Stout, D., 2005. Neural foundations of perception and action in stone knapping. In: Roux, V., Bril, B. (Eds.), Stone Knapping. The Necessary Conditions for a Uniquely Hominin Behaviour. McDonald Institute Monographs, Cambridge, pp. 273–286.
Stout, D., Quade, J., Semaw, S., Rogers, M.J., Levin, N.E., 2005. Raw material selectivity of the earliest stone toolmakers at Gona, Afar, Ethiopia. Journal of Human Evolution 48, 365–380.
Tactikos, J.C., 2005. A Landscape Perspective on the Oldowan from Olduvai Gorge, Tanzania. Unpublished Ph.D., Rutgers University.
Texier, P.-J., 1995. The Oldowan assemblage from NY 18 site at Nyabusosi (Toro-Uganda). Comptes Rendus de l'Academie des Sciences Paris 320, 647–653.
Texier, P.-J., Roche, H., 1995. The impact of predetermination on the development of some Acheulean chaînes opératoires. In: Bermúdez de Castro, J.M., Arsuaga, J.L., Carbonell, E. (Eds.), Evolución Humana en Europa y los Yacimientos de la Sierra de Atapuerca, Vol 2. Junta de Castilla y León, Valladolid, pp. 403–420.
Toth, N., 1982. The Stone Technologies of Early Hominids at Koobi Fora, Kenya; An Experimental Approach. Unpublished Ph. D., University of California, Berkeley.
Toth, N., 1985. The Oldowan reassessed: a close look at early stone artifacts. Journal of Archaeological Science 12, 101–120.
VVAA, 1967. Discussions on terminology. In: Bishop, W.W., Clark, J.D. (Eds.), Background to Evolution in Africa. University of Chicago Press, Chicago, pp. 861–875.
Willoughby, P.R., 1987. Spheroids and Battered Stones in the African Early Stone Age. BAR International Series 321. Oxford.
Wynn, T., 1979. The intelligence of later Acheulean hominids. Man 14, 371–391.
Wynn, T., 1981. The intelligence of Oldowan hominids. Journal of Human Evolution 10, 529–541.
Wynn, T., 1989. The Evolution of Spatial Competence. University of Illinois Press, Illinois.
Wynn, T., 1993. Two developments in the mind of early *Homo*. Journal of Anthropological Archaeology 12, 299–322.
Wynn, T., 2002. Archaeology and cognitive evolution. Behavioral and Brain Sciences 25, 389–438.
Wynn, T., McGrew, W.C., 1989. An ape's view of the Oldowan. Man 24, 383–398.
Young, D.E., Bonnichsen, R., 1984. Understanding Stone Tools: A Cognitive Approach. Center for the Study of Early Man, Peopling of the Americans series, Vol. 1, University of Maine at Orono, 2.

3. From Homogeneity to Multiplicity: A New Approach to the Study of Archaic Stone Tools

Eudald Carbonell*, Robert Sala
Institut Català de Paleoecologia Humana i Evolució Social
Àrea de Prehistòria Universitat Rovira i Virgili
Plaça Imperial Tarraco Tarragona
1 43005 Spain
eudald.carbonell@urv.net, robert.sala@urv.cat

Deborah Barsky , Vincenzo Celiberti
Centre Européen de Recherches Préhistoriques de Tautavel
Ave. Léon Jean-Grégory Tautavel
66720 Tautavel
dbarsky@hotmail.fr, vceliberti@hotmail.com

Keywords: Mode 1, Oldowan, variability, diversity, multiplicity, human dispersal

Abstract

Evidence from sites both in Africa and in Eurasia points towards an ever expanding archaeological record for archaic industries attributed largely to the Oldowan tradition. We have therefore found it necessary to re-examine data from some of these sites and to propose a new framework for the evolution of early technology to better express its variability. The variability observed in so-called Oldowan or Mode 1 lithic assemblages is described and analyzed in a three phase evolutionary sequence: variability, diversity and multiplicity. Finally, these three phases are described for the African and European archaeological record until the appearance of Mode 2.

3.1 Recent Discoveries of Early Technology Give Rise to a New Dilemma

The study of the rich lithic assemblages from Olduvai Gorge (Leakey 1971) laid the foundation for further studies of archaic stone-tools. Localities such as FLK and DK furnished, for the first time, lithic tools in association with hominin remains (*Paranthropus boisei, Homo habilis),* thus raising the important question, unanswered to this day, as to the authorship of these ancient artifacts. The early date (currently known to be around 1.8 Ma; Tamrat et al. 1995), and primitive aspect of the Olduvai artifacts led Mary Leakey to fully describe a new prehistoric culture, the "Oldowan". Following this work, other sites with lithic artifacts dating to the same period or older, such as some localities of the Lake Turkana basin, 1.9 Ma old, i.e. localities FxJj1, FxJj3, FxJj10 in Koobi Fora (Isaac 1973, 1975, 1984, 1997) or localities 71, 57 and 123 in Omo, 2.3 and 2.4 Ma old (Chavaillon 1970, 1975, 1976; Roche and Tiercelin 1977; Howell et al. 1987), were assigned to this cultural group. This cultural complex was called "Mode 1" by G. Clark (Clark 1977) in his original revision of archaic stone-tools in Africa. Since then several artifact-bearing sites in East Africa, older than 1.8 Ma, have been excavated (Kimbel et al. 1996; Semaw et al. 1997, 2003; Roche et al. 1999; Semaw 2000).

The discovery of the Gona assemblages at localities EG10, EG12, OGS6 and OGS7 (Semaw et al. 1997, 2003) pushed back the date for the first human technologies to 2.6 Ma, much older than previously believed, once again underlining the question as to the identity of the hominin(s) who made

* Address for correspondence: eudald.carbonell@urv.net

E. Hovers and D.R. Braun (eds.), *Interdisciplinary Approaches to the Oldowan,* DOI: 10.1007/978-1-4020-9059-2,

these tools. Given such new data, the "Oldowan" or "Mode 1" complex takes on an even broader perspective, both chronologically and technologically. As a result, some authors (Piperno 1989; Kibunjia et al. 1992; Kibunjia 1994; Roche 1996; Roche et al. 1999; de Lumley et al. 2004, 2005) suggested the term "Pre-Oldowan" or "Archaic Oldowan" for assemblages predating 2.0 Ma, thus subdividing the Mode 1 complex. Other authors (Semaw 2000) stress the homogeneity of the industrial record between 2.6 and 1.8 Ma. Either way, the study of these recently discovered assemblages has revealed the complexity within the Mode 1 complex. It therefore appears necessary to define and synthesize variations observed within this large complex in order to better understand the origins and evolution of early stone tool technology, and perhaps even to discover a true "Mode 0".

3.2 A Hypothesis for the First Phases of Tool Evolution

Apart from questions about the identity of the authors of these early industries, several other questions arise from the new discoveries.

1. The variability now recognized *within* Mode 1 assemblages, even earlier the 2.0 Ma border (Roche 1996), leads us to hypothesize the existence of an earlier phase, yet unknown, a true "Mode 0".
2. What might such "Mode 0" industries be like? Currently we have no assemblages earlier than 2.6 Ma. A possible flake in Member C of the Shungura Formation dates 2.8 Ma old (Howell et al. 1987), an isolated occurrence that is anything but firm evidence. Nothing can inform us of the technical system to which it belongs. Although de la Torre (2004) expressed doubts about the validity of the archaeological record of the Shungura E member, this find opens the possibility for further research in these sediments or others of similar ages.
3. Overall, it appears that there may have been a long period during which tool use evolved, followed by a period of tool-making which would have begun very early on in the evolution of prehuman behavior. The capacity for tool use among chimpanzees, widely described (Boesch and Boesch 1983; McGrew 1992; Boesch et al. 1994; Boesch-Achermann and Boesch 1994; Boesch and Tomasello 1998; Whiten et al. 1999, 2005; Panger et al. 2002), may be considered as an example of what ancestral hominin behavior might have been like before a hypothetical Mode 0 phase. Finally, we should consider the possibility that *Homo* was not the only tool-making genus.

Presently, early stone tool technology represented at sites older than 2.0 Ma is not homogeneous. Such a state of homogeneity must be reserved for the hypothetical Mode 0. To the contrary, Gona and other known archaic assemblages represent an initial phase of variability and differentiation. The subsequent evolution of lithic tools may then be seen as the constant loss of homogeneity of the technological components. Even if the very earliest, homogeneous lithic strategies remain undiscovered at present, it may be hypothesized that they have to be the products of lineal sequential reduction of rocks and that their structure shows no variability, diversity or multiplicity. Although such conditions seem stable, the evolutionary morphological potential of such industries could have been sufficient to produce change and variability due to environmental pressures or the necessity for adaptation to new geographical regions. Such external conditions could trigger the latent evolutionary potential for change, which might then be expressed in arising new technological tendencies. Once initiated, such tendencies would be further developed over time, leading to a constant loss of homogeneity and to a proportional gain of entropy within technological systems. Ultimately, the loss of homogeneity of the sequential technological structure is observed through the emergence of hierarchical episodes leading to diversity. In this scheme, the Gona assemblages are closest to a primeval homogeneous phase, and, perhaps, the first to show initial variability.

3.3 A Complex Evolutionary Framework

The discovery of 2.6 Ma old lithic assemblages at Gona raises questions about who the first tool-makers might have been, given that the earliest known occurrence of the genus *Homo* in the fossil record is only about 2.33 Ma old (Kimbel et al. 1996; Prat et al. 2005). This problem continues to be a central issue for our proposal of an initial, as yet undiscovered phase older than the one represented by the Gona assemblages. Using knowledge from primatology and ethology, it may be possible to solve this apparent inconsistency, as has been outlined by other authors (Panger et al. 2002). Occasional or even systematic use of wood or lithic implements is well documented among some birds and many mammals, apes being the group of special interest in this context. Capacity for tool use by gorillas, orangutans, macaques and capuchin monkeys has been demonstrated (McGrew 1992; van Schaik and Pradhan 2003; Fragaszy et al. 2004). However it is among chimpanzees and bonobos, both captive individuals and free ranging groups, that this behavior is particularly systematic and complex (Boesch and Boesch 1983; McGrew 1992; Boesch et al. 1994; Boesch-Achermann and Boesch 1994; Boesch and Tomasello 1998; Whiten et al. 1999, 2005; van Schaik and Pradhan 2003). Chimpanzees and bonobos currently use a variety of materials as tools in order to access certain foods otherwise not available to them. Beyond the classical ethological research on the use of vegetal material for accessing ants and of stone blocks to crack nuts, two recent works describe such behavior and its resultant remains. The first concerns an excavation of a chimp "archaeological site". At this site lithic waste, accidentally broken off blocks when used by chimpanzees as hammers and/or anvils was discovered (Mercader et al. 2002, 2007). The second concerns research about bonobo language and communication skills (Savage-Rumbaugh et al. 1998) and their capacity for tool production in a controlled experiment (Toth et al. 1993).

In spite of chimpanzees' and bonobos' capacity for using objects, they have never developed long and complex operational chains of production nor have they modified the properties of the materials they use in order to gain new capacities, e.g. the capacity for cutting. To the contrary, chimpanzees choose objects according to their actual capacities and properties, for example the capacity for cracking a cobble, which does not require any modification in order to be accomplished. The sharp edges observed in the earliest known hominin lithic tools show that hominin capacities far surpass those of chimpanzees and bonobos. It should be stressed that an evolutionary bifurcation exists between the capacity for using objects and that for their production. Such discontinuity could have emerged as two-phase process. Hominins may have inherited chimpanzee-like behavior from a common ancestor. The evolution of this inherited capacity enabled *Australopithecus* and *Paranthropus* to achieve an initial transformation, non-exclusively targeted towards use of lithic material. This phase would have led to a second one, a derived behavior typical of hominins: tool *production* that goes beyond simple use. Both behavioral transformations have, however, the same goal: accessing new food resources.

The example possibly closest in time to this first use of cutting tools among hominins may be the evidence from Bouri, where gazelle bones bearing cut marks were found in the same stratigraphic level as *Australopithecus garhi* fossils, albeit not in direct association with it (Asfaw et al. 1999; de Heinzelin et al. 1999). The tools themselves were not found.

In conclusion, systematic tool use, especially among chimpanzees, allows us to propose a possible evolutionary scheme whereby a bifurcation took place from an initial shared phase of lithic tool use (shared by chimpanzees and hominins) which then developed towards a second phase of lithic tool production only among the genera *Australopithecus* and *Paranthropus*. These two archaic phases in the evolution of technological capacities could be placed temporally between 3.0 and 2.6 Ma. This model, which can be empirically tested pending appropriate assemblages, provides an evolutionary scenario for hominin tool production, explaining similarities between chimpanzee, bonobo and human tool use behavior. More specifically, it explains the initial variability recorded at Gona by asserting the existence of a previous evolutionary stage. According to our model this first phase must be homogeneous.

3.4 The Evolutionary Development of the First Tool Strategies

We present a model for the evolution of tool production beginning with the industries represented by the Gona material until the appearance of the strategies characterizing the Acheulian (Mode 2). The model establishes phases of development of lithic technology within Mode 1 and offers the landmarks that could characterize each of these phases.

We envision that, once the basal homogeneity (our hypothetical "Mode 0") has been overcome, three phases of development for these first production strategies could appear. The first one is a phase of variability, displayed by Gona; a second one is a phase of diversity, and a final one of multiplicity, corresponding to the appearance of Mode 2 in East Africa. The Mode 1 variability and potential for diversity were thus the base for later evolution of tool-making. This variability had increased by the introduction of new adaptations and improvements on tool production.

In this study we define variability as the existence of more than one method of acquiring flakes. The claim for variability in the Gona assemblages is based on the existence of two relatively developed technical systems, both unifacial: one orthogonal and another discoidal. These two methods were selected due to ecological constraints.

For us the appearance of retouched tools and the beginning of standardization indicates both a major complexity of the operational chain as well as a new level of preplanning. The appearance of retouched tools corresponds to a new degree in the evolutionary model, which we call diversity: the existence of two main systems, each with its own variability, within the general conception of Mode 1. When the development of technical systems goes further and includes large flake economy and the better conception of standardized tools it is understood as the generation of multiplicity, with the coexistence of the two Modes, Mode 1 and 2.

3.4.1 The Evolution in Africa

The lithic sample from sites EG10 and EG12 in the Afar region of Ethiopia (Semaw et al. 1997; Semaw 2000, 2005) may serve as a model to outline the characteristics of early stone tool morphology. These assemblages, knapped mainly from locally available trachyte cobbles, are composed mostly of detached pieces (flakes, flake fragments and debris), that attain frequencies of 97.81% and 97.97%, respectively, in EG10 and EG12; and flaked pieces (cores or chopper/cores), occurring at frequencies of 2.19% and 2.3% at EG10 and EG12, respectively (Semaw 2000). Flakes and cores are relatively small (average whole flake length in both assemblages is close to 35 mm; Semaw 2000) and show no standardization in their morphology. However, the technology used by the

Gona hominins, mainly unidirectional and recurrent flaking (that is, repeatedly from the same edge of the core) was systematic and well-mastered. Because of the dominant use of these unifacial, unidirectional flaking methods, these early assemblages *seem* homogeneous while in fact they are quite variable. For us, as for Semaw (2000), the application of unidirectional recurrent flaking resulted at Gona in specific core morphologies that varied according to initial cobble shape. Core morphology and dimensions relative to the cobble natural characteristics seem to indicate that core reduction at Gona was performed through relatively long (but not exhaustive) sequences. Unifacial peripheral recurrent removals sometimes led to the production of discoidal cores. However, it is important to note that the resulting sinuous core edge was rarely subsequently exploited and appear to have been manufactured without any intention of exploiting it bifacially. This shape of the core edge could be either related to platform rearrangement or to final removals from a second surface, as has been proposed for Lokalalei 2C (Delagnes and Roche 2005).

Occasional exploitation of negatives of previous removals as striking platforms sometimes resulted in orthogonal type cores. Although the edge thus formed could have served as a guide for bifacial discoidal type flaking, this technological potential was not exploited. Systematic use of such platforms with frequent core rotation, leading to multidirectional flaking, was rarely used at Gona. The scarcity of the latter technique, which tends to produce globular or polyhedron cores, explains the lack of subspheroids or spheroids in these assemblages.

These technical systems do not show homogeneity. Rather they exhibit structural variability. In the case of Gona, we observe that unidirectional flaking is nascent to a number of other technological sequences (*chaînes opératoires*), such as orthogonal and even discoidal flaking, which are always unifacial. This variability does not include either the development of the bifacial knapping edge or the retouch of flakes, which will be introduced in the next evolutionary phase.This variability is also characteristic at Lokalalei 2C (2.34 Ma Nachukui Formation, West Turkana; Roche et al. 1999; Roche, 2000; Delagnes and Roche 2005). Here unidirectional recurrent reduction sequences on basalt and phonolite cobbles were extended, leading to the use of several striking platforms on a single core. Comparatively, at Lokalalei 1, only one kilometer away and located in the same stratigraphic position, lava cobbles were apparently less skillfully knapped. Differences between these two assemblages are most often attributed to the relatively poor quality of raw materials exploited at Lokalalei 1 (Kibunjia et al. 1992; Kibunjia 1994; Roche et al. 1999; see Harmand 2009). Other interpretive hypotheses such as ecological or activity differences are difficult to formulate due to lack of significant faunal remains at both sites. We emphasize however that both assemblages demonstrate mainly unidirectional systems with some degree of variability, as has been shown by Delagnes and Roche (2005) in the unidirectional and parallel flaking of two or more opposed and/or orthogonal edges, resulting in cores with unipolar, bipolar or orthogonal scar patterns. Cobble shape and length of operational schemas influenced the resultant core morphology. At Lokalalei 2C, hominins exhausted successive striking platforms, perhaps in order to maximize the use of fine quality raw material.

Several sites in the Omo Basin have yielded lithics dated between 2.4 and 2.3 Ma (localities Omo 57, Omo 71, Omo 123, FtJi1, FtJi2, FtJi5 in Members F and E of the Shungura Formation, Lake Turkana Basin, Ethiopia; Howell et al. 1987). These industries, comprising cores, flakes and debris knapped mainly from quartz pebbles, show a strong frequency of hammer on anvil flaking technique, well adapted to the reduction of small rounded cobbles of crystalline rocks. At these sites only unidirectional flaking was practiced. Once again, products were relatively small sized and flakes were not retouched into tools.

Early lithic technology after 2.0 Ma is well illustrated by the industry from the Fejej FJ-1a site in Ethiopia, dated 1.96 Ma (Bahain et al. 2000; de Lumley et al. 2004; Barsky et al. 2006), where numerous refits of flakes onto cores demonstrate clearly that hominins mastered several organized debitage schemes. The site, located on a residual fluvial sand ridge in a floodplain of the Turkana Lake Basin, only 10 km north of the border with Kenya, is rich in well preserved faunal and lithic artifacts. Raw materials used for producing tools, mainly quartz and basalt pebbles, were collected from nearby streams. Lithic and faunal refits demonstrate that tool manufacture and meat-processing took place at the site. Quartz was the preferred raw material for the different knapping procedures, but basalt cores show longer knapping sequences. The assemblage comprises mainly small and nonstandardized flakes (average flake length = 37 mm) and cores as well as some cobbles which show worked edges (mainly choppers or "chopper-cores"). We note that among the pebble tools, very few edges were bifacially worked (only 2.4% are "chopping-tools"). Bifacial or multifacial technology is also very rare at this site where nearly 60% of the cores were exploited using unifacial knapping methods. Retouched pieces are very rare (0.4% of the assemblage) and show no standardization. As in the older assemblages cited above, the Fejej industry demonstrates that hominins had not attained the technological threshold of systematic, bifacial manufacture of a worked edge. This characteristic appears to come later as the landmark of a new phase, which we call *diversity*.

Assemblages from Fejej FJ-1a show the same tendency observed in Gona. Unidirectional flaking is dominant in all its forms: linear recurrent, recurrent on an anvil, orthogonal and discoidal, showing its characteristic variability. Bidirectional and bifacial knapping were also applied, sometimes on an anvil. When practiced on an anvil, unidirectional flaking resembles a sort of shearing action, the energy of the blow traveling directly into the volume of the cobble. Resulting cores are truncated cobbles, one extremity of which shows negatives of anterior removals that may seem unidirectional or bidirectional, depending on the presence of *contre-coups*. A cobble that has been partially knapped by unidirectional flaking may be confounded with a 'chopper' but as in other archaic assemblages, it seems that flakes were the ultimate goal of the knappers as opposed to a manufactured cobble edge (pebble tool).

Certain elements of unidirectional flaking just barely suggested in the Gona assemblages find a stronger expression in the FJ-1a assemblage. For example, orthogonal flaking is relatively frequent and the presence of some globular and polyhedron cores (11% of the cores) bears witness to the development of the polyhedral system into multidirectional flaking. Note that spheroids are not yet present in this assemblage. Knapping sequences are relatively short and most elements maintain a more or less developed cortical cover. Intentionally retouched flakes or debris are not present.

Compared with earlier assemblages, occasional use of multidirectional flaking resulting in polyhedron shaped cores at FJ-1a appears thus to be an important technological development signaling the beginning of diversity.

The Fejej FJ-1a assemblage appears to be analogous to other industries of similar age discovered in several localities within the KBS Member at Koobi Fora, Kenya (FxJj1, FxJj3 and FxJj10, at ca. 1.9 Ma; Isaac 1997). Nevertheless, in the Oldowan assemblages of DK, FLK *Zinjanthropus* Floor and FLK NN, Levels 1, 2, 3 from Lower and Middle Bed I of Olduvai Gorge (around 1.8 Ma; Leakey 1971), at Kokiselei 5 (West Turkana, younger than 1.8 Ma; Texier et. al. 2006) and at Ain Hanech in Algeria, at about 1.8 Ma age (Sahnouni 1998; Sahnouni et al. 2002) development of the multidirectional flaking leads to the appearance of relatively numerous subspheroids or spheroids and/or polyhedron cores. At these sites bifacial flaking seems to be extensively practiced for the first time, resulting in "chopping-tools" and a few bifacial discoids. These flaking methods contributed from then on to the diversity of technical systems. We also observe here the beginnings of flake or debris edge configuration (small retouched tools), scarce in the older assemblages. All these sites correspond to the Mode 1 tradition but it is clear that compared to Gona, Lokalalei and Fejej they belong to a new phase, characterized not by variability but by diversity. This diversity focused on bifacial flaking that led to the development of edges as a striking platform, dividing the core into two surfaces. This structure was to further evolve into the technique of large flake production characteristic of Mode 2.

3.4.2 The Human Dispersal into Europe

This model of emerging entropy has been presented for the earliest lithic complexes in Africa but it may also be applied to the development of technology in Europe. Recent discoveries at sites such as Dmanisi (Gabunia et al. 2000; de Lumley et al. 2002, 2005), Atapuerca (Carbonell et al. 1995, 1999, 2001, 2005; Bermúdez de Castro et al. 1997; Parés et al. 2006; Carbonell et al. 2008), Orce (Oms et al. 2000a, b; Toro et al. 2003), Ceprano (Manzi et al. 2001) or Pirro Nord (Arzarello et al. 2006) have pushed the dates of human occupation in Europe back to at least 1.4 Ma and that of Eurasia back to 1.8 Ma. All these sites appear to show lithic artifacts belonging to the Mode 1 complex long before the arrival of the Acheulian in Eurasia and, in some cases, long after the rise of the Acheulian in Africa. Thus the development of lithic technology in Eurasia seems to repeat the general scheme of evolution seen in Africa. Although we have evidence of the Acheulian culture at Ubeidiya 1.4 Ma old (Bar-Yosef and Goren-Inbar 1993) there are no other Eurasian sites of similar age with Mode 2 lithic tools.

The origin and evolution of tool-making are not the only issues raised by recent discoveries in Eurasia. Sites such as Atapuerca and Dmanisi, with the archaic anatomies of their *Homo* fossils and accompanying Mode 1 technology, as well as the anatomy of *Homo floresiensis* (Brown et al. 2004; Morwood et al. 2004, 2005; Argue et al. 2006) suggest a much earlier dispersal than previously thought. This proposition is not consistent with previous views of the capacities of early hominin taxa and may explain the coexistence of an African Mode 2 with Eurasian Mode 1 technology after the first appearance of the Acheulian in Africa around 1.6 Ma (Asfaw et al. 1992). This leads us to propose that early hominins had behavioral and anatomical capacities to adapt to new situations with their Mode 1 technology. It is to be highlighted that Mode 1 was a more powerful technology and a more powerful adaptive tool than previously thought, the diversity of which enabled hominins to adapt to different ecological and geographical regions.

At Dmanisi in Georgia an archaic stone industry in association with a rich faunal assemblage and numerous remains of *Homo georgicus* (de Lumley 2006) bear witness to the presence of hominins outside of Africa at an early date. The Paleolithic levels discovered at the site (Celiberti et al. 2004; de Lumley et al. 2005) are dated between 1.81 and 1.7 Ma (de Lumley et al. 2002). The lithic assemblage of Dmanisi (2064 objects) is knapped from locally available raw materials (mainly volcanic) and composed by numerous flakes and angular fragments (81.5%), as well as some cores (5.1%) and numerous broken and worked pebbles (de Lumley et al. 2005). The assemblage varies little throughout the stratigraphy.

Although very similar to some of the African assemblages described above, the Dmanisi industry presents some characteristics signaling the consolidation of processes leading from variability to diversity. Flakes are small (average flake length = 41.7 mm) and without standardized morphology. Unifacial knapping and its variants (including unifacial discoidal flaking) is dominant (42.3% of the cores), yet bifacial (including orthogonal) and multifacial knapping is relatively common and a few cores show bifacial centripetal removal negatives. Globular or polyhedron cores are also present, although rare (less than 5% of the cores). Spheroids are altogether absent. Knapping sequences are sometimes prolonged but most products do show residual cortex.

Most interesting is the morphology of the "core tools" which, contrary to some of the earlier African assemblages, appear more clearly distinguishable from cores. Although largely dominated by choppers bearing negatives of a single removal (60.1% of the pebble tools), several of the tools show bifacially worked edges (chopping-tools). However, the Dmanisi assemblage is devoid of standardized large or small tools (handaxes or small retouched tools) and does not show systematic bifacial edge exploitation. It thus belongs very clearly to the Mode 1 complex. We may therefore consider the assemblage to be in a sort of transitional position where initial variability, represented by the unifacial and unidirectional components, remains dominant. The beginning of diversity, the subsequent stage of development according to our model, is expressed by the presence of nascent bifacial and multifacial forms in a slightly larger quantity.

The Orce Basin is situated in the southern Iberian Peninsula and corresponds to an ancient lacustrine Basin of Pliocene and Lower Pleistocene Age. There are three main sites in the region: Venta Micena, Barranco León and Fuente Nueva 3. Only the latter two sites revealed lithic artifacts flaked from local limestone and flint: 295 artifacts at Barranco León and 244 at Fuente Nueva 3 (Toro et al. 2003). These two sites were dated to a period from 1.3 to 1.2 Ma using paleomagnetic and biochronological criteria (Oms et al. 2000a, b). The industry from both sites is very similar and comprises mostly of debitage products (flakes, angular fragments) along with cores (Toro et al. 2003). Both hard-hammer and hard hammer on anvil techniques were applied for the production of small nonstandardized flakes. There appears to be no real large or small tool configuration at either site. The scarcity of pebbles in the region led hominins to exploit locally available angular stones. Cores are cubic or polyhedron with relatively long debitage sequences shown by the transformation of the debited edge and the reduction of the original format. The dominant technology appears to have been unidirectional recurrent. Core morphology was once again dependent on the initial form of the block.

Another important Mode 1 site in Spain is that of Atapuerca. The Atapuerca hills are located in the north of Spain, on the northeastern ridge of the Meseta. It is located in the Duero valley, very close to the Ebro from which some Mediterranean climatic influences reached Atapuerca during the Pleistocene. On the hills of Atapuerca the Lower Pleistocene sites are located on the Trinchera del Ferrocarril.

Within Trinchera the best known site is that of Gran Dolina. Its archaeological stratigraphy begins with the TD4 and TD5 levels in which have been found eight artifacts presenting unidirectional reduction. Above these levels the large human occupation of level TD6 has been discovered, presenting a rich association of human and faunal remains and artifacts. This basal half (TD4–TD6) of Gran Dolina has been dated by paleomagnetism, ESR and micromammals chronology to the late Matuyama magneto-chron (Carbonell et al. 1995; Falguères et al. 2000). The well-known Aurora Stratum, a clear horizon with the *Homo antecessor* fossil record, is located within level TD6.

The Atapuerca Lower Pleistocene record has been augmented by the discovery of another site in the Trinchera del Ferrocarril area, known as Sima del Elefante, which is currently under study. It preserved five levels with Lower Pleistocene Mode 1 lithic artifacts. Those are always flakes, made on both flint and limestone (Parés et al. 2006). Although rare, the latter are evidence of the exploitation of limestone fallen from the cave walls. The archaeology of the site indicates that human occupation was very short. The nature of the occupation and the use of the limestone from inside the cave demonstrate an economic system in which immediacy prevailed, as is typical for the archaic phases of Mode 1.

At Gran Dolina only the sample from TD6 is large enough for a qualitative and quantitative analysis (Figure 3.1). The main raw materials used at TD6 are Neogene flint and quartzite while other rocks such as quartz, limestone and Cretaceous flint were less often exploited. All of these rocks may be found within an area of less than 2 km from the sites. Flakes (54%) and retouched pieces (10%) dominate the assemblage. The knapping strategies are variable and diverse, their main characteristic being the sequences devoted to the exploitation of cores to obtain flakes. These sequences when applied to Neogene flint begin with the production and transport to the site of large flakes or fragments. The reduction sequence was then completed on site. In this way all elements of the knapping sequences are present at TD6. At this time, 0.8 Ma ago, the centripetal bifacial strategy is consolidated even if the orthogonal and unidirectional strategies, originating from the archaic technological substratum, are dominant. The important percentage of retouched flakes is also in accordance with an evolutionary trend within the parameters of Mode 1. In spite of that, there is no generation of more standardized morphologies of large tools, such as handaxes and cleavers or picks. We may thus place the TD6 ensemble within a phase of diversity inside the Mode 1 complex.

3.5 Discussion

In the Gona lithic assemblages, as well as some other archaic (older than 2.0 Ma) assemblages, we observe that the sequence of lithic reduction begins inevitably by unidirectional knapping. Stone reduction was practiced using either hard hammer and/ or hard hammer on anvil methods, depending on raw material and initial cobble shape. We hypothesize that all lithic reduction systems derive from this single basic method and that subsequent knapping methods seem to have arisen out of the unidirectional strategy. This strategy could have characterized the *homogeneity* of the hypothesized Mode 0. When unidirectional knapping is pressed further it opens up different potential knapping strategies which vary according to, on the one hand, the initial form of the cobble, and, on the

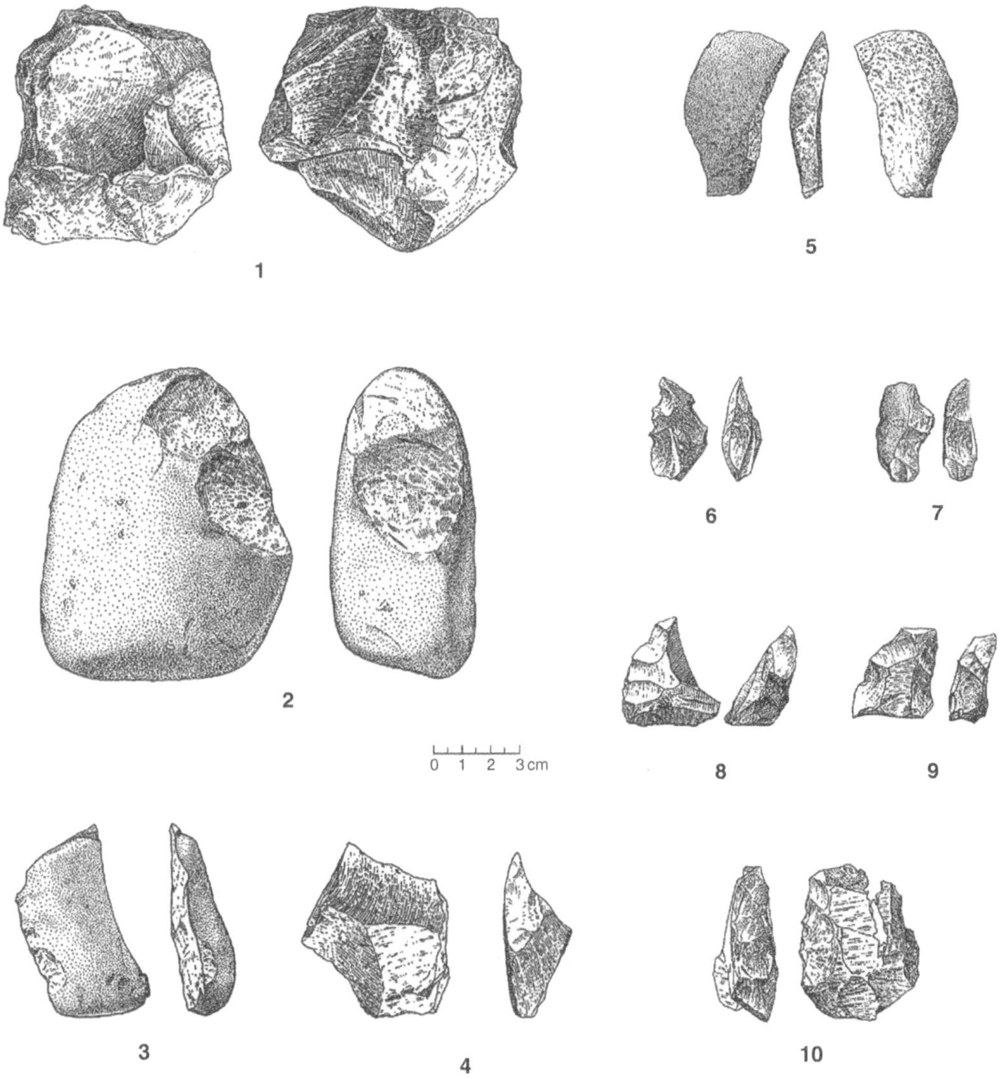

FIGURE 3.1. Atapuerca-TD6, one example of a phase of diversity, showing one multifacial core (1) with retouched flakes (6–10) along with one unifacial core (2) and flakes (3–5). Some flakes (for example #10) come from bifacial cores.

other hand, the edges formed by sequential removals. When pursued at any length on a single cobble, the unidirectional method leads to centripetal (unifacial discoidal) flaking. That is to say, the core will present centripetal removal negatives and an edge separating exploited from unexploited surfaces. Flakes, most often short, have cortical striking platforms and unidirectional (converging) anterior removal negatives. If however there is a change of striking platform, orthogonal flaking will result. In this case, an edge will be formed which may or may not be further exploited. We have observed that in early Mode 1 industries, the evolutionary potential of this edge was not exploited. These are the traits characterizing the *variability* in Early Mode 1 sites like Gona while true intentional exploitation of the bifacial edge does not take place until the appearance of bifacial flaking at the *diversity* phase in late Mode 1 complex where the first bifacial discoidal cores appear (Figure. 3.2 a–b).

Whereas the orthogonal tendency is developed at Gona, multidirectional flaking is not. This method, consisting of a change in direction after each removal and resulting in globular shaped cores is however present, although poorly represented, at sites such as Fejej FJ-1a. It is subsequently developed at the classical Oldowan localities of Bed I at Olduvai Gorge when the *diversity* phase arises.

This emergence and evolution of lithic tools demonstrates the constant loss of homogeneity of the components comprising the whole system and the appearance of new strategies such as multidirectional flaking and later the bifacial system. The initial variability shown by the Gona assemblage could have been triggered by environmental pressures in the sense of developing its latent evolutionary potential into the arising of those new technological tendencies.

The establishment of variability begins through the unipolar recurrent flaking that leads to centripetal flaking

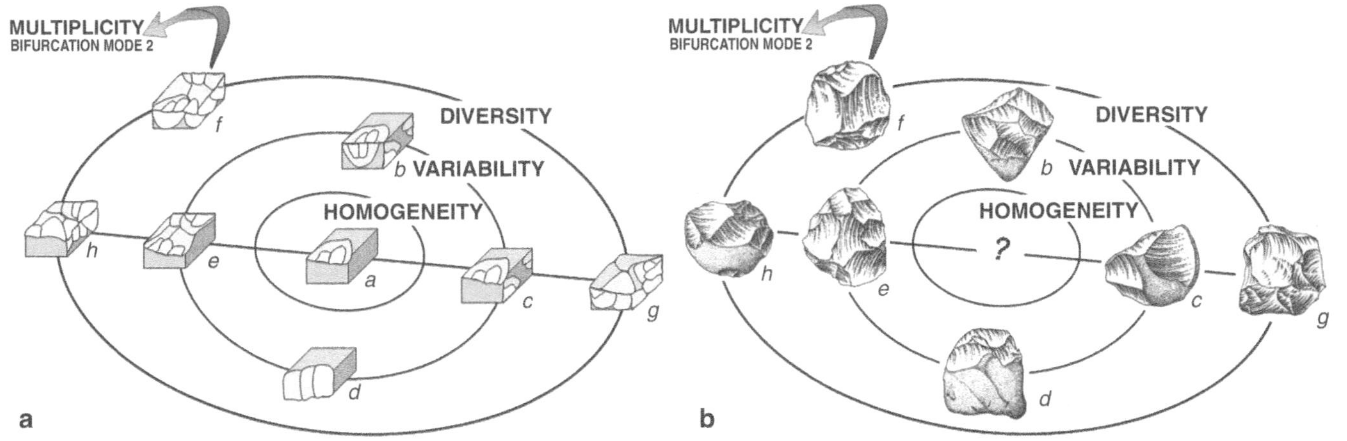

FIGURE 3.2. Evolution of Mode 1. The potential of the simple unipolar recurrent sequence to evolve into any other reduction sequence form makes it likely to be the nascent sequence and, then, the representative of an hypothetic Mode 0 (inner circle, a). Within a model of the evolution of Mode 1 the phase of variability (second circle) is represented by sites such as Gona and Fejej with reduction systems that are orthogonal (b, c, d) and discoidal (e). The phase of diversity (outer circle) adds to the previous strategies the bifacial (f, h) and multifacial (g) reduction sequences along with spheroids and configuration by retouch. It is represented by sites in Bed I of Olduvai, Koobi Fora, Ain Hanech and Atapuerca-TD6. The evolution of bifacial discoids within a new economy of large flakes produces the bifurcation through Mode 2. In Figure 3.2a we can see the models for the reduction sequences, while Figure 3.2b shows the actual artifacts from main sites: b and c from Gona (redrawn from Semaw 2000), d from Atapuerca-TD4, e from Fejej (redrawn from Lumley et al. 2004), f from Koobi Fora (redrawn from Isaac 1997), g from Atapuerca-TD6 and h from Ain Hanech (redrawn from Sahnouni 1998). There are no artifacts in the inner circle because the proposed Mode 0 is unknown.

and thus to the production of discoidal-type cores, even if unifacial. These cores present an edge which was only occasionally exploited at Gona and Fejej. Exploitation of this edge is however to become systematic at the classical Oldowan Bed I sites at Olduvai as well as at other sites dating to around 1.8 Ma. At Olduvai, discoids (interpreted by M. Leakey as another form of chopper or chopping-tool) most often maintain a cortical residue on one face and are therefore not yet completely bifacial. Bifacial debitage is however well expressed at these sites through the abundance of chopping-tools with truly bifacially worked edges, a concept absent from earlier assemblages and that could be the landmark for a new phase, that of diversity.

Whether or not this technological threshold was reached due to an increase in hominins' cognitive capacities, to environmental pressures causing behavioral changes in basic survival patterns (creation of new needs) or to other considerations remains to be discussed. However, new proliferation of this technique at Olduvai does not seem to be linked to raw material constraints. We may assume, as in genetic evolution, that the technological potential present in the initial homogenous system led inevitably to variability and diversity when triggered by the adequate conditions.

The operational schema observed at Gona presents thus several possibilities for hybridization, all operating around the formation of an edge which is still not systematic at the Gona stage of development. This edge presents in itself several potential technological schemas, such as orthogonal, bifacial and discoidal. Discoidal unifacial flaking generally resulted from simple recurrence of linear unidirectional removals around the periphery of more or less oval shaped pebbles and was thus a direct consequence of initial cobble shape. This was observed also by Leakey in her description of the DK discoids (Leakey 1971). It was however the discoidal method which was to follow a unique evolutionary system (Barsky and de Lumley 2004) opening up new technological potential and leading first to the development of a bifacial edge thus initiating the *diversity* phase and ultimately towards a new technological milestone that we call *multiplicity*. This stage has also benefited from the development of the large flake economy within the Mode 2 complex.

The rise of diversity in these ancient technological systems is essential to understanding how so-called Mode 1 assemblages evolved towards Mode 2 technology. At present the earliest known Mode 2 site is that of Konso Gardula (1.6 Ma, Ethiopia; Asfaw et al. 1992) but this technocomplex was to proliferate rapidly in Africa from around 1.5 Ma. At early Mode 2 sites, the unidirectional flaking method and its variants persist alongside the newly developed discoidal bifacial technology and first systematic production of large sized flakes. It would appear that it was the combination of the discoidal flaking method with bifacial conception and exploitation of its resulting edge which led progressively to systematic bifacial discoids and to the manufacture of handaxes and other standardized tools. With the evolution of Mode 2 industry, newly conceived worked edges, produced through configuration or generation of diversified forms, proliferate creating new diversity and multiplicity.

Summing up from the analysis of the African sites, we see some landmarks emerging that help distinguishing different phases of variability and diversity within Mode 1 (Table 3.1)

TABLE 3.1. Phases and milestones of the evolution of Mode 1.

Phase	Attribute	Characteristics	Milestones	Mode	Main sites
IV	Multiplicity	Development of discoidal knapping. Production of large flakes and large standardized tools.	Bifurcation of knapping techniques. Change in economical criteria.	2	Konso. Olduvai Gorge, Bed II .
III	Diversity	Bifacial knapping. Configuration of small flakes by retouch. Standardization of some tools like polyhedrons	Development of edge transformation as a central axis of knapping technology.	1	Olduvai Gorge Beds I and II. Ain Hanech
II	Variability	Transformation of one or more edges. Use of orthogonal and discoid methods. Mainly unifacial knapping.	Increased transformation.. Application of divers unidirectional knapping systems.	1	Kada Gona EG10, EG12. Lokalalei 1, Lokalalei 2C. Fejej FJ-1a
I	Homogeneity	Very rare edge transformation. Potential for change.		0	Hypothesised. Unknown

The unknown phase of homogeneity may be characterized by simple removals without systematic modification of an edge. The introduction of variability can be seen by the coexistence, at sites like Gona, of at least two technical methods: orthogonal and discoidal, both unifacial. The shift from variable assemblages, like Gona, Lokalalei 2C (Delagnes and Roche 2005) or Omo 57 (de la Torre 2004) into some that are invariant and unimodal could be explained by the adaptation to different environments or activities. The ecological or functional constrictions may have induced hominins' conscious selection for particular components of this technological variability, so we observe sites that are technically unimodal, like Lokalalei 1, while others show more aspects of the variability already observed at Gona. The existence of variability within these early assemblages bestows hominins with adaptive advantages that could help them survive in to a greater variety of situations, in contrast with the homogeneous previous phase.

Within Mode 1 the introduction of bifaciality and edge configuration on small tools by retouch as well as the first configured tools might be considered as the beginning of a new phase. As the former technical systems persist there is diversity. The bifurcation produced by the development of different procedures to obtain large flakes and standardized tools, within which the bifacial discoid technology, along with a new economy of resources and territory characterizes the rise of Mode 2. The persistence of Mode 1 in parallel with the development of Mode 2 populations at a regional and global level led to the multiplicity phase (Figure 3.3).

The process of generation of variability and diversity in Africa does not preclude its existence outside the continent. The recently discovered record at Dmanisi, Atapuerca, Orce and Pirro Nord can, conversely, be proposed as the base for a very similar evolutionary process. Dmanisi, Pirro Nord, Orce sites, TD4 and TD5 at Gran Dolina and possibly Sima del Elefante at Atapuerca could provide data for the reconstruction of the variability phase in Europe where orthogonal and unifacial knapping strategies dominated without the presence of truly bifacial cores.

The TD6 record may represent a phase of diversity where bifacial knapping is well developed and retouched artifacts are recorded along with other knapping strategies coming from

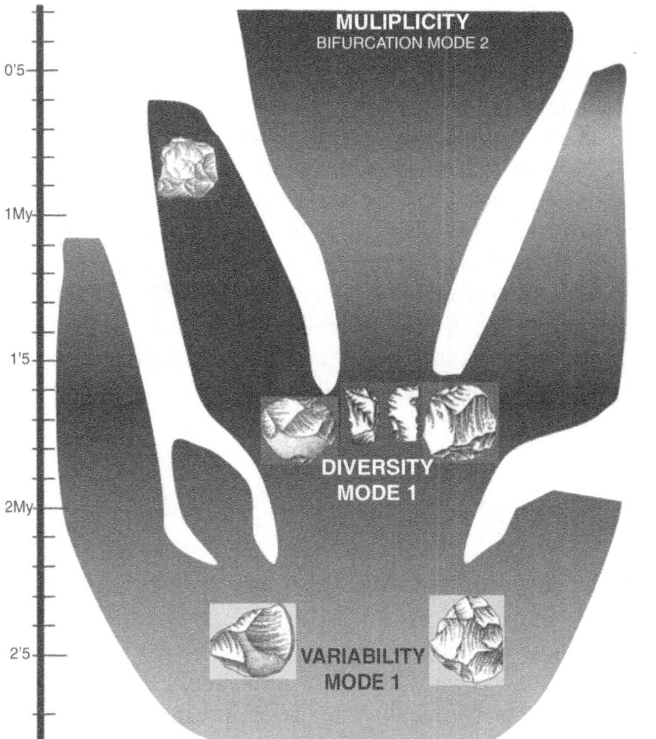

FIGURE 3.3. Evolutionary tree of Mode 1 showing the different phases from the hypothesized Mode 0. The moments of change are related to the origin of every phase but it does not preclude the end of the previous phase. At any stage residual populations maintaining the older strategies remain. Because of that Mode 1 should persist in reduced regional branches both in Africa and in Eurasia until about 0.6 Ma. The examples used illustrate the whole period to what they correspond and not only the strict span of time in which they are designed.

the core of the Mode 1 complex. Further development of bifacial centripetal techniques led to the creation of multiplicity through the arising of Mode 2 where centripetal techniques for obtaining large flakes dominated. Whether the first phase of this process has been provoked by an internal evolution or by acculturation might be the subject of further discussion but under our proposition it is clear that at least the potential development of Mode 1 variability into diversity is present in its internal and original structure and does not require any ulterior introduction of inventions from Africa.

Large flake economy, engaging in different activities in various parts of a group's territorial tract and the systematic fragmentation of the operative chain in different sites as an economic strategy, as well as the configuration of standardized and large artifacts, all typical of Mode 2 and characteristic of a multiplicity phase, must be the consequence of a true bifurcation between two very different modes, Mode 1 and Mode 2 and, more likely, have been introduced in Eurasia through population dispersal.

As we have stressed, the arrival of the Acheulian at Ubeidiya seems not to have undergone a radiation of that culture to other regions. Conversely, these Mode 2 populations may have produced ecological pressure inducing Mode 1 populations living in their periphery to disperse elsewhere in Eurasia (Carbonell et al. 1999). Current evidence points to the arrival of Acheulian cultures into the Far East around 0.8 Ma and into Europe at around 0.6 Ma ago (Hou et al. 2000; Piperno et al. 1998; Piperno 1999). Could these cultures represent the radiation of the developed Acheulian of Gesher Benot Ya'aqov (Goren-Inbar et al. 1992, 2000) or are they an extension of the primitive Ubeidiya Acheulian? In any case, this arrival of Mode 2 into Eurasia has to be considered as the milestone introducing multiplicity in Eurasian technical systems.

3.6 Conclusion

The tendency towards variability observed from the Gona lithic assemblage allows us to propose, accepting a progressive evolutionary model, the possibility for the existence of an earlier unknown industry of a technologically homogenous state, our "Mode 0". The variability shown in the Gona industry may explain the existence of some later assemblages of lesser apparent complexity such as Lokalalei. This state of variability also explains the existence of different orders of tendencies observed in the now numerous industries older than 2 million years in Africa. Consequently, from our point of view, the Gona industry is within Mode 1 (Oldowan) variability and does not represent a Mode 0 or founder phase. Previous hypotheses (Roche 1996; Roche et al. 1999) defend a homogenous state of industries older than 2.0 Ma without variability. However our analysis reveals the existence of potential and expressed variability in these assemblages. Furthermore it is no longer possible to accept the idea (Semaw 2000) of a single cultural identity prior to the emergence of Mode 2 given the diversity observed in industries after 1.8 Ma.

We then postulate that litho-technical evolution is characterized from its very beginnings by the constant loss of homogeneity and by the growing of entropy. The development of technological potential evolution within the state of variability at Gona led to diversity present in ensembles such as those of Bed I at Olduvai or Ain Hanech. And in a final development, the consolidation of this diversity led to the first multiplicity in human technology through the bifurcation of a part of the Mode 1 complex and eventually to the emergence of Mode 2 at sites like Konso Gardula. Multiplicity was thenceforth maintained by the universal coexistence of technological modes, each one with its own internal variability.

This process has also been proposed for the evolution of technical systems in Eurasia from a variability phase known at sites like Dmanisi, Orce, Atapuerca-Sima del Elefante and TD4-TD5 of Gran Dolina more than or around 1 Ma old. The development of these archaic systems towards diversity may be seen at Atapuerca — TD6, 0.8 Ma ago. Finally, the arrival of Mode 2 at around 0.6 or 0.8 Ma ago signals the appearance of multiplicity. From the study of European lithic artifacts it appears that in Europe 0.8 Ma ago the technical complex presents a stage of diversity characterized by some evolutionary trends such as: **1**. the consolidation of centripetal bifacial strategy; **2**. sometimes, like at TD6, this centripetal strategy is made on quite large flakes, that are never devoted to the elaboration of handaxes, picks or cleavers; and **3**. the importance of the configuration of small retouched flakes.

This stage is an evolution from what we know at sites like Dmanisi. The Mode 1 in Europe developed the same phases as those observed in Africa, with an earlier phase characterized by loss of homogeneity and the consolidation of variability as well as a second phase presenting the emergence of diversity. From a general and universal point of view, this phenomenon in Europe could be similar to the process we know from Africa at 1.6-1.8 Ma ago in which the variability and the gain of entropy are also consolidated.

Acknowledgments. Figure 3.1 has been drawn by Néstor Sanchís and illustrations of Figures 3.2 and 3.3 have been designed by Susana Alonso. We wish to thank them both. This work has been partially done thanks to the project "Ecología, biología y comportamiento social y técnico en el Pleistoceno y el Holoceno de la Sierra de Atapuerca" (reference number BXX2000-1258-C03) funded by the Spanish Commission for Science and Technology.

References

Argue, D., Donlon, D., Groves, C., Wright, R., 2006. *Homo floresiensis*: microcephalic, pygmoid, *Australopithecus*, or *Homo?* Journal of Human Evolution 51, 360–374.

Arzarello, M., Marcolini, F., Pavia, G., Pavia, M., Petronio, C., Petrucci, M., Rook, L. Sardella, R., 2006. Evidence of earliest human occurrence in Europe: the site of Pirro Nord (Southern Italy). Naturwissenchaften (electronic issue), 10.1007/s00114-006-0173-3. 24 October 2006.

Asfaw, B., Beyene, Y., Suwa, G., Walter, R.C., White, T.D., WoldeGabriel, G., Yemane, T., 1992. The earliest Acheulean from Konso-Gardula. Nature 360, 732–735.

Asfaw, B., White, T., Lovejoy, O., Latimer, B., Simpson, S. Suwa, G., 1999. *Australopithecus garhi*: a new species of early hominid from Ethiopia. Science 284, 629–635.

Bahain, J.-J., Barsky, D., Beyene, Y., Byrne, L., Echassoux, A., Laurent, M., de Lumley, H., de Lumley, M.-A., Marchal, F., Moullé, P.-E., Pleurdeau, D., Valensi, P., Desclaux, E., 2000. Le site de Fejej FJ-1, Sud Omo, Ethiopie. Cadre géochronologique et biostratigraphique, Paléoenvironnement. L'Homme et l'industrie. In: Brunet, M, Vignaud, P. (Eds.), Congrès International: Les Hominidés et leur environnement: histoire et interactions, Poster. Poitiers.

Barsky, D., de Lumley, H., 2004. Caractérisation des industries lithiques du Paléolithique Inférieur présentes dans les différents niveaux archéologiques de la Caune de l'Arago. In: Toussaint, M., Draily, C. Cordy, J.-M. (Eds), XIVème Congrès UISPP, Section 4. Premiers hommes et Paléolithique Inférieur. Sessions générales et posters, Actes du XIVème Congrès UISPP. BAR International Series 1272, Oxford, pp. 95–103.

Barsky, D., Cauche, D., Celiberti, V., Pleurdeau, D., Beyene, Y. de Lumley, H., 2006. Les industries lithiques archaïques du site de Fejej FJ-1 en Ethiopie. In: de Maret, P., Cornelissen, E., Ribot, I. (Eds.), XIVème Congrès UISPP, Section 15. Préhistoire en Afrique. Sessions générales et posters, Actes du XIVème Congrès UISPP. Liège, 2-8 September 2001. BAR International Series 1522, Oxford, pp. 1–9.

Bar-Yosef, O., Goren-Inbar, N., 1993. The Lithic Assemblages of Ubeidiya: A Lower Paleolithic Site in the Jordan Valley. Monographs of the Institute of Archaeology, 34. Jerusalem, Institute of Archaeology, The Hebrew University of Jerusalem.

Bermúdez de Castro, J.M., Arsuaga, J.L., Carbonell, E., Rosas, A., Martínez, I., Mosquera, M., 1997. A hominid from the Lower Pleistocene of Atapuerca, Spain: possible ancestor to Neandertals and modern humans. Science 276, 1392–1395.

Boesch, C., Boesch, H., 1983. Optimization of nut-cracking with natural hammers by wild chimpanzees. Behavior 83, 265–286.

Boesch, C., Marchesi, P., Marchesi, N., Fruth, B., Joulian, F., 1994. Is nut cracking in wild chimpanzees a cultural behaviour? Journal of Human Evolution 26, 325–338.

Boesch, C., Tomasello, M., 1998. Chimpanzee and human cultures. Current Anthropology 39, 591–614.

Boesch-Achermann, H., Boesch, C., 1994. Hominization in the rainforest: the chimpanzee's piece of the puzzle. Evolutionary Anthropology 3, 9–16.

Brown, P., Sutikna, T., Morwood, M.J., Soejono, R.P., Jatmiko, Wayhu Saptomo, E., Awe Due, R., 2004. A new small-bodied hominin from the Late Pleistocene of Flores, Indonesia. Nature 431, 1055–1061.

Carbonell, E., Bermúdez de Castro, J.M., Arsuaga, J.L., Allué, E., Bastir, M., Benito, A., Cáceres, I., Canals, A., Díez, C., van der Made, J., Mosquera, M., Ollé, A., Pérez-González, A., Rodríguez, J., Rodríguez, X.P., Rosas, A., Rosell, J., Sala, R., Vallverdú, J., Vergès, J.M., 2005. An Early Pleistocene hominin mandible from Atapuerca-TD6, Spain. Proceedings of the National Academy of Sciences 102, 5674–5678.

Carbonell, E., Bermúdez de Castro, J.M., Arsuaga, J.L., Díez, J.C., Rosas, A., Cuenca, G., Sala, R., Mosquera, M., Rodríguez Álvarez, X.-P., 1995. Lower Pleistocene hominids and artifacts from Atapuerca-TD6 (Spain). Science 269, 826–830.

Carbonell, E., García Antón, M.D., Mallol, C., Mosquera, M., Ollé, A., Rodríguez Álvarez, X.-P., Sahnouini, M., Sala, R., Vergés, J.M., 1999. The TD6 level lithic industry from Gran Dolina, Atapuerca (Burgos, Spain): production and use. Journal of Human Evolution 37, 653–693.

Carbonell, E., Bermúdez de Castro, J.M., Parés, J.M., Pérez González, A., Cuenca-Bescós, G., Ollé, A., Mosquera, M., Huguet, R., van der Made, J., Rosas, A., Sala, R., Vallverdú, J., García, N., Granger, D.E., Martinón-Torres, M., Rodríguez, X.P., Stock, G.M., Vergès, J.M., Allué, E., Burjachs, F., Cáceres, I., Canals, A., Benito, A., Díez, C., Lozano, M., Mateos, A., Navazo, M., Rodríguez, J., Rosell, J. & Arsuaga, J.L. 2008. The first hominin of Europe. Nature 452, 465–469.de

Carbonell, E., Mosquera, M., Olle, A., Rodriguez, X., Sahnouni, M., Sala, R., Verges, J. M., 2001. Structure morphotechnique de l'industrie lithique du Pléistocène inférieur et moyen d'Atapuerca (Burgos, Espagne). L'Anthropologie 105, 259–280.

Carbonell, E., Mosquera, M., Rodríguez, X.P., Sala, R., van der Made, J., 1999. Out of Africa: the dispersal of the earliest technical systems reconsidered. Journal of Anthropological Archaeology 18, 119–136.

Celiberti, V., Barsky, D., Cauche, D., Notter, O., Nioradze, M., Lordkipanidze, D., Gabunia, L., Lumley, H., 2004. Les industries lithiques archaïques du site de Dmanisi en Géorgie. In: XIV UISPP Congress. Liège, 2-8 september 2001. Oxford, BAR International Series 1272, pp. 29–36.

Chavaillon, J., 1970. Découverte d´un niveau Olduvayen dans la basse vallée de l´Omo (Ethiopie). Bulletin de la Société Préhistorique Française 67, 7–11.

Chavaillon, J., 1975. Le site Paléolithique ancien d´Omo 84 (Ethiopie). Documents pour servir à l´Histoire des civilisations Ethiopiennes 6, 9–18.

Chavaillon, J., 1976. Evidence for the technical practices of early Pleistocene hominids, Shungura Formation, Lower Omo Valley, Ethiopia. In: Coppens, Y., Howell, F.C., Isaac, G.Ll., Leakey, R.E.F, (Eds.), Earliest man and Environments in the East Rudolf Basin: stratigraphy, paleoecology, and evolution. Chicago University Press, Chicago, pp. 565–573.

Clark, G., 1977. World Prehistory in New Perspective. (3rd). Cambridge University Press, Cambridge.

de Heinzelin, J., Clark, J.D., White, T., Hart, W., Renne, P., WoldeGabriel, G., Beyene, Y., Vrba, E., 1999. Environment and behavior of 2.5 Million-year-old Bouri hominids. Science 284, 625–629.

de Lumley, H., Beyene, Y., Barsky, D., Byrne, L., Camara, A., Cauche, D., Celiberti, V., Fournier, A., Pleurdeau, D., 2004. L'industrie lithique préoldowayenne du site Fejej FJ-1, Ethiopie. In Lumley, H. d.andY. Beyene, Eds, Les sites préhistoriques de la région de Fejej, sud-Omo, Éthiopie, dans leur contexte stratigraphique et paléontologique. Éditions Recherche sur les civilisations. Association pour la diffusion de la pensée française (ADPF), Paris, pp. 391–635

de Lumley, H. d., Lordkipanidze, D., Féraud, G., García, T., Perrenoud, C., Falguères, C., Gagnepain, J., Saos, T., Voinchet, P.,2002. Datation par la méthode $^{40}Ar/^{39}Ar$ de la couche de cendres volcaniques (couche VI) de Dmanissi (Géorgie) qui a livré des restes d'hominidés fossiles de 1,81 Ma. Comptes Rendus de l'Académie de Sciences de Paris Palevol 1, 181–189.

de Lumley, H., Nioradze, M., Barsky, D., Cauche, D., Celiberti, V., Nioradze, G., Notter, O., Zvania, D., Lordkipanidze, D., 2005. Les industries lithiques préoldowayennes du début du Pléistocène inférieur du site de Dmanissi en Géorgie. L'Anthropologie 109(1), 1–182.

de Lumley, M.-A., 2006. L'Homme de Dmanissi (Homo georgicus), il y a 1.810.000 années. Comptes Rendus de l'Académie de Sciences de Paris Palevol 5, 273–281.

de la Torre, I., 2004. Omo revisited. Evaluating the technological skills of Pliocene Hominids. Current Anthopology 45, 439–465.

Delagnes, A., Roche, H., 2005. Late Pliocene hominid knapping skills: the case of Lokalalei 2C, West Turkana, Kenya. Journal of Human Evolution 48, 435–472.

Falguères, C., Bahain, J.-J., Bischoff, J., Arsuaga, J.L., Bermudez de Castro, J.M., Carbonell, E., 2000. U-series and ESR dating of Atapuerca sites: Sima de los Huesos, Gran Dolina and Trinchera Galeria. Mediterranean Prehistory Online 2, 14 pp.

Fragaszy, D., Izar, P., Visalberghi, E., Ottoni, E. B., Gomes de Oliveira, M., 2004. Wild capuchin monkeys (Cebus libidinosus) use anvils and stone pounding tools. American Journal of Primatology 64, 359–366.

Gabunia, L., Vekua, A., Lordkipanidze, D., Swisher, C.C., Ferring, R., Justus, A., Nioradze, M., Tvalchrelidze, M., Antón, S. C., Bosinski, G., Jöris, O., de Lumley, M.-A.., Majsuradze, G., Mouskhelishvili, A., 2000. Earliest Pleistocene hominid cranial remains from Dmanisi, Republic of Georgia: taxonomy, geological setting, and age. Science 288, 1019–1025.

Goren-Inbar, N., Belitzky, S., Goren, Y., Rabinovich, R., Saragusti, I., 1992. Gesher Benot Ya'aqov -the "Bar": an Acheulian assemblage. Geoarchaeology 7, 27–40.

Goren-Inbar, N., Feibel, C., Verosub, K.L., Melamed, Y., Kislev, M. E., Tchernov, E., Saragusti, I., 2000. Pleistocene milestones on the Out-of-Africa corridor at Gesher Benot Ya'aqov, Israel. Science 289, 944–947.

Harmand, S. 2009. Variability of raw material selectivity and techno-economic behavior in the early Oldowan: evidence from the Late Pliocene sites of Lokalalei. In: Hovers, E. and Braun, D.R. (Eds.), Interdisciplinary Approaches to the Oldowan. Springer, Dordrect, pp. 85–97.

Hou, Y., Potts, R., Yuan, B., Guo, Z., Deino, A., Wang, W., Clark, J., Xie, G., Huang, W., 2000. Mid-Pleistocene Acheulean-like stone technology of the Bose Basin, South China. Science 287, 1622–1626.

Howell, F.C., Haesaerts, P., de Hainzelin, J., 1987. Depositional environments, archaeological occurrences and hominids from members E and F of Shungura Formation (Omo Basin, Ethiopia). Journal of Human Evolution 16, 665–700.

Isaac, G.L., 1973. Notes on the archaeological evidence from the Koobi Fora Formation. Wenner-Gren Foundation for Anthropological Research, New York.

Isaac, G.L., 1975. Stratigraphy and cultural patterns in East Africa during the middle ranges of Pleistocene time. In Butzer, K. W., Isaac, G.L. (Eds.), After the Australopithecines. Stratigraphy, Ecology, and Culture Change in the Middle Pleistocene. Mouton Publishers, The Hague, pp. 495–542

Isaac, G.L., 1984. The archaeology of human origins: studies of the Lower Pleistocene in East Africa 1971–1981. In: Wendorf, F., Close, A. (Eds.), Advances in Old World Archaeology 3. Academic Press New York, pp. 1–87

Isaac, G.L., (Ed.), 1997. Plio-Pleistocene archaeology. Koobi Fora Research Project, Vol. 5. Clarendon Press, Oxford.

Kibunjia, M., 1994. Pliocene archaeological occurrences in the Lake Turkana basin. Journal of Human Evolution 27, 159–171.

Kibunjia, M., Roche, H., Brown, F.H., Leakey, R.E., 1992. Pliocene and Pleistocene archaeological sites west of Lake Turkana, Kenya. Journal of Human Evolution 23, 431–438.

Kimbel, W.H., Walter, R.C., Johanson, D.C., Reed, K.E., Aronson, J.L., Assefa, Z., Marean, C.W., Eck, G.G., Bobe, R., Hovers, E., Rak, Y., Vondra, C., Yemane, T., York, D., Chen, Y., Evensen, N.M., Smith, P.E., 1996. Late Pliocene *Homo* and Oldowan tools from the Hadar Formation (Kada Hadar Member), Ethiopia. Journal of Human Evolution 31, 549–561.

Leakey, M.D., 1971. Excavations in Beds I and II. 1960-1963. Olduvai Gorge, Vol. 3. Cambridge University Press, Cambridge.

Manzi, G., Mallegni, F., Ascenzi, A., 2001. A cranium for the earliest Europeans: phylogenetic position of the hominid from Ceprano, Italy. Proceeding of the National Academy of Sciences 98, 10011–10016.

McGrew, W.C., 1992. Chimpanzee Material Culture. Implications for Human Evolution. Cambridge University Press, Cambridge.

Mercader, J., Panger, M., Boesch, C., 2002. Excavation of a chimpanzee stone tool site in the African rainforest. Science 296, 1452–1455.

Mercader, J., Barton, H., Gillespie, J., Harris, J., Kuhn, S., Tyler, R., Boesch C., 2007. 4,300-Year-old chimpanzee sites and the origins of percussive stone technology. Proceedings of the National Academy of Science U.S.A. 104, 3043–3048.

Morwood, M.J., Brown, P., Jatmiko, Sutikna, T., Wahyu Saptomo, E., Westaway, K.E., Awe Due, R., Roberts, R.G., Maeda, T., Wasisto, S., Djubiantono, T., 2005. Further evidence for small-bodied hominins from the Late Pleistocene of Flores, Indonesia. Nature 437, 1012–1017.

Morwood, M.J., Soejono, R.P., Roberts, R.G., Sutikna, T., Turney, C.S.M., Westaway, K.E., Rink, W.J., Zhao, J.-X., van den Bergh, G.D., Due, R.A., Hobbs, D.R., Moore, M.W., Bird, M.I., Fifield, L.K., 2004. Archaeology and age of a new hominin from Flores in eastern Indonesia. Nature 431, 1087–1091.

Oms, O., Agustí, J., Gabàs, M., Anadón, P., 2000a. Lithostratigraphical correlation of micromammal sites and biostratigraphy of the Upper Pliocene to Lower Pleistocene in the Northeast Guadix-Baza Basin (southern Spain). Journal of Quaternary Science 15, 43–50.

Oms, O., Parés, J.M., Martínez-Navarro, B., Agustí, J., Toro, I., Martínez-Fernández, G., Turq, A., 2000b. Early human occupation of Western Europe: paleomagnetic dates for two paleolithic sites in Spain. Proceedings of the National Academy of Science 97, 10666–10670.

Panger, M., Brooks, A.S., Richmond, B. G., Wood, B., 2002. Older than the Oldowan? Rethinking the emergence of Hominid tool use. Evolutionary Anthropology 11, 235–245.

Parés, J.M., Pérez González, A., Rosas, A., Benito, A., Bermúdez de Castro, J.M., Carbonell, E., Huguet, R., 2006. Matuyama-age lithic tools from Sima del Elefante site, Atapuerca (northern Spain). Journal of Human Evolution 50, 163–169.

Piperno, M., 1989. Chronostratigraphic and cultural framework of the *Homo habilis* sites. In Giacobini, G. (Ed.), Hominidae. Proceedings of the 2nd International Congress of Human Paleontology. Jaca Book, Milan, pp. 189–195

Piperno, M. (Ed.), 1999. Notarchirico. Un sito del Pleistocene medio iniziale nel bacino di Venosa. Museo Nazionale Preistorico Etnografico «Luigi Pigorini»/Edizioni Osanna, Venosa.

Piperno, M., Lefèvre, D., Raynal, J.P., Tagliacozzo, A., 1998. Notarchirico. An Early Middle Pleistocene site in the Venosa Basin. Anthropologie XXXVI, 1-2, 85–90.

Prat, S., Brugal, J.-P., Tiercelin, J.-J., Barrat, J.-A., Bohn, M., Delagnes, A., Harmand, S., Kimeu, K., Kibunjia, M., Texier, P.-J., Roche, H., 2005. First occurrence of early *Homo* in the Nachukui Formation (West Turkana, Kenya) at 2.3-2.4 Myr. Journal of Human Evolution 49, 230–240.

Roche, H., 1996. Remarques sur les plus anciennes industries en Afrique et en Europe. In: Facchini, F. (Ed.), XIII International Congress of Prehistoric and Protohistoric Sciences, Colloquia, Vol. 4. The first humans and their cultural manifestations, pp. 55–68.

Roche, H., Delagnes, A., Brugal, J.-P., Feibel, C., Kibunjia, M., Mourre, V., Texier, P.-J.,1999. Early hominid stone tool production and technical skill 2.34 Myr ago in West Turkana, Kenya. Nature 399, 57–60.

Roche, H., Tiercelin, J.J., 1977. Découverte d'une industrie lithique ancienne in situ dans la formation d'Hadar, Afar Central, Éthiopie. Comptes Rendus de l'Académie de Sciences de Paris 284 (Série D), 1871–1874.

Sahnouni, M., 1998. The Lower Palaeolithic of the Maghreb: excavations and analyses at Ain Hanech, Algeria. Cambridge Monographs in African Archaeology 42. BAR International Series 689. Oxford.

Sahnouni, M., Hadjouis, D., de Made, J.V., Derradji, A.-e.-K., Canals, A., Medig, M., Belahrech, H., Harichane, Z., Rabhi, M., 2002. Further research at the Oldowan site of Ain Hanech, North-eastern Algeria. Journal of Human Evolution 43, 925–937.

Savage-Rumbaugh, S., Shanker, S.G., Taylor, T.J., 1998. Apes, Language, and the Human Mind. Oxford University Press, New York.

Semaw, S., 2000. The World's oldest stone artefacts from Gona, Ethiopia: their implications for understanding stone technology and patterns of human evolution between 2.6-1.5 million years ago. Journal of Archaeological Science 27, 1197–1214.

Semaw, S., Renne, P., Harris, J.W.K., Felbel, C.S., Bernor, R.L., Fesseha, N., Mowbray, K., 1997. 2.5-million-year-old stone tools from Gona, Ethiopia. Nature 385, 333–336.

Semaw, S., Rogers, M.J., Quade, J., Renne, P.R., Butler, R.F., Dominguez-Rodrigo, M., Stout, D., Hart, W.S., Pickering, T., Simpson, S.W., 2003. 2.6-Million-year-old stone tools and associated bones from OGS-6 and OGS-7, Gona, Afar, Ethiopia. Journal of Human Evolution 45, 169–177.

Tamrat, E., Thouveny, N., Taieb, M., Opdyke, N.D., 1995. Revised magnetostratigraphy of the Plio-Pleistocene sedimentary sequence of the Olduvai Formation (Tanzania). Palaeogeography, Palaeoclimatology, Palaeoecology 114, 273–283.

Texier, P.-J., Roche, H.,Harmand, S., 2006. Kokiselei 5, formation de Nachukui, West Turkana (Kenya): un témoignage de la variabilité ou de l'évolution des comportements techniques au Pléistocène ancien? In: Otte, M. (Ed.), XIVe Congrès de l'Union Internationale des Sciences Préhistoriques et Protohistoriques, African Prehistory, BAR International Series 1522, Oxford, pp. 11–22.

Toro, I., Agustí, J., Martínez-Navarro, B., (Eds.), 2003. El Pleistoceno inferior de Barranco León y Fuente Nueva 3, Orce (Granada). Memoria científica campañas 1999-2002. Arqueología. Monografías, 17. Junta de Andalucía, Sevilla.

Toth, N., Schick, K.D., Savage-Rumbaugh, E.S., Sevcik, R.A., Rumbaugh, D.M., 1993. *Pan* the tool-maker: investigations into the stone tool-making and tool-using capabilities of a bonobo *(Pan paniscus)*. Journal of Archaeological Science 20, 81–91.

van Schaik, C.P., Pradhan, G R., 2003. A model for tool-use traditions in primates: implications for the coevolution of culture and cognition. Journal of Human Evolution 44, 645–664.

Whiten, A., Goodall, J., McGrew, W.C., Nishida, T., Reynolds, V., Sugiyama, Y., Tutin, C.E.G., Wrangham, R.W., Boesch, C., 1999. Cultures in chimpanzees. Nature 399, 682–685.

Whiten, A., Horner, V., de Waal, F.B M., 2005. Conformity to cultural norms of tool use in chimpanzees. Nature 437, 737–740.

4. An Overview of Some African and Eurasian Oldowan Sites: Evaluation of Hominin Cognition Levels, Technological Advancement and Adaptive Skills

Deborah Barsky*
Centre Européen de Recherches Préhistoriques de Tautavel
Ave. Léon Jean-Grégory
66720 Tautavel France
dbarsky@hotmail.fr

Keywords: Oldowan, technology, stone tools, archaic assemblage, raw materials

Abstract

This article attempts to synthesize the growing amount of information available using concrete examples from some key sites in Africa and Europe to draw conclusions concerning hominin cognition levels, degree of technological advancement and adaptive skills. Sites with Mode 1 or Oldowan type industries are now known outside of Africa; around the Mediterranean Basin and also in central, southern and eastern Asia. Although Africa remains the most probable origin for the first hominin populations to occupy Eurasia, little is known of their paleodemography or cultural transmission modes. Technological studies may provide some answers to questions about raw material procurement patterns and tool manufacturing capacities for hominin groups occupying Eurasia between around 2.0 and 0.8 Ma. What can they tell us about hominin migration patterns? How might comparative lithic assemblage studies be used to discern cultural exchange?

4.1 Introduction

The so-called "Oldowan" typo-technological complex has taken on widespread geographical and chronological connotations after recent discoveries, in both Africa and Eurasia, of new sites yielding archaic stone tools. Additionally there has been new work on some already known sites, including the application, wherever possible, of relatively precise dating methods. Geographically, the Oldowan was once considered to be a uniquely African phenomenon as the "first" tool-makers, once presumed to be *Homo habilis*, seemed to be restricted to this continent and *Homo erectus*, inventor of the so called "Acheulian" (Mode 2, Clark 1970) tool kit, was thought to be the first hominin to venture out of Africa and eventually reach into Asia and Europe. A little over a decade ago, few authors advanced theories suggesting human presence in Europe earlier than about 0.5 Ma. A short chronology hypothesis was proposed to explain the majority of occupational evidence in Middle Pleistocene Europe (Roebroecks and Van Kolfschoten 1994). Much has changed since that time. Relatively recent introduction of technological (over typological) factors into the analysis of early stone tool assemblages has contributed to the recognition and validation of early Eurasian assemblages whose anthropic origin had sometimes been questioned. Among the latter are Vallonnet Cave (Alpes-Maritimes, France) Orce, (Guadix Baza Basin, Spain) and ancient river terraces around the Mediterranean Basin. Modern methodology has come to include an experimental phase where artefacts are reproduced using the same raw materials as those in the samples, thus allowing for a more precise understanding of assemblage morphology. However, it is the discovery of the Dmanisi site (Nioradze and Justus 1998; Gabunia et al. 2000; de Lumley et al. 2002, 2005; Celiberti et al. 2004) with its numerous hominin remains in association with a rich faunal assemblage and unquestionable stone artifacts, precisely dated to 1.81 Ma (Layer VI; de Lumley et al. 2002), which has recently changed our views of when was the first dispersal out of Africa and which hominin groups first left Africa.

* Address for correspondence: dbarsky@hotmail.fr

E. Hovers and D.R. Braun (eds.), *Interdisciplinary Approaches to the Oldowan,* DOI: 10.1007/978-1-4020-9059-2,

In Africa, work on the Gona localities EG10 and EG12 (Ethiopia, Hadar region; Semaw et al. 1997; Semaw 2000, 2005) has most particularly contributed to extending the lower chronological boundary of the African Oldowan to 2.6 Ma, well beyond its original limit of around 1.8 Ma (Olduvai Gorge and Leakey 1971; Tamrat et al. 1995). The appearance of assemblages comprising handaxes (so-called Mode 2 or Acheulian industries) continues to be the defining factor signaling the "end" of the Oldowan, as these tools are thought to represent the beginnings of systematization, predetermination and expression of the notion of symmetry; concepts absent from Oldowan (Mode 1) assemblages (Carbonell et al. 1999; de Lumley et al. 2004) Thus, recent evidence allows us to place the upper chronological limit for the African Oldowan industrial complex at around 1.65 Ma, the date for the oldest known handaxe industries from Konso Gardula (Ethiopia; Asfaw et al. 1992) and from Kokiselei 4 in Kenya (Texier et al. 2006). Although the nonlinear character of hominin techno-cultural evolution should be evoked here by the fact that Oldowan type industries do continue to exist even after the appearance of the Acheulian in many places, (including at Olduvai itself), for the purposes of this article we shall highlight major trends. As such, we note that around 1.5 Ma, Acheulian sites in Africa become more common (for example in Tanzania, at Olduvai Upper and Middle Bed II, Leakey 1971; in Ethiopia, at Gadeb, Clark and Kurashina 1976, 1979 and at Melka Kunture, Garba IV; Piperno et al. 1974; Chavaillon and Piperno 1975).

In Eurasia, upper and lower chronological limits for the Oldowan cultural complex have been largely extended by recent discoveries. The lower chronological boundary may be situated at around 2.0 Ma with growing evidence from several sites in China (Longgupo, Sichuan Province, Huang et al. 1995; Renzidong, Anhui Province, Longgudong, Hubei Province; Dong 2006), attesting to hominin presence in mainland Asia at this early date or even earlier, and raising questions about early migration patterns. However, dating and the anthropic origin for some of these accumulations are yet to be confirmed. It has been argued that *Homo erectus* may have evolved in Asia from an earlier form of *Homo* (Culotta 1995; Dennel and Roebroeks 2005). It does appear that a tool-making pre-*Homo erectus* hominin occupied mainland Asia as early as 2.0 Ma. Presence of an early tool-making *Homo* has also been claimed in Southern Asia as early as 1.9 Ma (Riwat, Potwar Plateau; Raynolds and Johnson 1985; Rendell et al. 1989; Gaillard 2006).

In this complex scenario of the peopling of Eurasia, the upper chronological limit for the "Oldowan" (or the Mode 1/Mode 2 transition) varies considerably, from as early as 1.4 Ma in the near East (Ubeidiya in the Levantine corridor; Bar-Yosef and Goren-Inbar 1993) to around 0.8 Ma in China (at Lantian, Shaanxi Province and at Yunxian, Hubei Province; Feng 2005; south China's Bose Basin; Hou et al. 2000) and to around 1 Ma in southern Asia (Isampour; Gaillard 2006). In western Europe stone assemblages including handaxes are presently known to appear later, around 0.6 Ma, for example in France (P levels of the Caune de l'Arago; Barsky and de Lumley 2004, 2005) and Italy (Notarchirico, Venosa; Piperno 1999).

The geographical and chronological extent of the Oldowan (Leakey 1971) or Mode 1 (Clark 1970) techno-cultural complex may thus be resumed as follows (Figure 4.1).

Can the term Oldowan really encompass this entire chronological and cultural framework?

This article examines different aspects of lithic production and use (assemblage composition, raw material procurement patterns, technological considerations), noting similarities and/or differences before drawing conclusions that may or may not be related to chronological or evolutionary progression. The observations rely primarily on the author's own study of assemblages from several African and Eurasian sites. Therefore, some recently discovered/published assemblages are not included in the discussion.

We have chosen to discuss assemblage composition, that is to say, types of products composing such early industries, and to outline results concerning raw material procurement patterns and technological aspects from the sample sites, in order to ascertain to what extent lithic analysis might be a factor in the evaluation of early hominin cognitive levels. Although the notion of an identifiable relationship between hominin cognition levels and stone tool assemblage morphology is highly debatable, it is interesting to explore whether or not certain technological criteria might be considered as factors revealing a more or less "evolved" behavior towards stone reduction. Once established, the reasons for variability in such criteria may be discussed taking into account environmental context, geographical situation, chronology and homi-

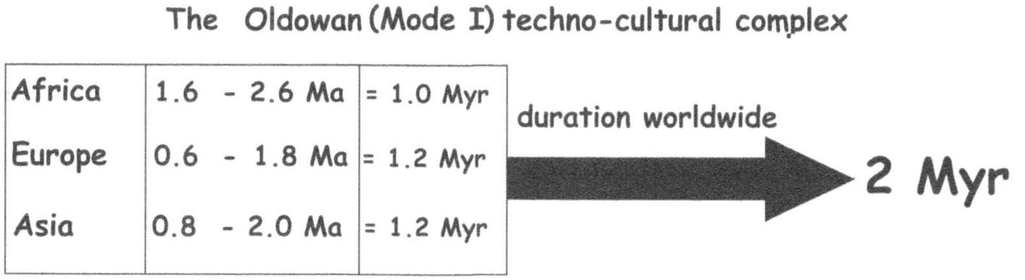

FIGURE 4.1. Age estimates of the Oldowan (Mode 1) technocomplex in various geographical regions and its duration worldwide.

nid type. As yet, the existence of a recognizable relationship between stone tool morphology and hominin cognition levels remains utopian.

4.2 Defining the Oldowan

In order to ascertain the scope of the Oldowan, it is necessary to recognize and define the cultural nature of assemblages dating therefore to between 2.6 Ma and 1.6 Ma in Africa, between (at least) 2.0 Ma and 0.8 Ma in South and East Asia and from around 1.8 Ma to 0.6 Ma in Europe (Figure 4.1). The fact is, direct comparisons of sites over such large geographical and/or chronological distances is problematic and there are difficulties in using broad definitions tying lithic assemblage similarities to a single cultural entity (Delagnes and Roche 2005). To further complicate this situation, recent studies confirm that lithic assemblage morphology varies according to site context and and raw material quality (Asfaw et al. 1991; Kibunjia 1994; Roche et al. 1999; Roche 2000). Further, anthropic elements should certainly be taken into consideration before comparing lithic assemblages across continents and eras. For example, the hominin species responsible for the Gona assemblages were certainly very different from the authors of the Dmanisi ones. This begs the question whether stone tool analysis can contribute to the evaluation of human cognitive levels and, if so, be applied as a determining factor for defining inter-specific differences?

We have outlined here some questions which complicate the concept of the term Oldowan as a reference to a coherent cultural unit. This is not to say that we are in favor of using a plethore of new terms dividing the Oldowan into different semantic categories based on multiple differences in morpho-technological details. Many of these differences are due to raw material variability or site function, as some authors have shown (Asfaw et al. 1991; Kibunjia 1994; Roche et al. 1999; Roche 2000). Given the new data from China, such as that from the Longgupo site, simplified schemas drawing a line from Africa, through the Levantine corridor, to Dmanisi in the Caucasus and then on either to the west, towards Europe or to the East, towards Asia, may no longer be used to illustrate early hominin migration patterns which, in reality, must have been far more complex. Although an initial exodus of early human populations out of Africa appears to remain a valid concept, the presumed date for these migrations is certainly much older than previously thought. Indeed, hominins seem to have reached China relatively early and exchanges between Asia and Europe after around 2.0 Ma ought to be more closely examined in relation to those of large faunal species for whom such migrations are well documented. This, of course, raises questions related to hominid evolutionary models and cultural transmission modes. Many of these early sites are unfortunately completely lacking or poor in hominin remains, thus contributing to problems in determining which hominin species was responsible for which stone assemblages.

In order to justify the broad connotationsof the term Oldowan as currently used, some authors speak of "cultural stasis"(the length of which varies depending on the way the upper and lower limits of the Oldowan are defined), whereas, as we shall see, evidence points towards real variability in the numerous assemblages presently encompassed within this concept. A new taxonomic definition thus appears necessary. Whether it be based on geographical, chronological, technological or even perhaps anthropological factors in relation to early hominin groups, remains problematic.

4.3 The "Oldowan" lithic assemblages from Olduvai Gorge (Leakey 1971)

In the attempt to establish a strict definition of the "Oldowan" cultural complex we refer to the eponymous lithic assemblages from Olduvai Gorge (Leakey 1971), localities DK, FLK NN levels 1–3, FLK "*Zinjanthropus*" Floor and FLK North, levels 1–6, from Lower and Middle Bed I (dating to around 1.8 Ma.). We then consider the same aspects in assemblages from our sample sites, referred to as "Pre-Oldowan" or "Oldowan", in order to test if and how they might differ from the original sample. We hope that this study will contribute, not to the semantic arguements dividing the Oldowan into different cultural entities, but rather towards understanding the vital elements contributing to the variability observed in early industries.

For a strict definition of Oldowan assemblage composition, let us refer to the eponymous case of lithic assemblages (cited above) from Olduvai Gorge. These assemblages are composed of material classed as "tools, used material and *débitage*". Global composition of "Oldowan" assemblages from Bed I, Olduvai Gorge is as follows (Leakey 1971) :

- whole, broken and worked pebbles (choppers and bifacially worked choppers or chopping-tools, "chopper cores")
- polyhedrons, discoides
- subspheroids and spheroids
- heavy-duty scrapers (these tools as defined by M. Leakey (1971) appear to be equivalent to *rostro-carénés* (de Lumley et al. 2004, 2005) or *nucléus racloirs* (Semaw 2005).
- light-duty scrapers and burins (cf. small retouched tools)
- anvils, hammerstones, cobblestones, nodules and blocks, light duty flakes and other fragments, *débitage*, manuports (?).

At the Olduvai Gorge localities considered here, hominins (*Paranthropus boisei*, *Homo habilis*) collected their raw materials (lava, quartz) no more than a few kilometers away from the site. Assemblage composition is dominated by *débitage* products : broken and whole flakes and angular fragments. These assemblages do not include handaxes or cleavers (configured tools : edges shaped by removals and/or retouch and modified intentionally after a mental template). Retouched (configured) edges are observed on some cores (heavy duty scrapers / *nucléus racloirs* or *rostro-carénés*). Some flakes or fragments were retouched to produce tools of relatively irregular morphology and with little typological diversity however the quantitative importance of such pieces varies according to different sources (Potts 1988; Kimura 2002; de la Torre and Mora 2005). One observes the frequent use of unidirectional recurrent knapping as well as unidirectional bifacial technology, producing sinuous cutting edges (chopping-tools). Knapping appears to be mostly unifacial and, depending on the length of débitage episodes, removals affect more or less of each pebble's periphery. Depending on initial pebble shape, the latter method sometimes resulted in the production of discoidal type "cores", referred to as "discoids" (Leakey 1971) Other flake producing technology was also applied at these sites, including multidirectional (multifacial) flaking producing globular (polyhedron) shaped "cores" and/or spheroids (configured tools). The overall small size of most of the *débitage* products is, as we shall see, another important characteristic of these assemblages.

4.4 Sample site comparisons: transformation or cultural stasis?

We have chosen to compare the lithics from some sample sites of different ages (but all globally attributed to Mode 1) in Africa and Europe, in order to observe how they may or may not be related to the above defined Oldowan tradition. Firstly, we shall examine data from the oldest stone tool assemblages presently known, those from localities EG10 and EG12 of Kada Gona (Hadar region, Ethiopia; Semaw et al. 1997; Semaw 2000, 2005). These sites have yielded lithic artifacts in a precisely dated context (2.6 Ma). Hominins settled near a stream that doubled also as a close-by raw material source, and cobbles were collected from a nearby conglomerate (100–200 m away; Semaw 2005). Raw material selection was clearly focused on trachyte (more than 75% of the artifacts from both sites), as opposed to rhyolite (20%), basalt or other available rock types (Semaw 2005). Sampling in the conglomerate revealed different rock type distribution (trachyte representing only 48%) showing that trachyte was preferentially chosen for knapping. Furthermore, trachyte pebbles present in the archaeological sample have a finer groundmass and smaller or less phenocrystals (Stout et al. 2005). Thus for these assemblages, the oldest presently known, hominins already possessed an intimate knowledge of the physical qualities of each rock type and were able to predict how they might react during percussion. We may conclude, in agreement with Semaw (2005), that very early on, hominins demonstrated selective behavior, showing concern for the quality of their raw materials.

The Gona assemblages, ascribed to the Oldowan cultural complex (Semaw 2005), are composed of numerous unmodified *débitage* products: flakes, flake fragments, debris and cores or chopper/cores. Worked pebbles with an intentionally modified edge are extremely rare and truly bifacially worked pebbles (chopping-tools) are absent. Further, the authors of these industries rarely practiced multidirectional (multifacial) flaking, which explains the absence of polyhedrons, spheroids and / or subspheroids (or cores with multiple directions of removal negatives). Unidirectional unifacial flaking was largely dominant and sometimes led to the production of discoid type cores, depending on initial pebble morphology and the length of the *débitage* episodes. So-called heavy-duty scrapers (*nucléus racloirs*; Semaw 2005; *rostro-carénés*; de Lumley et al. 2004, 2005) are present, and intentionally retouched products are absent (light-duty scrapers and burins). In the absence of intentionally worked tools, typological criteria are not applicable to these industries.

To summarize, at these sites, hominins practiced skilled use of unidirectional peripheral (or recurrent) flaking on pebbles whose shape was most probably carefully selected according to advantages linked to the presence of natural angles permitting flake removal, given the apparent scarcity of pieces showing prior preparation (such as intentionally breaking or decapping the pebble to create an appropriate striking platform) and the abundance of flakes with cortical striking platforms (Toth types II and III; Toth 1985; Semaw 2005). The Gona assemblages therefore share some aspects with Olduvai's Oldowan localities, such as the frequencies of small-sized, nonstandardized and unmodified knapping products (flakes and debris) and local raw material procurement. However, in spite of relatively fine quality raw materials, assemblage composition and technological aspects of the Kada Gona assemblages considered here differ essentially from the Olduvai Bed I sample in that :

- they do not show systematic bifacial or multifacial flaking techniques
- they do not comprise a large percentage of worked pebbles with unifacially or bifacially intentionally modified edges (clear choppers or chopping-tools)
- They do not comprise "configured" tools such as intentionally retouched flakes or debris, or spheroids.

Although Gona cores are relatively small, *débitage* episodes were frequently long enough to ascertain the morphologies related to dominant technological tendencies. The latter, resulting mainly from unifacial recurrent knapping (Semaw 2005), appear to be clearly distinct from Oldowan bifacial or multifacial products.

It is interesting to note that at this early period in hominin cognitive evolution, the technological characteristics observed from these industries reveal that hominins already possessed knowledge of the principles of conchoidal fracture and that they practised organized and well mastered knapping techniques (Semaw 2005). They were familiar with the basic mechanics involved in knapping stone such as choosing an appropriate angle for flake removal. Cobbles, chosen in consideration of the latter, thus required virtually no preparation and flakes were often removed from cortical striking platforms. The hominins responsible for these industries therefore adapted their technological capacities well to available raw materials which they worked in a systematic way.

Some 0.6 Ma later, at the Fejej FJ-1a site (1.96 Ma; de Lumley de et al. 2004; Barsky et al. 2006), hominins seemingly showed similar behavior towards raw material procurement, although their selection was focused on other rock types (mainly quartz). At this site, hominins processed carcasses on a fluviatile sand ridge, near a riverbank in a floodplain. Game was abundant and varied and rocks were collected from nearby alluvial deposits composed of quartz, basalt, granite, gneiss, quartzite sandstone and "other" pebbles. As at Gona, alluvial sampling revealed a completely different rock type distribution compared with the archaeological sample, with only 35% quartz pebbles as opposed to 91 % in the FJ-1a artifact sample. Basalt, representing 35 % of the alluvial sample, represents only 7% of the FJ-1a industry (de Lumley de et al. 2004). In preferentially selecting quartz pebbles over other available types the Fejej FJ-1a hominins practiced selective raw material procurement behavior similar to that observed at Gona. As we have seen at the Gona localities, this selective behavior was acquired very early on and may thus be considered a culturally transmitted behavior.

Generally speaking, the assemblage composition for this site seems to match the "Oldowan" scheme outlined above : frequent small sized flakes and debris with worked cobbles and/or cores (among the latter, core tools as such are difficult to distinguish from cores). However, the assemblage is completely lacking in light duty scrapers and burins (small, intentionally retouched pieces). Indeed, choppers are present, although most of them (almost half of the sample) show only a single removal negative (pebbles are not naturally present in the stratigraphy and removals are clearly of anthropic origin) and only 4 (out of 85) show a bifacially worked edge (most worked pebbles at our eponymous sites are bifacially worked). Heavy duty scrapers (*rostro-carénés*; de Lumley et al. 2004) are present.

Technological aspects of the Fejej FJ-1a have been particularly well defined thanks to numerous refits, mainly flakes onto cores (de Lumley et al. 2004; Barsky et al. 2006). Both hard-hammer and hard-hammer-on-anvil techniques for stone reduction have been clearly identified at this site. *Débitage* methods include unidirectional (linear), unidirectional on an anvil, bifacial bidirectional, orthogonal, centripetal (discoid), multifacial (rare).

This assemblage, referred to as "Pre-Oldowan" (de Lumley de et al. 2004), shows that hominins practiced raw material procurement patterns similar to those of the Oldowan scheme, yet assemblage composition and technological aspects of the Fejej industry differ considerably from those of our Oldowan sample from Olduvai Gorge. One might argue that observed differences may be due to the fact that most of the FJ-1a assemblage was knapped from quartz pebbles, a raw material generally known for its difficulties in performing controlled knapping. However, at Fejej this raw material was of fine quality and refits show that controlled knapping procedures were successfully performed. It is most interesting to note the presence of a few globular type cores (multidirectional, multifacial flaking) as they are technologically distinct from any of the variants of unidirectional flaking. Spheroids are absent, however, underlining the lack of true configured tools.

When compared to the older Gona assemblages, the Fejej hominins were capable of chosing the most appropriate knapping method from among several systematic methods. Presence of unidirectional knapping methods and their variants appears to suggest acquired know-how while the appearance of multifacial flaking may be considered to be a relatively advanced cognitive characteristic. At the same time, the absence of configured tools distinguishes the Fejej assemblage significantly from that of the Oldowan assemblages from Olduvai Gorge, suggesting that the capacity to create a predetermined form from a mental template has not yet been achieved.

Local raw material procurement and rock type discernment in accordance with physical properties is also common to many of the earliest European sites. However, as we shall see, over time additional concerns come into play, nuancing and complicating this behavioral aspect. A good example comes to us from the Dmanisi site, dated to 1.81 Ma (de Lumley et al. 2002), where hominins (*Homo georgicus*; M. A. de Lumley 2006) apparently came to scavenge drowned animal carcasses transported by river channels (de Lumley et al. 2005). Local alluvial sediments bordering the site provided a wide variety of rock types for the realisation of their industry (volcanic, metamorphic cristalline, silicieuous sedimentary and sedimentary). As in the older African and Asian Mode 1 sites cited above, selective behavior is noted at Dmanisi as hominins favored one rock type, in this case, fine grained tuff. Yet at Dmanisi this behavioral aspect takes on new meaning as it is not homogeneous in all facets of the assemblage. As we have seen, Mode 1 assemblages comprise a majority of *débitage* products (flakes and debris mostly, with some cores and/or pebble tools). In our previous examples from earlier African sites, one raw material was favored over other locally available types for all elements composing the assemblages (Gona and Fejej). At Dmanisi however different techno-typological elements were often made in different rock types (de Lumley et al. 2005). In other words, when considered, *débitage* products (such as flakes and cores), whole or broken pebbles and worked pebbles (pebble tools) show different raw material distribution patterns. This has also been shown for Olduvai Bed I assemblages (de la Torre and Mora 2005). Although fine-grained tuff was always preferentially used, its frequency in the various typological groups varies according to that of other rock types. For example, fine-grained tuff dominates largely among *débitage* products (flakes, debris, cores), whereas both fine-grained tuff and basalt pebbles are common among whole pebbles (with or without percussion marks). Cobbles were brought to the site as manuports from nearby river alluvial sediments. Pebble tools were worked from fine or rough grained tuff pebbles as well as and other volcanic rock types. We may conclude that rock type selection seems at Dmanisi to vary in accordance with assemblage components. This subtle change accompanies the transition towards the emergence of more clearly defined typo-technological groups, as discussed above.

The Dmanisi assemblage, attributed to the "Pre-Oldowan" (de Lumley et al. 2005), also differs from the eponymous Oldowan industries through the absence of spheroids, a seemingly African "tradition". Also absent are flakes or fragments intentionally configured by retouch. This site does however show strong similarities with Olduvai in raw material procurement patterns and in the relative frequencies of unidirectional bifacial flaking (chopping-tools). Multidirectional (or multifacial) flaking is, at Dmanisi, relatively rare. The hominins present at Dmanisi practiced opportunistic food procurement patterns while raw materials were selected with care. Their technology shows strong similarities to that of our African sample site of Fejej FJ-1a, with the possible exception of the slightly more standardized morphology of the pebble tools in the Dmanisi assemblage, where they are more clearly distinguished from the cores.

The more recent European sites of Barranco León and Fuente Nueva 3 in Orce, Andalusia, Spain (respectively 1.3 and 1.2 Ma) are both located in swampy areas close to the eastern shores of the Baza-paleo Lake. (Turq et al. 1996; Oms et al. 2000a, b; Toro-Moyano et al. 2003). In these sites hominins and hyenas competed for access to large herbivore carcasses abandoned by other carnivores. To make their tools, hominins collected nearby rocks such as limestone, sometimes as cobbles or, more often, as stones, or flint in the form of plates, nodules or sometimes pebbles. As at the African Oldowan sites, rocks were collected near to the site. Distinctive groupings of rock type with typo-technological elements become even clearer than at Dmanisi : flint is largely exploited for *débitage*, the dominant assemblage elements (flakes, debris and, to a lesser degree, cores) whereas limestone was reserved for percussion instruments and worked cobbles.

Unlike Olduvai, these Spanish sites do not comprise real configured tools. Knapping techniques were adapted according to different raw material constraints and other considerations mainly related to initial block form, explaining the relative frequency of hard-hammer-on-anvil technique that was often used to reduce small, cube-shaped flint blocks as well as many of the larger limestone pieces. Polyhedron-shaped cores were otherwise produced using hard hammer technique and requiring frequent core rotations. We therefore observe that differences noted in the general morphology of the Orce lithic assemblages compared with those from the Oldowan sites of Olduvai Gorge may be partially explained by differences in the initial shape and quality of raw materials used. However, as we have seen, other major behavioral factors tied apparently to technological know-how separate these industries.

4.5 Conclusions

We have seen that the Oldowan, as it is defined today, shows little coherence as a single cultural unit given its now overly enlarged geographical and chronological connotations. Using chosen attributes (assemblage composition, raw material procurement patterns and technology) we have compared lithic assemblages from some of the oldest known African and European sites with those from the Oldowan sites of Olduvai Gorge, demonstrating that, although these assemblages show strong similarities, they also present considerable differences, thus underlining some of the difficulties in maintaining the Oldowan as a single cultural complex and the need to describe its variability.

Behavior ascertained from raw material procurement patterns at our sample site shows that on-site or nearby raw material selection does not necessarily imply low cognitive capacities (see also Goldman-Neuman and Hovers 2009; Harmand 2009). On the contrary, raw material selection was the first step in fulfilling a given need and is evidence for what has come to be referred to as *planning*. In fact, quite a complex *chaîne opératoire* was required before specific needs could be concretely provided for by the choice of appropriate blocks. The earliest known industries, those from Gona, show that hominins were already capable of taking into consideration the physical qualities of stones available to them; their quality and also their shape. Hominins present at Gona already possessed knowledge of how to evaluate appropriate angles necessary to successfully detach flakes. This enabled them to optimize their choices and to avoid unnecessary preparative steps by selecting pebbles with naturally advantageous platforms as has been assessed from data published about flake and core morphology (Semaw 2005). We have seen that, over time, knowledge about different rock types was transmitted and refined, and that it was to become more complex in parallel with the growing importance of typological features. As tool types became more clearly defined, typological factors apparently came into play during raw material selection, whereby specific rocks were chosen for the manufacture of specific products. Can this more complex behavioral aspect be attributed to factors related to hominin anatomic evolution?

Very generally speaking, the archaic assemblages cited above include unmodified or slightly modified *débitage* products (flakes, debris and cores), with worked pebbles whose identification as tools intentionally manufactured to fit a mental template is more or less clear. Looking at typological factors, relatively fine quality raw materials available to Gona hominins would have permitted them to manufacture tools such as those present in the Oldowan assemblages defined by M. Leakey. It has been suggested that the reason for their absence in the Gona assemblages may be that hominins simply did not need to dispose of such functionally elaborate objects in their tool-kit (Semaw 2005). Yet, evidence for the absence of retouched or otherwise configured tools (subspheroids or spheroids) in the now growing number of assemblages predating the Oldowan assemblages of Olduvai Gorge, in Africa, leads us to consider the possibility that other factors may be responsible for this lack. European evidence also suggests that such items are absent or very rare in most of the early assemblages and that they become gradually more frequent prior to the appearance of Mode 2 industries. We may conclude that one outstanding feature of the eponymous Oldowan assemblages is the presence of configured tools (including retouched elements, subspheroids and spheroids) and that their manufacture may be considered to have been a relatively evolved characteristic, compared with assemblages without configured tools. It seems that the first step for tool-making hominins was simply that of producing flakes using opportunistic but well-mastered techniques and that their cognitive capacities (perhaps not yet transformed into traditions) did not initially include types. The production of "tools" in the sense of typologically definable elements, may therefore be considered as a step towards growing complexity in hominin cognitive evolution.

Another interesting criterion that may reveal a complex technological behavior is that of preparing blocks before detaching flakes in order to obtain products of a specific morphology. Such behavior, which is present early on in some archaic assemblages (Delagnes and Roche 2005), does not become systematic until the appearance of Acheulian-type industries and the standardization of intentionally modelled preestablished forms.

Other technological considerations may contribute to defining just what the different elements comprising archaic industries can tell us about the degree of complexity attained by hominins present at a given location at any one time. As we have seen, archaic assemblages, once considered homogenous and simple (the term "primitive" is often applied), are in reality relatively divers. Very early on (at Gona), hominins practiced systematic flake production methods, perfectly adapted to the raw materials available to them. Archaic industries from our sample sites in both Africa and Europe show that hominins were capable of applying two or more well-mastered technical systems for stone reduction. The example of the Fejej FJ-1a lithic assemblage clearly demonstrates that, nearly 2.0 Ma ago, hominins chose appropriately from a variety of known technological systems in order to optimize production. Might we consider that, compared with the older Gona assemblages, the larger variety of technical systems present at the Fejej site be attributed to a more "evolved" behavior or is this apparent increase in complexity merely a question of different site function?

We have also observed that the process of growing complexity of early hominin technology does not necessarily follow a progressive evolutionary trend although certain landmarks are brought to the fore, such as the appearance and development of multidirectional (multifacial) flaking leading to the production of polyhedron or globular type cores and sometimes further to the production of spheroids. Although unidirectional unifacial knapping remains largely dominant throughout this long chronological period, other methods do become progressively more frequent, in particular unidirec-

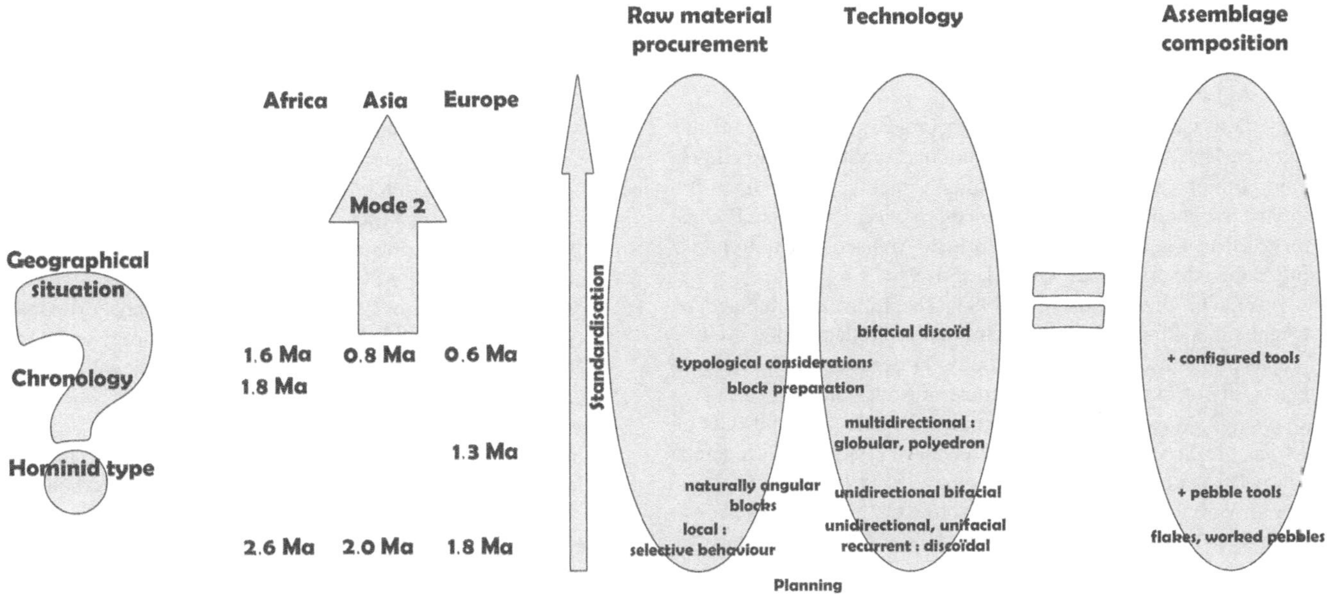

FIGURE 4.2. Major features contributiong to complexity and standardization in the technological behavior of the oldest known lithic assemblages in Africa, Asia and Europe.

tional bifacial and, of course, discoidal. The latter method, initially exclusively unifacial, was practiced bifacially from around 1.8 Ma (Leakey 1971; Barsky and de Lumley 2005; de Lumley et al. 2005), and then systematically from around 1.6 Ma (in Africa) (Asfaw et al. 1992). This method was to become emblematic of Mode 2 producing cultures, along with that of large flakes and systematized tools.

To summarize, these criteria may contribute to more precise definitions of the oldest known hominin cultural traditions (Figure 4.2). These features may be used to assess technological achievement levels and, indirectly, hominin cognitive advancement. We suggest that the following, when considered in relation to differences in : chronology, geography and hominin type, may constitute elements for comparing archaic lithic assemblages, and perhaps even contribute to the determination of cultural transmission modes and ancient migration routes.

References

Asfaw, B., Beyene, Y., Suwa, G., Walter, R.C., White, T.D., WoldeGabriel, G., Yemane, T., 1992. The earliest Acheulean from Konso-Gardula. Nature 360, 732–735.

Barsky, D., Lumley, H. de, 2004. Caractérisation des industries lithiques du Paléolithique inférieur présentes dans les différents niveaux archéologiques de la Caune de l'Arago. In: XIVth International Congress of the Union for Prehistoric and Protohistoric Sciences, Section 4: Human Origins and the Lower Paleolithic. Liège (Septembre 2001), British Archaeological Reports International Series 1272, Oxford, pp. 95–103.

Barsky, D., de Lumley, H., 2005. Comportement technologique des premiers niveaux à bifaces du nord de la Méditerranée. In: Molines N., Moncel, M-H., Monnier, J. –L. (Eds.), Les premiers peuplements en Europe. Colloque international: Données récentes sur les modalités de peuplement et sur le cadre chronostratigraphique, géologique et paléogéographique des industries du Paléolithique ancien et moyen en Europe (Rennes, 22-25 septembre 2003). British Archaeological Reports International Series 1364. Oxford, pp. 135–148.

Barsky, D, Cauche, D, Celiberti, V., Pleurdeau, D., Beyene, Y., de Lumley, H., 2006. Les industries lithiques archaïques du site de Fejej, Ethiopie. In: XIVth International Congress of the Union for Prehistoric and Protohistoric Sciences, Section 15:African Prehistory. Liège (Septembre 2001), British Archaeological Reports, International Series 1522, Oxford, pp.1–9.

Bar-Yosef, O., Goren-Inbar, N., 1993. The Lithic Assemblage of Ubeidiya: A Lower Paleolithic Site in the Jordan Valley. Monographs of the Institute of Archaeology, 34. Jerusalem, The Hebrew University of Jerusalem, The Institute of Archaeology.

Carbonell, E., Mosquera, M., Rodriguez, X.P., Sala, R., 1999. Out of Africa: the dispersal of the earliest technical systems reconsidered. Journal of Anthropological Archaeology 18, 119–136.

Celiberti, V., Barsky, D., Cauche, D., Notter, O., Nioradze, M., Lordkipanidze, D., Gabunia, L., Lumley, H. de, 2004. Les industries lithiques archaïques du site de Dmanissi en Géorgie. In: XIVth International Congress of the Union for Prehistoric and Protohistoric Sciences, Section 4: Human Origins and the Lower Paleolithic. Liège (Septembre 2001), British Archaeological Reports International Series 1272, Oxford, pp. 29–36.

Chavaillon,J., Piperno, M., 1975. Garba IV, site paléolithique ancien de Melka-Konturé. Bulletin de la Société Préhistorique Française 72, 134–138.

Clark, G., 1970. Aspects of Prehistory. University of California Press, Berkeley.

Clark, J.D., Kurashina, H., 1976. New Plio-Pleistocene archeological occurences from the Plain of Gadeb, Upper Webi Shebele Basin, Ethiopia, and statistical comparison of the Gadeb sites with other Early Stone Age assemblages. In: Clark, J.D., Isaac, G. (Eds.), The Earliest Industries of Africa. IXth International Congress of the Union for Prehistoric and Protohistoric Sciences , Colloque V, pp. 158–216.

Clark, J.D., Kurashina, H., 1979. Hominid occupation of the East-Central highlands of Ethiopia in the Plio-Pleistocene. Nature 282, 33–38.

Culotta, E., 1995. Asian hominids grow older. Science 270, 1116–1117.

de la Torre, I., Mora, R., 2005. Technological Strategies in the Lower Pleistocene at Olduvai Beds I and II. ERAUL 112, Liège.

de Lumley, H. , Lordkipanidze, D., Féraud, G., García, T., Perrenoud, C., Falguères, C., Gagnepain, J., Saos, T., Voinchet, P., 2002. Datation par la méthode $^{40}Ar/^{39}Ar$ de la couche de cendres volcaniques (couche VI) de Dmanissi (Géorgie) qui a livré des restes d'hominidés fossiles de 1,81 Ma. Comptes Rendus Palevol 1, 181–189.

de Lumley, H., Nioradzé, M., Barsky, D., Cauche, D., Celiberti, V., Notter, O., Zvania, D., Lordkipanidzé, D., 2005. Les industries lithiques pré-oldowayennes du début du Pléistocène inférieur du site de Dmanissi en Géorgie. L'Anthropologie 109, 1–182.

de Lumley, H., Beyene, Y., Barsky, D., Byrne, L., Camara, A., Cauche, D., Celiberti, V., Fournier, A., Pleurdeau, D., 2004. L'industrie lithique préoldowayenne du site de Fejej FJ-1, Ethiopie. In: Lumley H. de, Beyene, Y. (Eds.), Les sites préhistoriques de la région de Fejej, Sud Omo, Éthiopie, dans leur contexte stratigraphique et paléontologique. Editions recherche sur les civilisations. Association pour la diffusion de la pensée française, Paris, pp. 391–563.

de Lumley, M.-A., 2006. L'Homme de Dmanissi (*Homo georgicus*), il y a 1.810.000 ans. Comptes Rendus Palevol 5, 273–281.

Delagnes, A., Roche, H., 2005. Late Pliocene hominid knapping skills : the case of Lokalakei 2C, West Turkana, Kenya. Journal of Human Evolution 48, 435–472.

Dennell, R., Roebroeks, W., 2005. The Asian perspective on early human dispersal from Africa. Nature 438, 1099–1104.

Dong, W., 2006. Recent discoveries of hominid fossils in southern China. L'Anthropologie 110, 241–257.

Feng, X., 2005. Stratégie de débitage et mode de façonnage des industries du Paléolithique inférieur en Chine et en Europe entre 1 Ma et 400 000 ans. Thèse de Doctorat. Muséum National d'Histoire Naturelle, Paris.

Gabunia, L., Vekua, A., Lordkipanidze, D., Swisher, C.C., Ferring, R., Justus, A., Nioradze, M., Tvalchrelidze, M., Antón, S.C., Bosinski, G., Jöris, O., de Lumley, M.-A., Majsuradze, G., Mouskhelishvili, A., 2000. Earliest Pleistocene hominid cranial remains from Dmanisi, Republic of Georgia: taxonomy, geological setting, and age. Science 288, 1019–1025.

Gaillard, C., 2006. Les premiers peuplements d'Asie du Sud : vestiges culturels. Comptes Rendus Palevol 5, 359–369.

Goldman-Neuman, T., Hovers, E. 2009. Methodological issues in the study of Oldowan raw material selectivity: insights from A.L. 894 (Hadar, Ethiopia). In: Hovers, E. and Braun, D.R. (Eds.), Interdisciplinary Approaches to the Oldowan. Springer, Dordrect, pp. 71–84.

Goren-Inbar, N., Belitzky, S., Goren, Y., Rabinovich, R., Saragusti, I., 1992. Gesher Benot Ya'aqov - the "Bar": an Acheulian assemblage. Geoarchaeology 7, 27–40.

Harmand, S. 2009. Variability of raw material selectivity and techno-economic behavior in the early Oldowan: evidence from the Late Pliocene sites of Lokalalei. In: Hovers, E. and Braun, D.R. (Eds.), Interdisciplinary Approaches to the Oldowan. Springer, Dordrecht, pp. 85–97.

Hou, Y., Potts, R., Baoyin, Y., Zhengtang, G., Deino, A., Wei, W., Clark, J., Guangmao, X., Weiwen, H., 2000. Mid-Pleistocene Acheulian-like stone technology of the Bose Basin, South China, Science 287, 1622-1626. Huang, W., Ciochon, R., Gu, Y., Larick, R., Fang, Q., Schwarcz, H., Yonge, C., de Vos, J., Rink, W., 1995. Early *Homo* and associated artefacts from Asia. Nature 378, 275–78.

Kibunjia, M., 1994. Pliocene archaeological occurrences in the Lake Turkana basin. Journal of Human Evolution 27, 159–171.

Kimura, Y., 2002. Examining time trends in the Oldowan technology at Beds I and II, Olduvai Gorge. Journal of Human Evolution 43, 291–321.

Leakey, M.D., 1971. Olduvai Gorge, Excavations in Bed I and Bed II, 1960–1963. vol. 3. Cambridge University Press, Cambridge .

Nioradze, M., Justus, A., 1998. Stone tools of the ancient Palaeolithic site Dmanisi. In: Kopaliani, J. (Ed.), Dmanisi. Metsniereba, Tbilisi, pp. 140–159.

Oms, O., Agustí, J., Gabàs, M., Anadón, P., 2000a. Lithostratigraphical correlation of micromammal sites and biostratigraphy of the Upper Pliocene to Lower Pleistocene in the Northeast Guadix-Baza Basin (southern Spain). Journal of Quaternary Science 15, 43–50.

Oms, O., Parés, J.M., Martínez-Navarro, B., Agustí, J., Toro, I., Martínez-Fernández, G., Turq, A., 2000b. Early human occupation of western Europe: paleomagnetic dates for two paleolithic sites in Spain. Proceedings of the National Academy of Science 97, 10666–10670.

Piperno, M., 1999. Notarchirico. Un sito del Pleistocene medio iniziale nel bacino di Venosa. Osanna, Venosa.

Piperno, M. Bulgarelli-Piperno, G.M., 1974. First approach to the ecological and cultural significance of the Early Paleolithic occupation site of Garba IV at Melka Kunturé (Ethiopia). Quaternaria XVIII, 347–382.

Potts, R., 1988. Early Hominid Activities at Olduvai. Aldine and Gruyter, New York.

Raynolds, R.G.H., Johnson, G.D., 1985. Rates of Neogene depositional and deformational processes, north-west Himalayan Foredeep margin, Pakistan. In: Snelling, J.J. (Eds.), The Chronology of the Geological Record. Blackwell Scientific Publications, Oxford, pp. 297–311.

Rendell, H.M., Dennell, R.W., Halim, M.A., 1989. Pleistocene and Paleolithic Investigations in the Soan Valley, Northern Pakistan. British Archaeological Reports, International Series 544, Oxford.

Roche, H., 2000. Variability of lithic productions in East Africa. Acta Anthropologica Sinica, Supplement 19, 98–103.

Roche, H., Delagnes, A., Brugal, J. P., Feibel, C., Kibunjia, M., Mourre, V., Texier, P.-J., 1999. Early hominid stone tool production and technical skill 2.34 Myr ago in West Turkana, Kenya. Nature 399, 57–60.

Roebroecks, W., Van Kolfschoten, T., 1994. The earliest occupants of Europe: a short chronology. Antiquity 68, 489–503.

Semaw, S., 2000. The world's oldest stone artefacts from Gona, Ethiopia: their implications for understanding stone technology and patterns of human evolution between 2.6-1.5 million years ago. Journal of Archaeological Science 27, 1197–1214.

Semaw, S., 2005. Les plus anciens artefacts lithiques (2,6 – 2,5 millions d'années) des sites archéologiques du Pliocène final de EG-10 et EG-12 à Gona Est, Afar, Éthiopie. In: Sahnouni, M., Howell, F. C., Pigeaud, R. (Eds), Le Paléolithique en Afrique. L'histoire la plus longue. Guides de la Préhistoire mondiale, Collection Paléontologie humaine, Artcom'/Errance, Paris, pp. 13–52.

Semaw, S., Renne, P., Harris, J.W.K., Feibel, C.S., Bernor, R.L., Fesseha, N., Mowbray, K., 1997. 2.5-million-year-old stone tools from Gona, Ethiopia. Nature 385, 333–336.

Stout, D., Quade, J., Semaw, S., Rogers, M.J., Levin, N.E., 2005. Raw material selectivity of the stone toolmakers at Gona, Afar, Ethiopia. Journal of Human Evolution 48, 365–380.

Tamrat, E., Thouveny, N., Taieb, M., Opdyke, N.D., 1995. Revised magnetostratigraphy of the Plio-Pleistocene sedimentary sequence of the Olduvai Formation (Tanzania). Palaeogeography, Palaeoclimatology, Palaeoecology 114, 273–283.

Texier, P.-J., Roche, H., Harmand, S., 2006. Kokiselei 5, Formation de Nachukui, West Turkana (Kenya): un témoignange de la variablilité ou de l'évolution des comportements techniques au Pléistocène ancien? In: XIVth International Congress of the Union for Prehistoric and Protohistoric Sciences, Liège (septembre 2001). British Archaeological Reports International Series 1522, Oxford, pp. 11–22.

Toro-Moyano, I., Agustí, J, Martínez-Navarro, B. (Eds.) 2003. El Pleistoceno inferior de Barranco León y Fuente Nueva 3, Orce (Granada). Memoria científica campañas 1999-2002. Arqueología. Monografías, 17. Sevilla.

Toro-Moyano, I., Lumley, H de, Barsky, D., Celiberti, V., Cauche, D., Moncel, M.-H., Fajardo, B., Toro, M., 2003. Las industria líticas de Barranco León y Fuenta Nueva 3 de Orce. Estudio técnico y typológico. Análisis traceológico. Resultados preliminaires. In: Toro-Moyano, I., Agustí, J, Martínez-Navarro, B. (Eds.), El Pleistoceno inferior de Barranco León y Fuenta Nueva 3, Orce (Granada). Memoria científica campañas 1999 – 2002. Archeología monografías 17, pp. 183–206.

Toth, N., 1985. The Oldowan reassessed: a close look at early stone artefacts. Journal of Archeological Science 12, 101–120.

Turq, A., Martínez-Navarro, B., Palmqvist, P., Arribas, A., Agusti, J., Rodriguez-Vidal, J., 1996. Le Plio-Pléistocène de la région d'Orce, province de Grenade, Espagne: bilan et perspectives de recherches. Paléo 8, 161–204.

5. Early *Homo* Occupation Near the *Gate of Tears*: Examining the Paleoanthropological Records of Djibouti and Yemen

Parth R. Chauhan*
Stone Age Institute
Bloomington, IN 47407-5097
USA
pchauhan@indiana.edu

Keywords: Early Homo, Dispersal routes, Plio-Pleistocene, Bab al-Mandab, Yemen, Djibouti

Abstract

The Bab al-Mandab region has often been considered a primary crossing point for early hominins following a southern coastal route from East Africa to South and Southeast Asia. However, surprisingly little work has been done in the countries of Djibouti and Yemen, both of which hold the key to our understanding of the chronological, paleoenvironmental and adaptive contexts of such early movements. As a result, detailed and accurate information about hominin subsistence, raw material exploitation, climatic adaptations, and the rate and success of early dispersals in such regions still remain poorly understood. Being a part of the Rift Valley, Djibouti shows great potential for paleoanthropological research in parity with the rest of East Africa. Only one Oldowan site, near Lake Abbé, is currently known and dated to between 1.6 and 1.3 Ma by ESR, with presumably butchered remains of *Elephas recki ileretensis* and hundreds of artifacts on lavas. In addition, a complete articulated skeleton of *Elephas recki recki* was found in clays of the comparatively younger Gobaad Formation. Previous investigators have also reported a fragmentary maxilla, attributed to an older form of *Homo sapiens* and dated to ~250 Ka, from the valley of the Dagadlé Wadi. In Yemen and other parts of the Arabian Peninsula, archaeological investigations by numerous workers have yielded an abundance of Lower Paleolithic sites near the mountains and on fan surfaces, particularly in the Hadramaut area and the Tihama Plains, including the Al-Guza cave site with possible Oldowan artifacts. Surveys 25 to 40 km inland from the Gulf of Aden, South of Yemen, have yielded almost 40 Lower Paleolithic sites, including several Oldowan sites. Despite these commendable efforts, however, vast parts of both Djibouti and Yemen remain largely unexplored and much of the known evidence from both regions has not been absolutely dated or excavated. Until this is done, such data lend little support to early dispersal models that incorporate a southern coastal route to Southeast Asia during the Late Pliocene. This paper attempts to highlight and assess the earliest-known Mode 1 and Mode 2 evidence from Djibouti and Yemen, and correlate them with the available Plio-Pleistocene environmental records of the Bab al-Mandab region. Another objective is to provide a detailed synthesis of the original French publications on the paleoanthropological evidence from Djibouti, thus making it more widely available for comparative purposes.

5.1 Introduction

After the Oldowan industry's initial appearance in East Africa at ca. 2.6 Ma (Semaw et al. 1997), this technology began expanding rapidly to other parts of Africa (Leakey 1971; Kibunjia et al. 1992; Isaac 1997; Plummer 2004; Schick and Toth 2006; Sahnouni and Derradji 2007). Soon thereafter between 2.2 and 1.8 Ma, early *Homo* migrated into Eurasia for the first time (Larick and Ciochon 1996; Antón and Swisher 2004; Dennell and Roebroeks 2005). The earliest paleoanthropological evidence representing such an event comprises early *Homo* fossils and/or Oldowan-type assemblages from Dmanisi in West Asia (~1.8 Ma), Riwat and the Pabbi Hills in Pakistan (~2.0 Ma), Sangiran and Modjokerto in Southeast Asia (~1.8 Ma), and sites in central China (~1.6 Ma) (Dennell et al. 1988; Swisher et al. 1994; Gabunia et al . 2000; Zhu et al. 2003; Hurcombe 2004; de Lumley et al. 2005). The Acheulian industry appears later in East Africa, presumably between 1.7 and 1.6 Ma (Roche et al. 2003; Beyene 2003 cited in Suwa et al. 2007), before dispersing within and out of

* Address for correspondence: pchauhan@indiana.edu

E. Hovers and D.R. Braun (eds.), *Interdisciplinary Approaches to the Oldowan*, DOI: 10.1007/978-1-4020-9059-2,

Africa at ca 1.4 Ma (Clark 1994). A host of factors probably facilitated and/or affected these initial dispersals: ecological changes/selective pressures (Anton et al. 2002; Vrba 2007), stone tool technology (Anton and Swisher 2004), the tempo and direction(s) of faunal migrations (Martínez Navarro 2004; O'Regan et al. 2006), geo-topographic barriers and corridors (Dennell 2004), tropical diseases (Bar-Yosef and Belfer-Cohen 2001), and existing levels of resource competition with other hominin groups. The precise routes taken and the number of such migrations also currently form a major topic of research in human evolutionary studies (Anton and Swisher 2004; Langbroek 2004).

Although current interpretations regarding early human dispersals appear to be generally convincing, the paleoanthropological records of some of the most geographically critical regions are still poorly-understood. Two such regions that require further investigation are the northern regions of Eastern Africa such as Eritrea, Djibouti, and Somalia (WoldeGabriel et al. 2000) and the Arabian Peninsula (Petraglia 2003), particularly the southern portion. Their respective behavioral and paleontological records have significant implications for understanding the timing, context, routes, extents and frequencies of early hominin and faunal movements between continents (northern or through central Asia vs. southern or through southern Asia) (Bar-Yosef and Belfer-Cohen 2001; Dennell 2003, 2004). Given its geographic location, the Sinai Peninsula is often assumed to have been the primary migratory corridor between East Africa and West Asia for early hominin movement to explain the evidence known from Dmanisi, 'Ubeidiya and other early sites (e.g. Derricourt 2006). However, no convincing paleoanthropological evidence of appropriate age has been reported from associated regions such as the Nile Valley, the Sinai Peninsula and the Jordan Rift Valley, to reflect such a dispersal (e.g. Turner 1999). Likewise, the Straits of Bab al-Mandab (SBM henceforth) which have also been highlighted often (e.g. Whalen et al. 1989; Larick and Ciochon 1996; Mithen and Reed 2002; Petraglia 2003), have never been investigated adequately in order to *confirm* them as a major route during the Late Pliocene – Early Pleistocene. Our current knowledge of an existing land bridge at various temporal intervals (regardless of the short distance) at the SBM remains equally unclear (Petraglia 2005). During the Plio-Pleistocene, however, rates of sedimentation in the southern parts of the Red Sea were relatively high, indicating ongoing tectonic and erosional processes (Coleman 1974) near the SBM.

Although the currently available dataset from both sides of the SBM are inadequate (i.e. a lack of absolute dates and the low number of well-excavated sites), it is necessary to highlight the regions and outline some future research goals to increase the resolution of our current state of knowledge and test existing dispersal theories and environmental hypotheses related to human evolution (e.g. deMenocal 2004; Vrba 2007). Indeed, the earliest archaeological records of Djibouti and Yemen may not only have an important bearing on the entire Asian body of evidence, but also on the novel concept of *Homo erectus* evolving in Asia and subsequently entering Africa (Swisher et al. 1994; Asfaw et al. 2002; Dennell and Roebroeks 2005; Turner and O'Regan 2007). Such types of data sets can also be combined with genetic studies and the archaeological evidence to answer similar eco-geographic and contextual questions regarding the subsequent expansion of modern human populations into Eurasia, also presumably across the SBM as

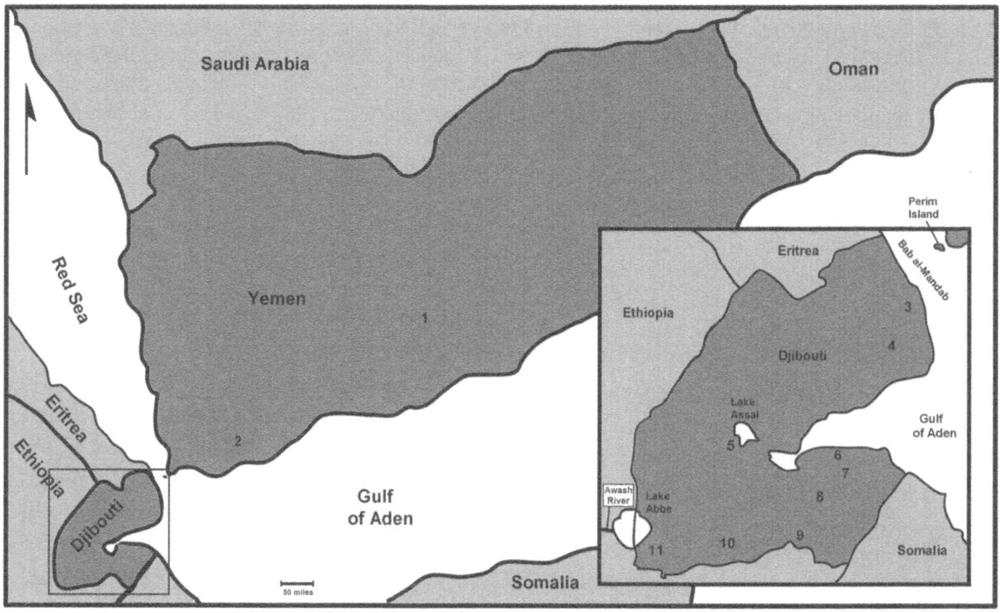

FIGURE 5.1. The Bab al-Mandab region. Paleoanthropological sites shown are: 1) Al Guza, 2) Jebel Tala, 3) Goh, 4) As Alé, 5) Sagantole, 6) Oflé, 7) Gombour As, 8) Armakato and Douré, 9) Ali Sabieh, 10) Anabok-koma, 11) Barogali and As Eyla. (Sources: Amirkhanov 1994; Berthelet 2002)

one route during the Upper Pleistocene (Rose 2004; Petraglia and Alsharekh 2004; Field and Lahr 2005; Beyin 2006).

The aim of the present paper is to briefly review the earliest paleoanthropological evidence from Djibouti and Yemen, countries geographically closest to the SBM. Much of the Early Paleolithic data presented here for Djibouti represents a synthesis of the extensive French literature on this region. The Yemeni evidence is discussed in less comparative detail since numerous publications in English are available. Some of the paleoanthropological sites from both Djibouti and Yemen as discussed in this paper are illustrated in Figure 5.1.

5.2 Djibouti

Originally a part of Ethiopia, Djibouti is situated on the western shore of the SBM directly across Yemen, forming the northeastern-most extension of the East African Rift Valley or the Afar triangle (WoldeGabriel et al. 2000). Several mountain ranges are found in the northern and central part of the country, also known for representing the lowest spot in Africa below mean sea level. The southwestern corner is dominated by Lake Abbé which is where the Awash River ends. Lake Assal, fed by groundwater, is found in the central part of the country and both lakes collectively form the end of a chain of lacustrine settings in this region of East Africa (Hughes and Beydoun 1992). It has been previously suggested that paleoanthropological sites similar to those found in Eritrea and Ethiopia (e.g. Abbate et al. 1998; Martínez Navarro et. al. 2004) are likely to be found in this area (Chavaillon 1987; also see Caton-Thompson [1957] for similar expectations on the Yemen side).

Tertiary sediments in Djibouti are restricted to linear patches that generally extend in a NW-SE direction with some exceptions (Zumbo et al. 1995: Figure 5.1). Plio-Pleistocene volcanic activity is evinced by extreme but intermittent magnetism from 2 to 0.25 Ma in the Stratoid series in the Afar depression and the Asal and Manda Inakir rifts, respectively (Zumbo et al. 1995; WoldeGabriel et al. 2000). Volcanic deposits are generally basaltic and rhyolitic and thought to range in age from 3.4 to 1 Ma and includes the Bahlo-Gamarri and Oummouna Formation. "The Asal Rift is the most active "axial zone" of the Afar Depression" (Zumbo et. al 1995:285) and these lavas are generally thought to be younger than 0.7 Ma (Courtillot et al. 184 cited in Zumbo et. al. 1995). The Manda Inakir Basalt is younger than 0.5 Ma. Trachyte from various parts of the Moussa Ali volcano gave variable K/Ar ages ranging from younger than 0.15 Ma to about 0.9 Ma. Depending on the flaking and functional qualities of these rocks, their respective age brackets may help in pinpointing and understanding the tempo of hominin occupation in the region in relation to raw material availability.

In Djibouti, field expeditions were first initiated in French Somalia and Abyssinia (modern day Ethiopia) by Pierre Teilhard de Chardin and P. Lamare in the late 1920s and later joined by H. de Monfreid, Abbe H. Breuil, and P. Werment during an expedition in the early 1930s. Most of the work at that time comprised surface collections; detailed investigations were not carried out. Amateur archaeologists visited the region after the mid-1940s, yet goal-oriented multidisciplinary paleoanthropological research was initiated only in the 1970s by French scientists such as Cl. Thibault and C. Roubet in Gobaad and Oued Chekheyti (Chavaillon 1987). Following the political independence of Djibouti, the majority of the work (albeit intermittent) has been carried out under the auspices of Institut Superieur d'Etudes et de Reserches Scientifique et Techniques (ISERST) (Berthelet et al. 1992; Berthelet 2002). Under the administrative direction of M. Anis Abdallah Kamra (founder of ISERST), several geological and archaeological expeditions were carried out by French scientists throughout different parts of Djibouti, especially by L. de Bonis, J. Chavaillon, X. Gutherz, and R. Joussaume. The expedition that was led by J. Chavaillon surveyed the Gobaad Basin in the Dikhil region (in the SW) and other areas such as Ali Sabîh (in the SE), Tajoura (central-western), and Obock (in the NE) (Berthelet 2002). One of the most prominent and well-known Oldowan sites in the Gobaad region in the Lake Abbé area is Barogali, excavated from 1985 to 1987 (Berthelet and Chavaillon 2001). In addition to Barogali, the Gobaad Formation yielded well-preserved vertebrate fossil material including a nearly-complete articulated skeleton of *E. recki recki* (Chavaillon and Berthelet 2001).

Barogali is located near the southern border with Ethiopia where lavas of indeterminate age are interstratified with sandy and clayey horizons. The sediments suggest an ancient marsh context that has preserved, in well-consolidated sediments, faunal remains of *Elephas recki ileretensis* in alleged behavioral association with Oldowan stone tool types (Berthelet et al. 1992; Berthelet 2001, 2002). The pachyderms at Barogali and comparable faunal localities in Djibouti include elephants and hippos. There is, however, a major problem of chronologically and taxonomically separating subspecies of *Elephas recki* because of significant temporal and metrical overlap between them (Todd 2005).

The artifacts at Barogali were produced on alkaline basalt of mediocre quality which is generally found on the periphery of the lacustrine basin (based on petrographic studies). In addition to the Last Appearance Datum (LAD) of 1.3–1.2 Ma then given for *E. recki ileretensis*, the investigators offer an age of 1.6–1.3 Ma for the site, estimated through preliminary ESR efforts on the lower third molar of the *Elephas* by R. Bouchez at the University of Grenoble (Berthelet and Chavaillon 2001; Berthelet 2002).[i] According to the investigators, the site fits well with classic Oldowan or Developed Oldowan sites. The excavation yielded 569 artifacts, variably worked and described as being technologically archaic and often shattered. Flaked pieces (*sensu* Isaac 1997) are represented by over 100 specimens. Broken lava pebbles represent 21.3% of the assemblage. At least five hammerstones were identified based on pit or percussion marks on the cortical surfaces. Broken pebbles and clast fragments number 116 and range in size between 5 and

150 mm. However, only 14% of these are between 60 and 150 mm. The majority is 20–50 mm in length and exhibit multiple fracture patterns. These patterns were identified as longitudinal, inclined and flat or concave, based on the position of the break and its morphological aspects on specimens over 20 mm in maximum length. The size or comparatively diminutive dimensions of the core specimens and their overall frequency may suggest that they were derived from unsuitable pebbles or even polyhedrons fractured during flaking or heavy-duty utilization. The authors stress that this evidence appears to indicate that the hominins favored a distinct material to pound bone with and use this function to identify the locality as a butchery site. Over 5% of the assemblage is pebble tools including 5 different types of bimarginal choppers (see Chauhan 2007 for this new chopper terminology), polyhedrons, spheroid, and core-scrapers (the latter again used to support butchery at the site).

Almost half of the choppers are side specimens and other chopper types include chisel, truncated, pointed, and peripherally-worked choppers. Polygonal and pentagonal, are the most common choppers (64%), and about seven choppers exhibit minimal flaking or less than five flake scars. The investigators appear to have counted flake scars on two faces on the flaked specimens (Faces A & B) and observe that choppers that have been worked peripherally retain about 9–13 flake scars, recalling discoid types from Olduvai Gorge. With regard to length and thickness, seven choppers are worked or retouched alternatively and three are single-sided.

The cores or flaked specimens (n = 14) are unipolar, centripetal and polyhedric and range between 40–100 mm in length. The polyhedric cores are equated in size and general morphology with those from Gombore IB and Melka Kunture. Polyhedrons form 22% of the pebble tools at Barogali and comprise a variety of shapes (e.g. pointed, prismatic, spheroids) and only three specimens are heavier than 850 g. The description of the facetted spheroid or "bola" stone is confusing, including information regarding the utilization and striking marks.

Furthermore, the authors state that such spheroids have been recovered from Olduvai Beds I and II, Ain Hanech, Sterkfontein Member 5, Melka Kunture and Garba IV. In addition, 3 heavy-duty scrapers and 5 core-scrapers are reported from Barogali, their maximum length varying between 60 and 80 mm. Both types of specimens show some morphological and technological diversity — the heavy-duty scrapers possess an edge angle of 70–80° while the core-scrapers have high and vertical retouch and an edge-angle closer to 90°. While the former are hexagonal and pentagonal, the latter are quarter-elliptical and semi-circular. The heavy-duty scrapers have 4 to 8 flake scars and the core-scrapers have 2 to 6 flake scars. In general, these specimens from Barogali were thought to have been used as picks or scrapers on animal remains (i.e. butchery, hide-working?). Unlike the formal tools, debitage is abundant and the number of detached pieces is particularly high (71% or 402 specimens). The investigators state that these may derive from tools and cores found at the site though no refits are mentioned or illustrated. They also argue that the butchery site had its own knapping workshop and possible functional purposes as well (these are not mentioned). More than half the flakes are broken and flake tools and "utilized flakes" comprise 15% of the debitage. The abundance of the flakes and flake types possibly demonstrates that core reduction took place at the site. Flake tools and casually retouched flakes only represent 6.3% of the assemblage and no small flakes appear to be retouched or utilized, showing a preference for larger flakes.

The best represented tool types are backed knives, notches, and denticulates and scrapers (side or end) are deemed exceptional. The knives (n = 4) are possibly naturally-backed, and the notched tools have been fashioned on flake fragments, though the notches are not very prominent. Only three denticulates are reported and the investigators recognize 24 flake specimens that may have been utilized but for which very little evidence is visible (i.e. they may be naturally damaged or represent interrupted flaking episodes). Indeed the investigators argue that the low number of flake tools in each category suggests the site represents a temporary habitation of a small group that the length of occupation was limited to a few days and that part of the tool-kit may have been transported by the hominins.

The majority of the *Elephas* remains were recovered within a 25 m^2 area of the excavation. The rest of the specimens attributed to *Elephas* sp. were found within an area of 35 m^2. The *Elephas recki* cranium was found away from the rest of the post-cranial skeleton. Molar and tusk fragments were also recovered away from the center of the proboscidean bone concentration. The highest concentration of artifacts is found near the cranium. In particular, a centripetal core, several flakes, and a chopper were found in association with the *Elephas* cranium. Other bones such as the mandible, vertebrae, ribs, scapulae and so forth, are scattered and located away from the cranium, again, in association with stone tools — both cores and flakes. It was hypothesized by the excavators that a single individual may have been displaced or scattered by scavengers and they also stress that the associated lithic material may not be in completely primary context (Berthelet 2001, 2002). Nonetheless, the investigators surmise that the kill site is a temporary camp of short duration and compared to other elephant butchery sites, Barogali is particularly important due to the abundance of stone tools and well-preserved vertebrate faunal remains. The absence of butchery marks may be due to taphonomic processes or simply represent the occasional paucity of such evidence on butchered pachyderm carcasses (see Delagnes et al. 2006). The authors of the original work on Barogali stress that the site represents a workshop site due to the nature of the archaeological materials (again, the investigators have not elaborated on the evidence to support this interpretation and thus, the associated material needs to be reanalyzed).

Another site with well-preserved elephant remains is Haïdalo, 8 km from Barogali. Here no artifacts were found.

Elsewhere in the Gobaad, at Gafolo (bordering Oued Dagadle), Faure and Guérin (1997) excavated *Elephas* (placed by them into *Palaeoloxodon*) *recki ileretensis* and *Hippopotamus amphibius* specimens. These paleontological sites also did not yield any artifacts. Three kilometers South of Barogali at the foot of outcrops at l'Oued Dagadle, a bimarginal chopper and one core-scraper were recovered. Between Dikhil and As Eyla, Cl. Thibault and his team noted the presence of worked quartzite clasts including bimarginal choppers, cores, and flakes. As Eyla has also yielded a spheroid-like artifact but no Acheulian bifaces (Chavaillon 1987). Based on stratigraphic correlation of geological formations, these artifact bearing localities may be as old as the Early Pleistocene or simply older than 1 Ma (Berthelet 2001, 2002). At Anabokkoma, the team recovered two choppers in a gravel context, as well as a bec, a large flake, and faunal fragments including elephant. At Armakato, Thibault carried out test trenching which yielded three fossiliferous layers as well as 172 artifacts and which is thought to be about 0.6 – 0.7 Ma in age (Chavaillon 1987).

In the region of Sangatole, to the SW of Lake Assal and near l'Oued Kalou, a polyhedric core was recovered and thought to represent the oldest *in situ* archaeological occurrence in that region. In the Oued Doure Basin of the Ali Sabîh region, investigations between 1990 and 1992 yielded surface and stratified archaeological material. Near rhyolite outcrops in the region, 345 large Acheulian artifacts (designated as Middle or Upper Acheulian) of the same material were collected including side and end choppers (n = 18), one polyhedron, and eight heavy-scrapers. The bifaces include cleavers and picks and there are numerous cores (n = 92) including Levallois, centripetal, unipolar, bipolar, and polyhedral types as well as flakes and flake tools (n = 118) including notch tools, denticulates, end-scrapers, burin, borer, backed knives and minimally retouched flakes. Additional Early Acheulian occurrences come from Gombourta near Arta and C. Roubet has reported the site of Ofle near the Gulf of Tadjoura and surveyed for sites at Sadday and Obock, where Late Acheulian assemblages were found (Chavaillon 1987). Younger lithic assemblages, belonging to the MSA and LSA, are not well-known but have been reported from Obock, Tables de Godoria, Ali Sabih, Arta, and Ouea, all of which have yielded flake-based assemblages including Levallois attributes such as prepared-core technology and small bifaces (assigned to the Fauresmith industry) as well as about 400 microlithic specimens collected from several sites by Boissaubert and Chavaillon (Chaivaillon 1987).

De Bonis et al. (1988) have described several Plio-Pleistocene vertebrate fossil localities and associated specimens from Djibouti and in addition to the intermittent geological research done by other teams they carried out paleoanthropological surveys in Anabokoma and Hara Idé near As Eyla. The former site comprises sections totaling to over 100 m in vertical thickness including strata of sand, gravel, clays, silts, and sandstone with occasional paleosol horizons. An episode of lava extrusion is recorded in the sequence as well. De Bonis and colleagues suggest ecological continuity in the faunal communities found above and below this marker horizon (the lava). The sites at Hara Idé including Barogali have yielded a variety of Plio-Pleistocene fauna including bones and dental material (Table 5.1). Paleomagnetic analyses of sediments associated with the vertebrate fossils show them to be of reverse polarity, indicating an age older than 780 Ka. Based on a comparative perspective, Ana Koma or Anabokoma is the richest and oldest site, the fauna suggesting an Early Pleistocene age, partially supported by the absence of *Metridiochoerus compactus* (de Bonis et al. 1988).[ii] Hara Idé I is thought to be younger than the Brunhes-Matuyama boundary but based on lithic typology, de Bonis et al. (1988) conversely place it at no older 0.9 Ma.

Most importantly, de Bonis et al. (1984) reported a locality near Oued Dagedle (Hara Idé Mx) that produced a hominin

TABLE 5.1. Key Plio-Pleistocene mammalian taxa from localities closest to both sides of the Bab al-Mandab*

	Djibouti	Saudi Arabia
Carnivora		
Crocuta crocuta		+
Panthera cf. *gombaszoegensis*		+
Vulpes cf. *vulpes*		+
Proboscidea		
gen. et sp. indet. (cf. *Elephas recki*)	+	+
Elephas recki recki	+	
Elephas recki ileretensis	+	
Perrisodactyla		
Equus sp.	+	+
Ceratotherium sp.	+	
Hipparion cf. *cornelianum*	+	
Artiodactyla		
Hippopotamus (cf.) *amphibius*	(+)	+
Hexaprotodon sp.?		+
gen. et sp. indet. (Camelidae)		+
Sivatherium maurusium	+	
Kolpochoerus cf. *phacocheroides*	+	
? *Metridiochoerus andrewsi*	+	
Metridiochoerus cf. *hopwoodi*	+	
Metridiochoerus compactus	+	
Bovidae		
Pelorovis cf. *oldowayensis*	+	+
Oryx sp.		+
gen. et sp. indet (Alcelaphini)		+
gen. et sp. indet. 1		+
gen. et sp. indet. 2		+
Syncerus sp.	+	
? *Kobus* cf. *kob*	+	
Numidocapra crassicornis	+	
? *Damaliscus* sp.	+	
Megalotragus kattwinkeli	+	
? *Aepyceros* sp.	+	
Antidorcas cf. *recki*	+	
Primates		
Homo sp.	+	

*Sources: de Bonis et al. 1988:324; Faure and Guérin 1997; Thomas et al. 1998:149.

maxillary fragment that is attributed to advanced *Homo erectus* or archaic or primitive *Homo sapiens* based on the dental evidence (de Bonis et al. 1988); several teeth are preserved. This locality is estimated to be about 250 Ka old from preliminary U, K, and Th compositions (Chaivaillon 1987). However, de Bonis et al. (1988) have suggested a conflicting age bracket of 0.6 ± 0.3 Ma to 1 Ma for this locality. From the fauna and general observations on the sediments and landscape, the investigators state that the region was more humid and wooded than today and supported Oldowan hominin populations during the Lower Pleistocene (Berthelet 2002). At other localities discussed above, de Bonis et al. (1988) conclude that the presence of hippo and reduncini suggest an open habitat with some humidity.

Following the work by French scientists, Djibouti has not received further scientific attention by Western or African paleoanthropologists. When considering their potential significance, these aforementioned sites clearly betray a broad temporal range (Lower to Middle Pleistocene) with conflicting age estimates and paleoenvironmental interpretations. On that basis, the hominin fossil site and other vertebrate fossil sites in Djibouti clearly merit renewed geochronological and stratigraphic investigations to resolve current confusion as evident in the literature. For instance, the alleged butchery activities at these locales can only be supported by *unequivocal* evidence of hominin-produced percussion or cut marks in comparable contextual integrities.

5.3 Yemen

Located in the southwestern part of the Arabian Peninsula, Yemen represents the eastern terrestrial boundary of the SBM. Although very little multidisciplinary paleoanthropological research has taken place in this region, it is evident from the known sites and its geographic location that the region may very well yield evidence representing the biological predecessors of the earliest-known hominin populations in Asia. The South Arabian and East African regions shared similar depositional histories during the Paleocene and Eocene (Hughes and Beydoun 1992). A continuous topographic landscape and basement geology can be found on either side of the SBM as both regions share the boundary between the Asian and African tectonic plates along the Red Sea (Coleman 1974; Chapman 1978). Large portions of Arabia, including the Rub' al Khali, were mostly savannah grassland with prominent drainage systems during the Late Pliocene as indicated by higher isotope values and C_4 grassland in north eastern Africa between 3.4 and 1.5 Ma (see Dennell 2004; Feakins et al. 2005; Edgell 2006) and comparable isotopic evidence from Early Pleistocene herbivore teeth in northern Arabia (Thomas et al. 1998). The general environment in both regions probably remained similar during the late Pliocene and Quaternary time frames. During the Early Pliocene, the Red Sea region witnessed deep-marine and normal salinity conditions, possibly a result of the lack of a land-bridge between the Red Sea and the Gulf of Aden, and Early Pleistocene volcanism is also known (Edgell 2006). These conditions continued to the present and are exemplified by the Wardhan Formation in Yemen which is composed of Plio-Pleistocene marine clastic sediments. There are correlative geological strata on the East African side of the SBM known as the Desset Formation. The Wardhan Formation is overlain by the Shagara Formation — a carbonate facies — which is synonymous with the Dharishab Formation in Ethiopia (Hughes and Beydoun 1992; also see Beydoun 1970). Coral reef deposits of limestone along the coastal zones are generally buried rapidly from colluvial and fluvial processes of these sedimentary structures located further inland and at higher elevations (Coleman 1974) such as the Hejaz Asir Mountains. Although a eustatic rise in sea-level in relation to a warmer climate has been proposed between 1.9 and 1.8 Ma (Edgell 2006), it is generally accepted that a land-bridge existed immediately prior to this time (Turner 1999; Turner and O'Regan 2007) and which may have been more intermittent since then due to fluctuating sea-levels and the rifting of the Red Sea area (see Whitmarsh et al. 1974; Werner and Mokady 2004). Paleolithic sites buried in such terrestrial and marine deposits (e.g., Bates et al. 2007) may enable in accurately identifying evidence for some of the earliest coastal dispersals through this region.

Vertebrate faunal material has been reported from throughout the peninsula (see Whybrow and Hill 1999) and one of the most important localities is located in the An Nafud Desert in northern Saudi Arabia (Thomas et al. 1998). The evidence here come from three occurrences associated with lacustrine deposits and comprise a number of large vertebrate taxa (Table 5.1) that suggest an Early Pleistocene Age and African affinities. Except for some pre-Pliocene evidence (Tattersall et al. 1995), no comparable faunal assemblages are currently known from Yemen.

The known Early Paleolithic evidence in the Arabian Peninsula records both Oldowan and Early Acheulean occupation, indicative of multiple early dispersal events (de Maigret 1983; Whalen et al. 1983; Bar-Yosef and Belfer-Cohen 2001; Anton and Swisher 2004; Amirkhanov 2006; see Petraglia 2003, 2005 for overviews). The majority of the surveys have been conducted in the southwestern and central areas, particularly the Tihama and Hadrawmaut regions and very little work has been done in the eastern and northeastern sectors (Van Beek et al. 1964). Paleolithic sites are known to occur in a variety of ecological and topographic settings, often on the surface of terraces and volcanic landscapes, but occasionally in stratified contexts. However, none of these lithic occurrences throughout the region have reliable absolute dates and only a few sites have been systematically excavated; some localities represented deflated surface occurrences and maintain very low artifact densities (e.g., Bulgarelli 1986; Zimmerman 2000).

Some of the earliest surveys were conducted by Caton-Thompson and Gardner (1939) in the 1930s and Caton-Thompson (1957) in southern Arabia near the SBM, who was indeed one of the first researchers to explicitly address and test the SBM as a dispersal corridor from East Africa (also see Goodenough et al. 1939). They reported lithic occurrences in surface context and of low artifactual density, with the exception of some stratified Mode 1 assemblages documented between the Wadi 'Amd and Tarim by Caton-Thompson (1953; Doe 1971) and in the Hadramawt region by Amirkhanov (1994, 1999, 2006). The Hadramawt region is located in the central part of the country and north of the Gulf of Aden, where about 53 Paleolithic occurrences have been reported (Amirkhanov 1994) including cave and open-air contexts. The best known early Mode 1 or Oldowan-like material allegedly comes from the cave of Al Guza where a step-trench of 2 m x 20 m (or 39 m^2) yielded 972 cores, choppers, polyhedrons, side scrapers, and flakes in stratified contexts (Amirkhanov (1994, 1999). Although Amirkhanov (1994) interprets this material as being Oldowan or of pre-Acheulian Age, Petraglia (2003) does state the possibility of naturally fractured specimens as many exhibit marginal cortex removal. In addition, the dating of the site remains ambiguous. For example, an age range of 0.6 to 1.0 Ma through paleomagnetic and TL methods on the cave deposits and potassium-argon elsewhere in the valley is reported (Amirkhanov 1999); based on the work of Amirkhanov (1994), Edgell (2006) assigns an age bracket of 1.4 – 1.0 Ma for the Oldowan evidence in the Hadramawt region. However, the uranium-thorium result is not elaborated upon by Amirkhanov as it was thought to be beyond the accepted range of this method and the TL date (0.6 Ka) is not accepted as it had been a controversial method at the time. Regarding the paleomagnetic dating, Amirkhanov (1999:49) states that "…the Lower Bed *probably* [emphasis mine] corresponds to Brunhes/Matuyama boundary. There is a remote possibility however, that the paleomagnetic data reflect the Jaramillo or an even earlier episode of normal polarity." As detailed methodological information is missing, the location of sediment sampling is unclear in relation to the stratigraphic profile comprising limestone bedrock and fragments, rock debris, scree, rubble, fine sediments, and artifacts (Amirkhanov 1994:220; 1999:47). In other words, a reinvestigation of the Al Guza stratigraphic context, geochronology, and the associated archaeological material is highly merited, as well as additional surveys to locate sites with comparable evidence with robust contextual integrity.

Compared to the Oldowan, the Acheulian evidence in southern Arabia appears to be more prominent and has been reported as surface concentrations and from stratified contexts (Petraglia 2003). There are 22 reported localities from the Hadramawt region, some of which as reported by Caton-Thompson (1957) are associated with loess deposits, thus possibly being dateable in the future. The Russian work was followed by the efforts of Whalen and Pease (1992) who reported 28 Lower Paleolithic occurrences in western Yemen in the Tihama region near the Bab al-Mandab, including Acheulian sites on alluvial fans, thought to reflect a peripheral geographic expansion of East African hominins (e.g. Whalen et al. 1989). The entire landscape here comprises alluvial fans, volcanic badlands, and coastal plains, all with a host of diverse raw material sources. Some of the deposits are of Lower Pleistocene age (Whalen and Pease 1992) and include clay and aeolian sand (Whalen and Schatte 1997). Some of the lithic evidence has been compared to the Acheulian variant from the Levant (i.e. the Acheulo-Yabrudian) because bifaces were not abundant and choppers were more prominent. Sites with polyhedrons and Levallois technology in this area were presumed to be Middle Acheulian (Petraglia 2003). Interestingly, Whalen and Pease (1989) did not observe any Oldowan or Developed Oldowan evidence in their study area. While this may appear surprising at first, it is important to appreciate that their study area represented only a fraction of a vast promising landscape that clearly deserves renewed survey efforts, particularly in light of the possible Oldowan evidence to the north at Shuwayhitiyah and Najran in Saudi Arabia (e.g., Whalen and Pease 1990) — although these may be a result of movements through the Sinai peninsula rather than the SBM. Later, Whalen and Schatte (1997) carried out surveys 25 to 40 km inland from the Gulf of Aden and reported sites in volcanic landscapes and on flat plains, which yielded 37 Lower Paleolithic sites, often in association with raw material sources. Several sites were identified as being Oldowan and the rest were identified as Acheulian or younger typo-technological assemblages. Artifact densities and site size vary throughout these localities and Acheulian bifaces in this region were not found to be very abundant. Recovered artifacts were produced from a variety of raw materials including quartzite, rhyolite, basalt, as well as more siliceous material such as chert. At the Oldowan sites, cores and core-based tools dominated the assemblages whereas the Acheulian sites did not yield abundant bifaces, the flake tools and choppers being more common. Unfortunately, all these were in surface context on granite basement rock with a marked absence of fine sediments. Recent surveys in Southwest Yemen by the author and his colleagues have yielded the first-known Oldowan-like lithic assemblages on Perim Island (located between Djibouti and mainland Yemen).This evidence still requires proper investigation; if corroborated, it will demonstrate for the first time that hominin groups were present at this specific location in the "middle" of the SBM.

5.4 Conclusion

The current evidence from Djibouti and Yemen suggests that Oldowan and Early Acheulian occupations took place in a variety of ecological and topographic contexts such as fluvial terraces and floodplains, alluvial fans, extensive rocky outcrops, caves, coastal plains, lake margins, and hilly terrain. While the majority of the sites in both countries are in surface context, several stratified occurrences have also been reported and show some potential for absolute dating and paleoenvironment reconstructions. At the same time, several major issues remain unresolved and require multidisciplinary investigations, such as the precise location of certain lithic occurrences, associated stratigraphic information, the nature of deposition of the lithic assemblages and fossils (if any), and the general paucity of hominin remains. Most importantly, the known Oldowan evidence (e.g. Barogali, Al Guza) from both respective regions does not allow robust correlations for an African dispersal through the Bab al-Mandab. This is primarily due to conflicting or unreliable geochronological information in addition to ambiguous archaeological interpretations. As a result of these lacunae, such information as hominin subsistence, technological abilities, raw material exploitation, climatic adaptations to local conditions, and the rate and success of dispersals into these areas is poorly understood. A greater challenge, however, will be to correlate the data from both countries to test if and when early African hominin populations passed through these regions or went *around* the Red Sea through the Levant when colonizing Asia. One possible approach may be to target and identify regional behavioral and adaptive signatures (e.g. discrete assemblage compositions of specific tool types, a preference for lacustrine and floodplain paleoenvironmental conditions, presence of faunal assemblages with African affinities), such as those from West Asia (Dmanisi), the near East ('Ubeidiya), and East Africa (e.g. Gona, Konso Gardula) and compare these with the evidence from Djibouti and Yemen. For example, some of the vertebrate fossil evidence (at least four mammalian species) occurs on both sides of the Bab al-Mandab, possibly demonstrating faunal migrations through this corridor. Such data sets can also be supplemented by geological data from the large number of volcanoes and associated sediments in both countries, thus facilitating the correlation of tephra deposits in association with Paleolithic and paleontological localities. Long term projects may prove to be useful in also identifying new adaptive patterns such as changing resource exploitation and subsistence strategies from West to East (i.e. Africa to Arabia) as hominins gradually moved from fluvio-lacustrine and forested or grassland zones into arid and semi-arid areas. Conversely, the evidence of absence of Late Pliocene-Early Pleistocene paleoanthropological data in northern Djibouti and/or SW Yemen may even subsequently eliminate the SBM as a dispersal route during this time. In any case, the ecogeographic context of the Bab al-Mandab region and its regional Paleolithic and paleontological records promise exciting discoveries in the future in relation to hominin movements through this zone during the Plio-Pleistocene. Until this region where Africa meets Asia is properly tested, our perceptions of the directions and environmental contexts of early dispersals and associated adaptive strategies will continue to remain hypothetical.

Acknowledgments. I thank Erella Hovers and David Braun for the opportunity to contribute to this important and fascinating topic. Mohamed Sahnouni and August Costa generously translated the French papers on the Djibouti evidence, without which this review would have been impossible; I also thank Frank Brown for providing the reference to Whitmarsh et al. (1974). The two anonymous reviewers provided very constructive and helpful comments which enabled a considerable improvement of the original draft. Encouragement and useful information provided by the following people increased my knowledge and appreciation for the Bab-al-Mandab region and its paleoanthropological record: Hizri A. Amirkhanov, Miriam Belmaker, Robin W. Dennell, Joy McCorriston, Jeffrey I. Rose, Mohamed Sahnouni, Sileshi Semaw, and Paul Zimmerman. I also extend my thanks to Kathy Schick and Nick Toth, codirectors of the Stone Age Institute & CRAFT Research Center (Indiana University), for the postdoctoral position and related support. Current research on the topic discussed in this paper is funded by the American Institute of Yemeni Studies and the Leakey Foundation. All interpretations and errors are my own.

References

Abbate, E., Albianelli, A., Azzaroli, A., Benvenuti, M., Tesfamariam, B., Bruni, P., Cipriani, N., Clarke, R.J., Ficcarelli, G., Macchiarelli, R., Napoleone, G., Papini, M., Rook, L., Sagri, M., Tecle, T.M., Torre, D., Villa, I. 1998. A one-million-year-old *Homo* cranium from the Danakil (Afar) Depression of Eritrea. Nature 393, 458–460.

Amirkhanov, H.A., 1994. Research on the Palaeolithic and Neolithic of Hadhramaut and Mahra. Arabian Archaeology and Epigraphy 5, 217–228.

Amirkhanov, H.A. 1999. Archaeological Evidence for Early Hominid Migration to the Arabian Peninsula. Anthropologie 37, 45–50.

Amirkhanov, H.A., 2006. Stone Age of South Arabia. Nauka, Moscow.

Anton, S.C., Leonard, W.R., Robertson, M.L., 2002. An ecomorphological model of the initial hominid dispersal from Africa. Journal of Human Evolution 43, 773–785.

Anton, S.C., Swisher, C.C., 2004. Early dispersals of *Homo* from Africa. Annual review of Anthropology 33, 271–296.

Asfaw, B., Beyene, Y., Suwa, G., Walter, R.C., White, T.D., WoldeGabriel, G., Yemane, T., 1992. The earliest Acheulean from Konso-Gardula. Nature 360, 732–735.

Asfaw, B., Gilbert, W.H., Beyene, Y., Hart, W.K., Renne, P.R., WoldeGabriel, G., Vrba, E.S., White, T.D. 2002. Remains of *Homo erectus* from Bouri, Middle Awash, Ethiopia. Nature 416, 317–320

Bar-Yosef, O., Belfer-Cohen, A., 2001. From Africa to Eurasia – early dispersals. Quaternary International 75, 19–28.

Bates, M.R., Bates, C.R., Briant, R.M., 2007. Bridging the gap: a terrestrial view of shallow marine sequences and the importance of the transition zone. Journal of Archaeological Science 34, 1537–1551.

Berthelet, A., 2001. L'outillage lithique du site de dépeçage à *Elephas recki ileretensis* de Barogali (république de Djibouti). Comptes Rendus de l'Academie des Sciences Series IIA Earth and Planetary Science 332, 411–416.

Berthelet, A., 2002. Barogali et l'Oued Doure. Deux gisements représentatifs du Paléolithique ancient en Republique de Djibouti. L'Anthropologie 106, 1–39.

Berthelet, A., Boisaubert, J.-L., Chavaillon, J., 1992. Le Paléolithique en république de Djibouti: nouvelles recherches. Bulletin de la Société Préhistorique Française 89, 238–246.

Berthelet, A., Chavaillon, J., 2001. The Early Palaeolithic butchery site of Barogali (Republic of Djibouti). In: Cavaretta, G., Gioia, P., Mussi, M., Palombo, M.R. (Eds.), *The World of Elephants*. International Congress Proceedings, Consiglio Nazionale delle Ricerche, Rome, pp. 176–179.

Beydoun, Z.R., 1970. Southern Arabia and northern Somalia: comparative geology. Philosophical Transactions of the Royal Society London A, 267, 267–292.

Beyin, A., 2006. The Bab al Mandab vs. the Nile–Levant: an appraisal of the two dispersal routes for early modern humans Out of Africa. African Archaeological Review 23, 5–30.

Bulgarelli, G.M., 1986. Archaeological activities in the Arab Republic, 1986: Paleolithic culture. East and West 36, 419–422.

Caton-Thompson, G., 1953. Some Palaeoliths from South Arabia. Proceedings of the Prehistoric Society 19, 189–218.

Caton-Thompson, G., 1957. The evidence of South Arabian Palaeoliths in question of Pleistocene land connection with Africa. In: Clark, J.D., Cole, S. (Eds.), Proceedings of the 3rd Pan-African Congress on Prehistory, Livingstone *1955*. Chatto and Windus, London, pp. 380–384.

Caton-Thompson, G., Gardner, E.W., 1939. Climate, irrigation, and Early Man in the Hadhramaut. Geographical Journal 93, 18–38.

Chapman, R.W., 1978. General Information on the Arabian Peninsula. In: Al-Sayari, S.S. Zötl, J.G. (Eds.), Quaternary Period in Saudi Arabia, Vol. I. Springer-Verlag, New York, pp. 4–18.

Chauhan, P.R., 2007. Soanian cores and core tools from Toka, northern India: towards a new typo-technological organization. Journal of Anthropological Archaeology 26, 412–441.

Chavaillon, J., 1987. Cinquante ans de préhistoire en République de Djibouti. *Arqueologia* 15, 64–67.

Chaivaillon, J., Berthelet, A., 2001. *The Elephas recki* site of HaÁdalo (Republic of Djibouti). In: Cavaretta, G., Gioia, P., Mussi, M., Palombo, M.R. (Eds.), The World of Elephants. International Congress Proceedings, Consiglio Nazionale delle Ricerche, Rome, pp. 191–193.

Clark, J.D., 1994. The Acheulian industrial complex in Africa and Elsewhere. In: Corruccini, R.S., Ciochon, R.L. (Eds.), Integrative Paths to the Past. Prentice-Hall, New Jersey, pp. 451–469.

Clark, J.D., de Heinzelin, J., Schick, K., Hart, W.K., White, T.D., WoldeGabriel, G., Walter, R.C., Suwa, G., Asfaw, B., Vrba, E. Haile-Selassie, Y., 1994. African *Homo erectus*: old radiometric dates and young Oldowan assemblages in the Middle Awash Valley, Ethiopia. Science 264, 1907–1909.

Coleman, R.G., 1974. Geological background of the Red Sea. In: Whitmarsh, R.B., Weser, O.E., Ross, D.A. (Eds.), Initial Reports of the Deep Sea Drilling Project, Volume 23. US Government Printing Office, Washington, pp. 813–819.

Courtillot, V., Achache, J., Landre, F., Bonhommet, N., Montigny R.G., Ferraud, G., 1984. Episodic spreading and rift propagation: new palaeomagnetic and geochronologic data from the Afar nascent passive margin. Journal of Geophysical Research 89, 3315–3333.

de Bonis, L., Geraads, D., Guerin, G., Haga, A., Jaeger, J.-J., Sen, S., 1984. Découverte d'un Hominidé fossile dans le Pléistocène de la République de Djibouti. Comptes Rendus de l'Academie de Sciences Paris 299 : 1097–1100.

de Bonis, L., Geraads, D., Jaeger, J.-J., Sen, S., 1988. Vertébrés du Pléistocène de Djibouti. Bulletin de la Société Geologique Française 4, 323–334.

de Lumley, H., Nioradzé, M., Barsky, D., Cauche, D., Celiberti, V., Nioradzé, G., Notter, O., Zhvania, D., Lordkipanidze, D., 2005. Les industries lithiques préoldowayennes du début du Pléistocène inférieur du site de Dmanissi en Géorgie. L'anthropologie 109, 1–182.

De Maigret, A., 1983. ISMEO activities: Arab Republic of Yemen. East and West 33, 342–343.

Delagnes, A., Lenoble A., Harmand, S., Brugal, J.-P., Prat, S., Tiercelin, J.-J., Roche, H., 2006. Interpreting pachyderm single carcass sites in the African Lower and Early Middle Pleistocene record: A multidisciplinary approach to the site of Nadung'a 4 (Kenya). Journal of Anthropological Archaeology 25, 448–465.

deMenocal, P.B., 2004. African climate change and faunal evolution during the Plio-Pleistocene. Earth and Planetary Science Letters 220, 3–24.

Dennell, R.W., 2003. Dispersal and colonisation, long and short chronologies: how continuous is the Early Pleistocene record for hominids outside East Africa? Journal of Human Evolution 45, 421–440.

Dennell, R.W., 2004. Hominin dispersals and Asian biogeography during the Lower and Early Middle Pleistocene, c. 2.0 – 0.5 Mya. Asian Perspectives 43, 205–226.

Dennell R.W., Roebroeks, W., 2005. An Asian perspective on early human dispersal from Africa. Nature 438, 1099–1104.

Dennell, R.W., Rendell, H., Hailwood, E., 1988. Early tool-making in Asia: two-million-year-old artefacts in Pakistan. Antiquity 62, 98–106.

Derricourt, R., 2006. Getting "Out of Africa": sea crossings, land crossings and culture in the hominin migrations. Journal of World Prehistory 19, 119–132.

Doe, B., 1971. Southern Arabia. McGraw-Hill Book Company, New York.

Edgell, H.S., 2006. Arabian Deserts- Nature, Origin and Evolution. Springer, New York.

Faure, M., Guérin, C., 1997. Gafolo: nouveau site à *Palaeoloxodon recki et Hippopotamus ambhibius* (République de Djibouti). Comptes Rendus Academie des sciences Paris 324–II2, 1017–1030

Feakins, S.J., deMenocal, P.B., Elington, T.I., 2005. Biomarker records of late Neogene changes in northeast African vegetation. Geology 33, 977–980.

Field, J.S., Mirazon-Lahr, M., 2005. Assessment of the southern dispersal: GIS-based analyses of potential routes at oxygen isotopic stage 4. Journal of World Prehistory 19, 1–45.

Gabunia, L., Vekua, A., Lordkipanidze, D., Swisher, C.C., Ferring, R., Justus, A., Nioradze, M., Tvalchrelidze, M., Anton, S.C., Bosinski, G., Joris, O, de Lumley, M-A., Majsuradze, G., Mouskhelishvili, A., 2000. Earliest Pleistocene cranial remains from Dmanisi, Republic of Georgia: taxonomy, geological setting, and age. *Science* 288, 1019–1025.

Goodenough, W., Caton-Thompson, G., Sclater, W.L., Lees, G.M., J.V. Harrison, A., 1939. Climate, irrigation, and early man in the Hadhramaut: discussion. The Geographical Journal 93, 36–38.

Hughes, G.W., Beydoun, Z.R., 1992. The Red Sea-Gulf of Aden: biostratigraphy, lithostratigraphy and palaeoenvironments. Journal of Petroleum Geology 15, 135–156.

Hurcombe, L., 2004. The stone artefacts from the Pabbi Hills. In: Dennell, R.W. (Ed.), Early Hominin Landscapes in Northern Pakistan: Investigations in the Pabbi Hills. British Archaeological Reports International Series 1265, Oxford, pp. 222–292.

Isaac, G. (Ed.). 1997. Koobi Fora Research Project, Volume 5: Plio-Pleistocene archaeology. Clarendon Press, Oxford.

Kibunjia, M., Roche, H., Brown, F.H., Leakey, R.E., 1992. Pliocene and Pleistocene archaeological sites west of Lake Turkana, Kenya. Journal of Human Evolution 23, 431–438.

Langbroek, M., 2004. 'Out of Africa': An investigation into the Earliest Occupation of the Old World. Archaeological Reports International Ser 1244, Oxford.

Larick, R., Ciochon, R.L., 1996. The African emergence and early Asian dispersals of the genus Homo. American Scientist 84, 538–551.

Leakey, M., 1971. Olduvai Gorge, Volume 3: Excavations in Beds I and II, 1960–1963. Cambridge University Press, Cambridge.

Martínez-Navarro, B., 2004. Hippos, pigs, bovids, sabertoothed tigers, monkeys and hominids: dispersals during Late Pliocene and Early Pleistocene times through the Levantine Corridor. In: Goren-Inbar, N., Speth, J.D. (Eds.), Human Paleoecology in the Levantine Corridor. Oxbow Books, Oxford, pp. 37–51.

Martínez-Navarro, B., Rook, L., Segid, A., Yosif, D., Ferretti, M.P., Shoshani, J., Tecle, T.M., Libsekal, Y., 2004. The large mammals from Buia (Eritrea): systematics, biochronology, and palaeoenvironments. Rivista Italiana di Paleontologia e Stratigafia 110, 61–88.

Mithen, S., Reed, M., 2002 Stepping out: a computer simulation of hominid dispersal from Africa. Journal of Human Evolution 43, 433–462.

O'Regan, H.J., Bishop, L.C., Elton, S., Lamb, A., Turner, A., 2006. Afro-Eurasian mammalian dispersal routes of the late Pliocene and early Pleistocene and their bearings on earliest hominin movements. Courier Forschungsinstitut Senckenberg 256, 305–314.

Petraglia, M.D., 2003. The Lower Paleolithic of the Arabian Peninsula: occupations, adaptations, and dispersals. Journal of World Prehistory 17, 141–179.

Petraglia, M.D., 2005. Hominin responses to Pleistocene environmental change in Arabia and South Asia. In: Head, M.J., Gibbard, P.L. (Eds.), Early–Middle Pleistocene Transitions: The Land–Ocean Evidence. Geological Society Special Publications 247, London, pp. 305–319.

Petraglia, M.D., Alsharekh, A., 2004. The Middle Palaeolithic of Arabia: implications for modern human origins, behavior, and dispersals. Antiquity 78, 671–684.

Plummer, T., 2004. Flaked stones and old bones: biological and cultural evolution at the dawn of technology. Yearbook of Physical Anthropology 47, 118–164.

Roche H., Brugal J.-P., Delagnes A., Feibel C., Harmand S., Kibunjia M., Prat S., Texier, P.J., 2003. Les sites archéologiques plio-pléistocènes de la Formation de Nachukui (Ouest Turkana, Kenya): bilan préliminaire 1996–2000. Comptes Rendus Palévol 2, 663–673.

Rose, J., 2004. The question of Upper Pleistocene connections between East Africa and South Arabia. Current Anthropology 45, 551–555.

Sahnouni, M. Derradji, A., 2007. The Lower Palaeolithic of the Maghreb: current state of knowledge. Adumatu 15, 7–44.

Schick, K., Toth, N., 2006. An Overview of the Oldowan industrial complex: the sites and the nature of their evidence. In: Toth, N., Schick, K. (Eds.), The Oldowan: Case Studies into the Earliest Stone Age. Gosport, Stone Age Institute Press, pp. 3–42.

Semaw, S., Renne, P., Harris, J.W.K., Feibel, C.S., Bernor, R.L., Fesseha, N., Mowbray, K., 1997. 2.5-million-year-old stone tools from Gona, Ethiopia. Nature 385, 333–336.

Swisher, C. III, Curtis, G.H., Jacob, T., Getty, A.G., Suprijo, A., Widiasmoro, 1994. Age of the earliest known hominids in Java, Indonesia. Science 263, 1118–1121.

Tattersall, I., Clark, J.M., Whybrow, P., 1995. Paleontological Reconnaissance in Yemen. Yemen Update 37, 21–24.

Thomas, H., Geraads, D., Janjou, D., Vaslet, D., Memseh, A., Billiou, D., Bocherens, H., Dobigny, G., Eisenmann, V., Gayet, M., de Lapparent de Broin, F., Petter, G, Halawani, M. 1998. First Pleistocene faunas from the Arabian peninsula: An Nafud desert, Saudi Arabia. Comptes Rendus Academie des Sciences 326, 145–152.

Todd, N.E., 2005. Reanalysis of African *Elephas recki:* implications for time, space, and taxonomy. Quaternary International 126–128, 65–72.

Turner, A., 1999. Assessing earliest human settlement of Eurasia: Late Pliocene dispersions from Africa. Antiquity 73, 563–570.

Turner, A., O'Regan, H., 2007. Zoogeography: primate and early hominin distribution and migration patterns. In: Henke, W. Tattersall, I. (Eds.), Handbook of Paleoanthropology: Principles, Methods, and Approaches. Springer, New York, pp. 421–440.

Van Beek, G., Cole, G., Jamme, W.F., 1964. An archaeological reconnaissance in Hadhramaut, South Arabia – a preliminary report. Annual Report of the Board of Regents of the Smithsonian Institution, The Smithsonian Report for 1963, Publication 4587: 521–555.

Vrba, E., 2007. Role of Environmental Stimuli in Hominid Origins. In: Henke, W. Tattersall, I. (Eds.), Handbook of Paleoanthropology: Principles, Methods, and Approaches. Springer, New York, pp. 1441–1482.

Werner, N.Y., Mokady. O., 2004. Swimming out of Africa: mitochondrial DNA evidence for late Pliocene dispersal of a cichlid from Central Africa to the Levant. Biological Journal of the Linnean Society 82, 103–109.

Whalen, N. M., Pease, D.W., 1990. Variability in Developed Oldowan and Acheulean bifaces of Saudi Arabia. Atlal 13, 43–48.

Whalen, N. M., Pease, D.W., 1992. Archaeological survey in Southwest Yemen, 1990. Paléorient 17, 129–133.

Whalen, N., Davis, W. Pease, D., 1989. Early Pleistocene migrations into Saudi Arabia. Atlal 12, 59–75.

Whalen, N. M., Schatte, K.E., 1997. Pleistocene sites in southern Yemen. Arabian Archaeology and Epigraphy 8, 1–10.

Whalen, N. M., Sindi, H., Wahida, G., Siraj-ali, J. S. 1983. 1—Excavation of Acheulean sites near Saffaqah in ad-Daw dmi 1402–1982. Atlal 7, 9–21.

Whitmarsh, R.B., Weser, O.E., Ross, D.A. et al. (Eds.), 1974. Initial Reports of the Deep Sea Drilling Project, Volume 23. Washington, US Government Printing Office, pp. 5–1180.

Whybrow, P. J., Hill, A., (Eds.). 1999. Fossil Vertebrates of Arabia. Yale University Press, New Haven, CT.

WoldeGabriel, G., Heiken, G., White, T.D., Asfaw, B., Hart, W.K. Renne, P.R., 2000. Volcanism, tectonism, sedimentation, and the paleoanthropological record in the Ethiopian Rift System. Geological Society of America, Special Paper 345, 83–99.

Zhu, R., An Z., Potts, R., Hoffman, K.A., 2003. Magnetostratigraphic dating of early humans in China. Earth Science Reviews 61, 341–359.

Zimmerman, P., 2000. Middle Hadramawt Archaeological Survey-Preliminary results, October 1999 season. Yemen Update 42, 27–29.

Zumbo, V., Feraud, G., Vellutini, P., Piguet, P., Vincent, J., 1995. First $^{40}Ar/^{39}Ar$ dating on Early Pliocene to Plio-Pleistocene magmatic events of the Afar – Republic of Djibouti. Journal of Volcanology and Geothermal Research 65, 281–295.

6. *Homo floresiensis* and the African Oldowan

Mark W. Moore*
Archaeology and Palaeoanthropology
University of New England, Armidale, New South Wales
2351 Australia
mmoore2@une.edu.au

Adam Brumm
McDonald Institute for Archaeological Research,
Downing Street, University of Cambridge,
Cambridge CB2 3ER,
United Kingdom
arb74@cam.ac.uk

Keywords: Stone tools, Liang Bua, Mata Menge, Hominin cognition

Abstract

The small-bodied hominin *Homo floresiensis* was recently identified at Liang Bua, Flores, Indonesia. Some researchers have argued that *H. floresiensis* represents pathological individuals from a behaviorally modern *Homo sapiens* population, arguing in part that the stone-tools found in association are too "advanced" to have been manufactured by a nonmodern hominin. Here we show that the Pleistocene stone-tools from Flores, including Liang Bua, are technologically and morphologically similar to the 1.2–1.9 Mya Oldowan/Developed Oldowan tools from Olduvai Gorge in Africa. The Pleistocene lithic technology on Flores was therefore within the capabilities of small-brained, nonmodern hominins.

6.1 Introduction

The remains of *Homo floresiensis*, a small-bodied hominin with a cranial capacity of 417 cm^3, were recently discovered together with abundant stone-tools at Liang Bua on the Indonesian island of Flores (Brown et al. 2004; Morwood et al. 2004; Falk et al. 2007). All diagnostic hominin skeletal elements from deposits dated between 95 and 16.6 Ka at Liang Bua are consistent in size and morphology with *H. floresiensis* (Morwood et al. 2004, 2005), whereas the hominin remains excavated from more recent levels of the site are all *Homo sapiens* of normal stature and brain size. On this evidence, the most parsimonious inference is that *H. floresiensis* manufactured the associated tools (Morwood et al. 2004:1089; Reynolds 2004:4; Moore et al. in press).

A cursory description of the stone-tools appeared in the initial report on Liang Bua, accompanied by an illustration of seven artefacts (Morwood et al. 2004). The tools were described as predominately "simple flakes" struck bifacially from small radial cores but including blades/microblades and retouched flakes known as "perforators" (Morwood et al. 2004:1089, Figure 6.5). Morwood et al. (2004:1091) noted that stone tool-making by *H. floresiensis* was in keeping with the relatively advanced cognitive capacities of *Homo* generally, despite the small brain size. Lahr and Foley (2004:1044), in an accompanying commentary, stated that "elsewhere such implements are associated with *H. sapiens*, and their contrast with tools found anywhere with *H. erectus* is very striking". While the tools might indicate "baffling evidence for complex, supposedly 'sapient'" behaviors among archaic hominins, the authors also speculate that the tools were made by modern humans dispersing across southern Asia (Lahr and Foley 2004:1044).

* Address for correspondence: mmoore2@une.edu.au

E. Hovers and D.R. Braun (eds.), *Interdisciplinary Approaches to the Oldowan*, DOI: 10.1007/978-1-4020-9059-2,

Lahr and Foley's scenarios figured prominently in an ensuing debate about the legitimacy of *H. floresiensis*. The characterization of the Liang Bua tools as "advanced" was cited to support the view that the holotype (Brown et al. 2004) and similar individuals (Morwood et al. 2005) were modern *H. sapiens* suffering from microcephaly (Hennenberg and Thorne 2004; Jacob et al. 2006; Martin et al. 2006a, b; Richards 2006). In this scenario, nonpathological members of a modern *H. sapiens* population must have been responsible for the Liang Bua tools because: 1) the manufacture of such tools is beyond the cognitive capabilities of a small-brained hominin like *H. floresiensis* (e.g. Martin in Culotta 2005:209); and/or 2) tools like those at Liang Bua are associated only with modern humans and *H. neanderthalensis* elsewhere in the world (e.g. Jacob et al. 2006:13421; Martin et al. 2006b:1128).

Here we challenge these points by comparing the Pleistocene stone artifacts from Liang Bua (after Moore 2005, 2008; Moore and Brumm 2007; Moore et al. in press) and the nearby 800 Ka site of Mata Menge (after Brumm et al. 2005, Moore and Brumm 2007) to Oldowan and Developed Oldowan artifacts from Olduvai Gorge in Africa (after Leakey 1971). Although hominins from more than one genus may have been responsible for the Olduvai tools, the tool-making species certainly had a smaller cranial capacity and simpler cognitive development than either *H. neanderthalensis* or modern *H. sapiens*. The comparison shows that the stone artifacts on Flores are formally and technologically similar to those from Olduvai Gorge. Since tools like those on Flores are associated with nonmodern hominins in Africa, we infer that the Flores stone tool assemblages were within the cognitive capabilities of a small-brained hominin such as *Homo floresiensis*.

6.2 Pleistocene Stone Knapping on Flores

Liang Bua is a 30 metre wide limestone cave located on the Wae Racang River in western Flores (Morwood et al. 2004, 2005). The cave was exposed ca. 190 Ka ago by the Wae Racang, an event that deposited an artifact-containing conglomerate into the former subterranean chamber (Westaway et al. 2007). The river is now located about 30 metres below and 200 metres distant from Liang Bua. The cave subsequently infilled with silts and clays, creating a stratified archaeological sequence up to 12 metres in thickness. The sequence is divisible into nine major stratigraphic units dating from 130 to 3 Ka (Westaway 2006). A full description of the Liang Bua lithic analysis is available in Moore (2005; Moore et al. in press).

The Pleistocene stone-workers at Liang Bua procured stone from water-deposited gravels. The stone-tools were

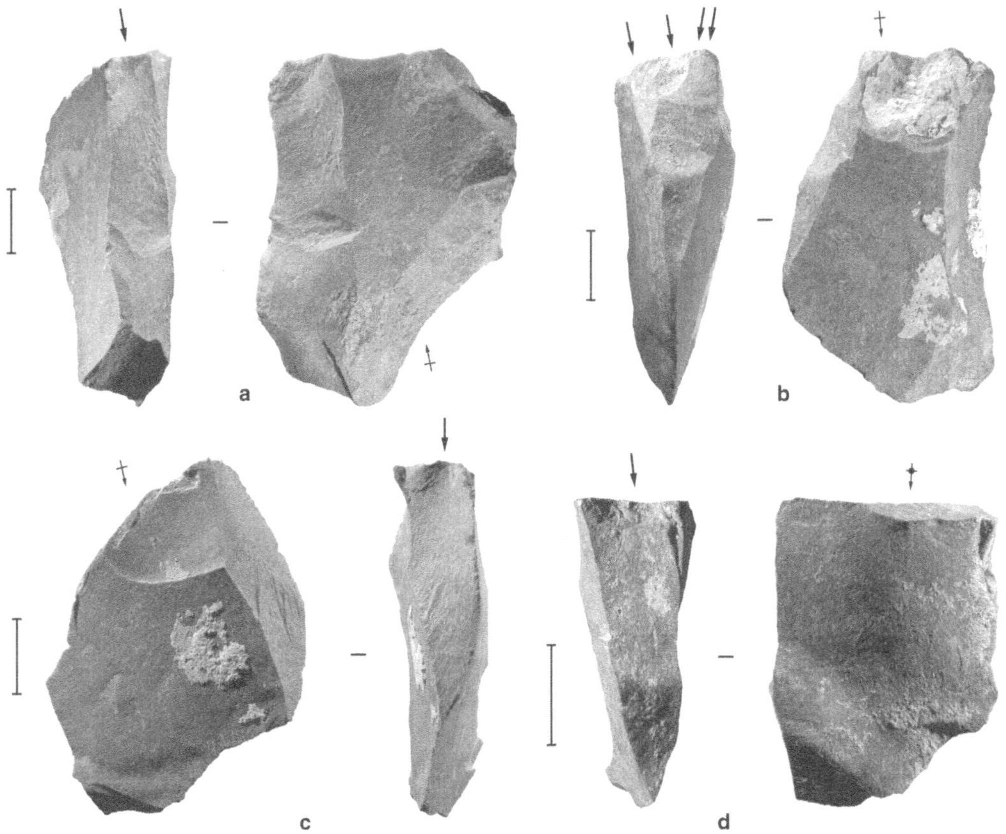

FIGURE 6.1. Burinated flake blank cores recovered from the Liang Bua Pleistocene deposits. The burin scars are indicated by arrows. Burin flakes were removed directly down the edge on artifacts b–d. The orientation of the scar on A is tranchet-like (after Debénath and Dibble 1994:36). Scales — 10 mm.

mostly made from volcanic and metavolcanic rocks of medium to high flaking quality (ca. 3.5–4.0 on Callahan's [1979:16] scale), along with minor amounts of chert and chalcedony. Coarse-grained volcanic river cobbles were collected from terrace or riverbed deposits and carried into the cave to use as hammerstones. Flakes were struck from cobbles at similar stone sources outside the cave and the cores were left behind (Moore and Brumm 2007). These flake blanks were large (more than 40 mm in maximum dimension), relative to the flakes subsequently struck from them. The large flake blanks and, sometimes, small cobbles were carried into the cave and reduced into small flakes. A few cobbles were discarded in the cave after initial assaying (after Wilke and Schroth 1989).

Four knapping techniques were practiced inside the cave:

(1) The "freehand oblique" technique involved striking flakes from flattish core surfaces.
(2) The "burin" technique involved striking small flakes down the narrow edges of cores (Moore et al. in press; after Debénath and Dibble 994:34-36; Tomásková 2005:85) (Figure 6.1).
(3) The "cross-axis truncation" technique involved placing a core flat on an anvil and striking near the middle of a face, fragmenting it (after Crabtree 1973:49; Bergman et al. 1987).
(4) The "bipolar" technique involved securing a flake edgewise on an anvil and striking the uppermost edge, splitting it (after White 1968, Hayden 1980; Figure 6.2).

Flakes were often struck from both faces of flake blanks and from all or part of the blank's periphery. They were classified as "flake blank cores" if the reduction scars were mostly invasive; and "retouched flakes" if the scars were mostly noninvasive (after Odell 2004:74). Retouched flakes are morphologically irregular with the exception of "perforators" (Figure 6.3). "Radial cores" were reduced centripetally and bifacially but reduction eliminated traces of the original blank (Figure 6.4). "Multiplatform cores"— usually of cuboidal or globular morphology — were created from the reduction of small cobbles or from radial/flake blank cores that were rotated from the dominant reduction plane, establishing new reduction planes (Figure 6.5). This sometimes

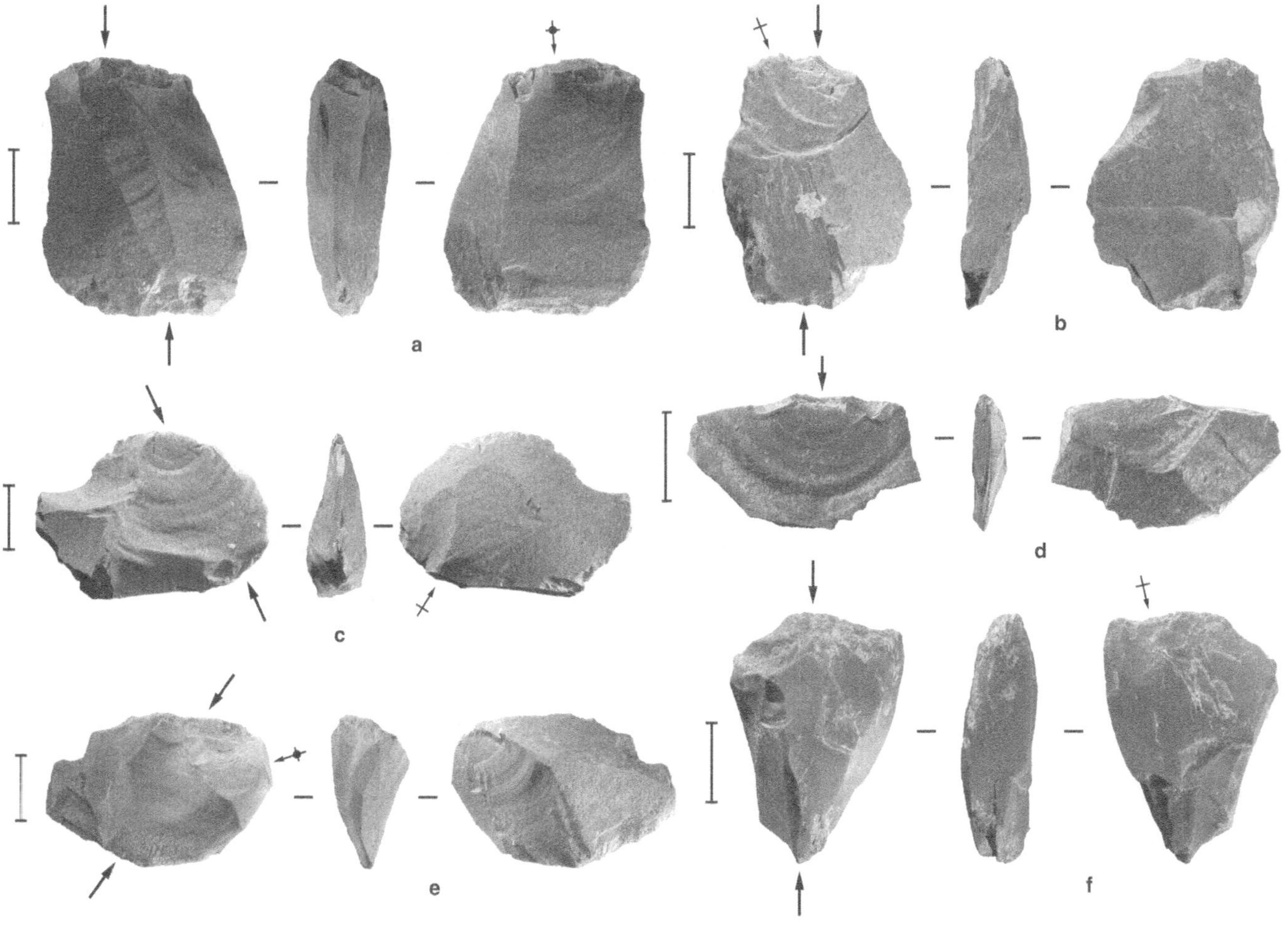

FIGURE 6.2. Bipolar artifacts recovered from the Liang Bua Pleistocene deposits. Artifacts a, c, e, and f were made on flake blanks. Arrows indicate opposed areas of crushing. Scales — 10 mm.

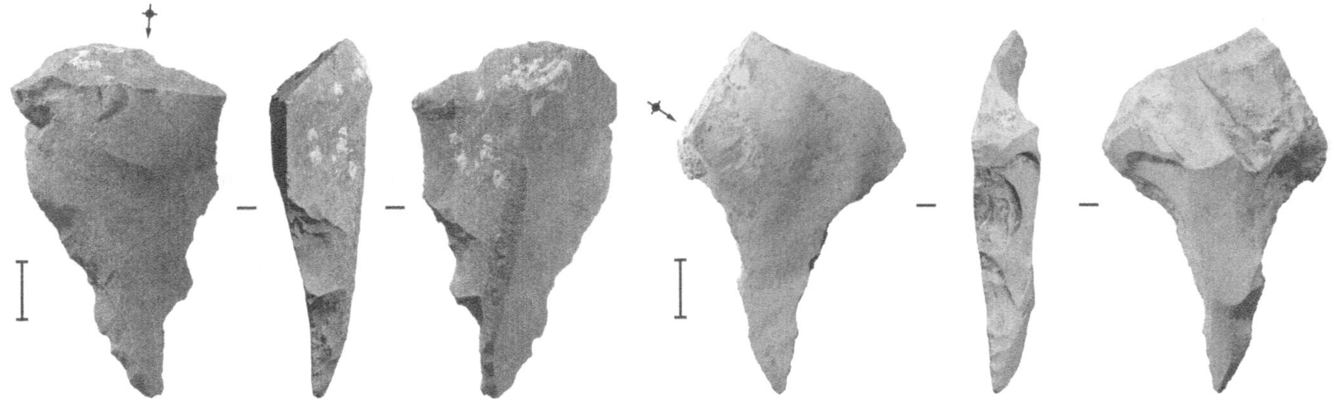

FIGURE 6.3. Retouched flake "perforators" recovered from the Liang Bua Pleistocene deposits. Scales — 10 mm.

occurred through the bridging techniques of burination and/or cross-axis truncation. Burin flakes struck down former platform surfaces produced crest-like objects (after Inizan et al. 1999:137-138) called "redirecting flakes" (after McCarthy 1976:22). Multiplatform cores were occasionally used as hammerstones (Figure 6.6).

The bipolar technique was always applied to different cores than those reduced by the freehand planar, burin, and cross-axis truncation techniques. Blanks for bipolar flaking were small-to-medium-sized flakes. Some 9.7% of the Liang Bua flakes are "blades", flakes measuring twice as long as wide with parallel dorsal ridges. A blade-like flake is produced if the core mass down which the flake is struck happens to be elongated. An elongated core mass can be created deliberately by carefully-staged flake removals, but can also occur incidentally. The small proportion of blades at Liang Bua and the lack of cores with consistently elongated scars suggest that the knappers created these artifacts incidentally (Moore 2005, 2007).

Mata Menge is a much older site located in the Soa Basin 50 km east of Liang Bua (Brumm et al. 2006; Brumm 2007; Moore and Brumm 2007). The Soa Basin has revealed a number of stratified open-air archaeological sites that contain stone artifacts in association with fossilized remains of *Stegodon florensis*. Mata Menge occurs in the Ola Bula Formation, deposited by one or more distal volcanic fans entering a series of lakes between 960 and 700 Ka ago (O'Sullivan et al. 2001). The artifact- and bone-bearing deposits at Mata Menge lie between two tuffaceous silts with ascending zircon fission-track ages of 880±70 Ka and 800±70 Ka (Morwood et al. 1998; O'Sullivan et al. 2001). Some 507 artifacts have been excavated from this dated context (Brumm et al. 2006; Brumm 2007).

There are strong similarities between the stone artifact assemblages from Mata Menge and Liang Bua (Brumm et al. 2006). Both assemblages show an emphasis on the use of volcanic/metavolcanic fluvial cobbles as raw materials and the transport of flakes as blanks for cores. The reduction

FIGURE 6.4. Radial cores recovered from the Liang Bua Pleistocene deposits. Scales — 10 mm.

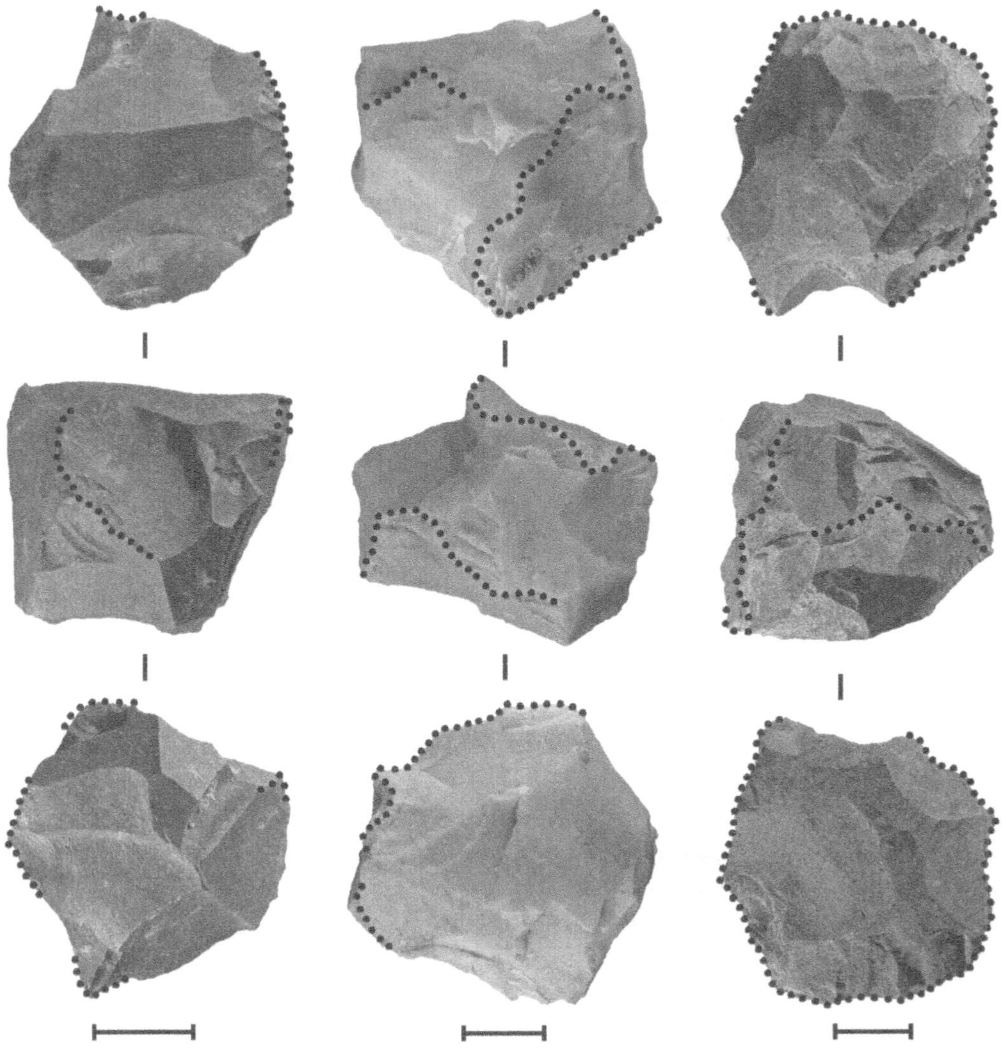

FIGURE 6.5. Multiplatform cores recovered from the Liang Bua Pleistocene deposits. Multiplatform cores were knapped in relation to more than one reduction plane, creating two or more platform edges (indicated by dotted lines). Scales — 10 mm.

sequences at Mata Menge and Liang Bua both emphasize freehand bifacial radial reduction of cobbles and flake blanks. Small, invasively reduced bifacial radial cores from the two sites are virtually indistinguishable. In addition, multiplatform cores, cores with burination scars, truncated flakes and cores, and perforators are found in both assemblages. One possible example of bipolar reduction — a relatively common technique at Liang Bua — was recorded at Mata Menge (Brumm et al. 2006; Brumm 2007). Brumm et al. (2006:628) conclude that "a long-term technological theme, involving the reduction of cores bifacially and radially, and the manufacture of a suite of technically and morphologically distinctive artifacts, is evident on Flores from at least 840 Kyr ago right up to the disappearance of *H. floresiensis* 12 Kyr ago."

6.3 Olduvai Gorge and Flores Compared

The approach to stone reduction on Pleistocene Flores is very similar to that described for 1.2–1.9 Ma Oldowan/Developed Oldowan assemblages from sites DK, FLK North, HWK East, and TK in Beds I and II at Olduvai Gorge (Leakey 1971) (Figure 6.7). The suite of tool-producing techniques at both places was relatively uncomplicated. Cobbles, cobble pieces, or flake blanks were reduced by direct hard-hammer percussion. Flaking was often arranged centripetally and bifacially on Flores, the same pattern that produced "discoids" at Olduvai Gorge. The partially flaked Mata Menge radial cores made on cobbles closely resemble Leakey's (1971) classic Olduvai "choppers". Core reduction produced small amorphous flakes. These were usually discarded without modification, although

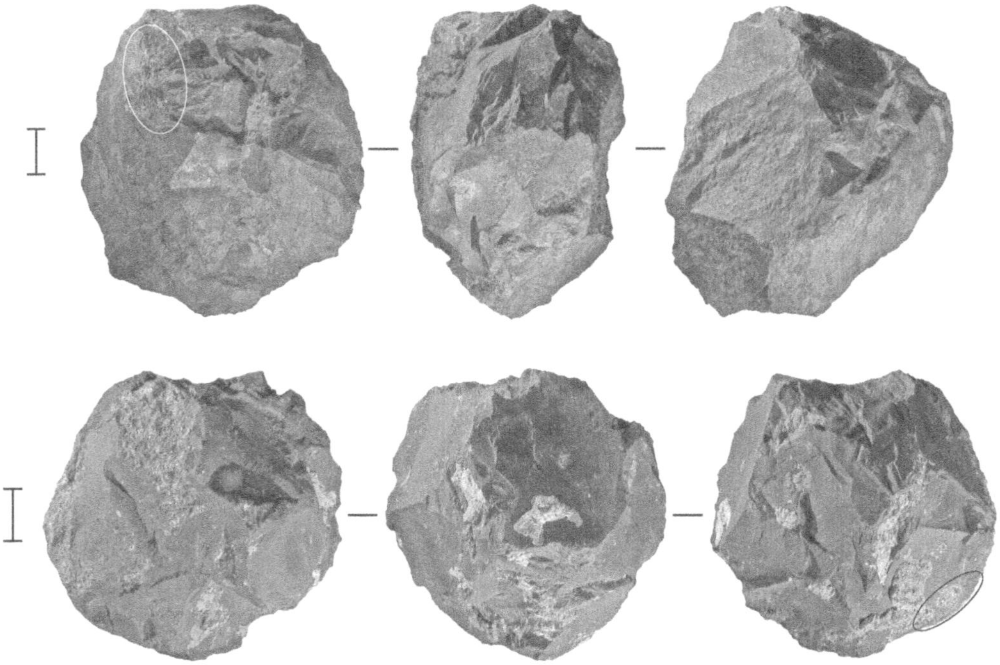

FIGURE 6.6. Multiplatform cores used as hammerstones, recovered from the Liang Bua Pleistocene deposits. Bashed areas are circled. Scales — 10 mm.

in both regions flakes were sometimes modified by minimal unifacial or bifacial noninvasive flaking. This produced retouched flakes on Flores or "scrapers" at Olduvai Gorge. Some flakes were intensively retouched on opposing margins, producing perforators on Flores and "awls" at Olduvai Gorge. Anvil-supported percussion is recorded at Liang Bua, and similar evidence occurs at Olduvai in the form of "*outils écaillés*" and "truncations" (after Clark 1971:xvi). Burination of core and flake edges is a notable feature of the Pleistocene Flores technologies, a technique Leakey (1971) also observed at Olduvai Gorge. Multiplatform cores at Liang Bua and Mata Menge were sometimes used as hammerstones, resembling Olduvai "polyhedrons" or "subspheroids" in morphology and method of creation (Leakey 1971).

In sum, the Pleistocene artifact assemblages from Flores are technologically and typologically similar to Oldowan/Developed Oldowan assemblages from Olduvai Gorge. The same stone-working techniques — freehand planar, burin, cross-axis truncation, and bipolar — compose the suite of stone-working techniques at both localities. Shared artifact types include bifacial radial cores (Leakey's [1971] "discoids" and "choppers"), retouched flakes, ("scrapers"), perforators ("awls"), bifaces, cobble hammerstones, redirecting flakes ("resharpening flakes"), bipolar cores ("*outils écaillés*"), assayed cobbles ("modified blocks") and multiplatform cores ("polyhedrons"). Leakey notes that "awls" and "*outils écaillés*" appear at Olduvai in the Developed Oldowan after ca. 1.5 Ma. Flakes with the attributes of blades are not reported at Olduvai, but are present 2.6 Ma ago at OGS-7 (Semaw et al. 2003).

6.4 Conclusion

The Liang Bua stone-tools have been cited as evidence that the *H. floresiensis* remains are pathological individuals from a population of modern humans (Hennenberg and Thorne 2004; Jacob et al. 2006; Martin et al. 2006a, b; Richards 2006). Our review has shown remarkable formal and technological similarity between the 1.2–1.9 Ma stone artifact assemblage from Olduvai Gorge — made by hominin(s) with less cranial capacity and cognitive development than modern humans — and the stone artifact assemblage from Mata Menge and Liang Bua on Flores. The techniques and types described above are the complete range documented to date for Pleistocene Flores. Therefore, assertions that tools like those at Liang Bua are elsewhere only associated with *H. sapiens* or *H. neanderthalensis* are incorrect (see also Clark 1971:xvi) and the belief that such tools are necessarily outside the capabilities of a small-brained hominin like *H. floresiensis* is unfounded. By these criteria, there is presently no behavioral evidence for modern *H. sapiens* in the Liang Bua deposits that contain the remains of *H. floresiensis*. This negates the archaeological evidence cited for reclassifying the *H. floresiensis* specimens as pathological modern humans. Conversely, some researchers have accepted the attribution of *H. floresiensis* as a new tool-making species and suggested that the "sophistication" of the stone-tools highlights our poor understanding of hominin brain and behavioral evolution, particularly on islands (Diamond 2004; Flannery 2004; Taçon 2005; see also Falk

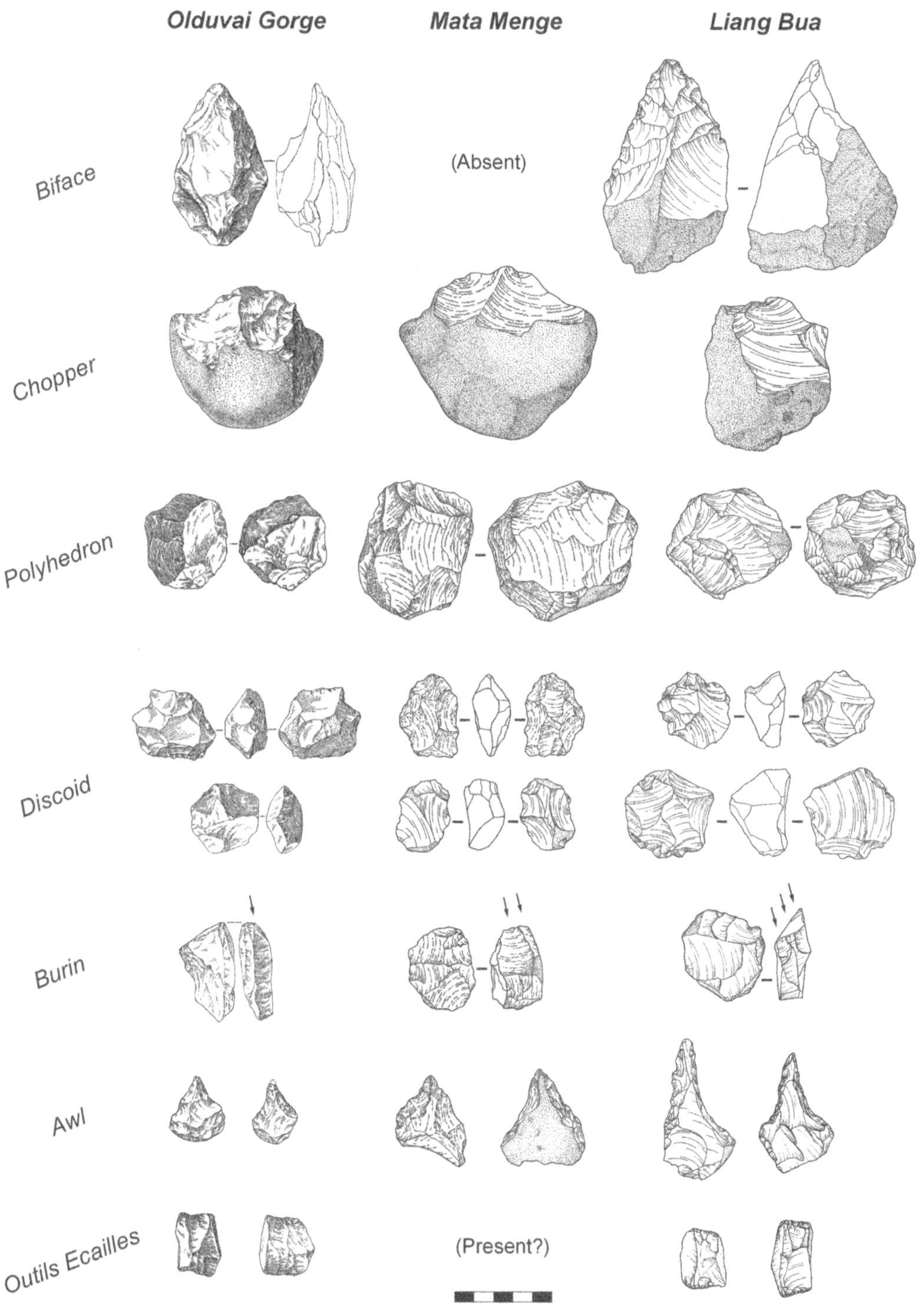

FIGURE 6.7. Comparison of artefacts from Olduvai Gorge, Mata Menge, and the Liang Bua Pleistocene deposits. Categories after Leakey (1971). Scale — 50 mm.

et al. 2005, 2007; Kohn 2005; Wong 2005; Niven 2007). The observation that the Flores assemblages are technically simple and Oldowan-like alters the behavioral context of these discussions.

Reynolds (2004), in counterpoint to Lahr and Foley (2004), observed that the Liang Bua assemblage represents a "minimal investment" technology with "the presence of some blades, both large and small, and a few retouched tools" (2004:4), and as such is typical of Southeast Asian Paleolithic assemblages. Technological simplicity and continuity are long recognized themes in Southeast Asian prehistory (White 1977). Seen in regional context (Moore and Brumm 2007; Moore et al. in press), the Pleistocene technological patterns identified on Flores are unexceptional. Given that a broadly similar technological pattern persists in Southeast Asia into later prehistory (Moore (2005:695-703; Moore and Brumm 2007; Moore et al. in press), ostensibly "universal" Old World-based models of technological change as proxy measures of hominin phylogenetics and evolution (e.g. Foley and Lahr 1997, Mellars 2006) do not apply here.

Acknowledgments. Funding for the Liang Bua excavations was provided by an Australian Research Council (ARC) grant to M.J. Morwood. The 2003–2004 excavations at Liang Bua Cave were conducted by the Indonesian National Centre for Archaeology (ARKENAS), with R.P. Soejono (counterpart), M.J. Morwood (counterpart), T. Sutikna, E. Wahyu Saptomo, Jatmiko, S. Wasisto, Rokus Awe Due and D. Hobbs. The Liang Bua work was authorized by Dr T. Djubiantono (ARKENAS Director). The 2004–2006 excavations at Mata Menge were conducted by the Geological Research and Development Centre (GRDC), with F. Aziz (director), G.D. van den Bergh, M. J. Morwood (counterpart), I. Kurniawan, S. Sudjarwardi, Dadang, Ngaliman, I. Botha, J. Tode, D. Hobbs and 20 Ngadha people led by K. Podhi. The Mata Menge work was authorized by Dr Djadjang (GRDC Director). M.W.M. was funded by an ARC Australian Postgraduate Award and an ARC Australian Postdoctoral Fellowship. A.B. was funded by an ANU PhD cofunded stipend scholarship.

References

Bergman, C.A., Barton, R.N.E., Collcutt, S.N., Morris, G., 1987. Intentional breakage in a Late Upper Palaeolithic assemblage from Southern England. In: Sieveking, G., Newcomer, M.H. (Eds.), The Human Uses of Flint and Chert: Proceedings of the Fourth International Flint Symposium. Cambridge University Press, Cambridge, pp. 21–32.

Brown, P., Sutikna, T., Morwood, M. J., Soejono, R. P., Jatmiko, Wayhu Saptomo, E., Rokus Awe Due., 2004. A new small-bodied hominin from the Late Pleistocene of Flores, Indonesia. Nature 431, 1055–1061.

Brumm, A., 2007. The End of the Movius Line: Palaeolithic Stone Technology from a Southeast Asian Perspective. Ph.D. Dissertation, Australian National University.

Brumm, A, Aziz, F., van den Bergh, G.D., Morwood, M.J., Moore, M.W., Hobbs, D.R., Kurniawan, I., Fullagar, R., 2006. Early stone technology on Flores and its implications for *Homo floresiensis*. Nature 441, 624–628.

Callahan, E, 1979. The basics of biface knapping in the eastern fluted point tradition: a manual for flintknappers and lithic analysts. Archaeology of Eastern North America 7, 1–180.

Clark, J.D., 1971. Foreword. In: Leaky, M.D., Olduvai Gorge, Volume 3: Excavations in Beds I and II, 1960–1963. Cambridge University Press, Cambridge, pp. xv–xvii.

Crabtree, D.E., 1973. The obtuse angle as a functional edge. Tebiwa 16, 46–53.

Debénath, A., Dibble, H.L., 1994. Handbook of Paleolithic Typology. Volume One: Lower and Middle Paleolithic of Europe. University Museum, University of Pennsylvania, Philadelphia.

Diamond, J., 2004. The astonishing micropygmies. Science 306, 2047–2048.

Flannery, T., 2004. Of dwarves and dragons. The Bulletin, 9 November, 32–34.

Foley, R. Lahr, M.M., 1997. Mode 3 technologies and the evolution of modern humans. Cambridge Archaeological Journal 7, 3–36.

Hayden, B., 1980. Confusion in the bipolar world: bashed pebbles and splintered pieces. Lithic Technology 9, 2–6.

Henneberg, M., Thorne, A., 2004. Flores human may be pathological *Homo sapiens*. Before Farming 4, 2–4.

Inizan, M.-L., Reduron-Ballinger, M., Roche, H., Tixier, J., 1999. Technology and Terminology of Knapped Stone. Préhistoire de la Pierre Taillée 5. Cercle de Recherches et d'Etudes Préhistoriques, Paris.

Jacob, T., Indriati, E., Soejono, R.P., Hsü, K., Frayer, D.W., Eckhardt, R.B., Kuperavage, A.J., Thorne, A., Henneberg, M., 2006. Pygmoid Australomelanesian *Homo sapiens* skeletal remains from Liang Bua, Flores: population affinities and pathological abnormalities. Proceedings of the National Academy of Sciences 103, 13421–13426.

Kohn, M., 2005. The little troublemaker. New Scientist 186, 41–45.

Leakey, M.D., 1971. Olduvai Gorge, Volume 3: Excavations in Beds I and II, 1960–1963. Cambridge University Press, Cambridge.

Martin, R.D., MacLarnon, A.M., Phillips, J.L., Dussubieux, L., Williams, P.R., Dobyns, W.B., 2006a. Comment on 'the brain of LB1, *Homo floresiensis*.' Science 312, 999b.

Martin, R.D., MacLarnon, A.M., Phillips, J.L., Dobyns, W.B., 2006b. Flores hominid: new species or microcephalic dwarf? The Anatomical Record 288A, 1123–1145.

McCarthy, F.D., 1976. Australian Aboriginal Stone Implements. Second edition. Australian Museum Trust, Sydney.

Mellars, P., 2006. Going east: new genetic and archaeological perspectives on the modern human colonization of Eurasia. Science 313, 796–800.

Moore, M.W., 2005. The Design Space of Lithic Technology. Ph.D. Dissertation, University of New England.

Moore, M.W., 2007. Lithic design space modelling and cognition in *Homo floresiensis*. In: Shalley, A., Khlentzos, D. (Eds.), Mental States: Nature, Function and Evolution. John Benjamins, Amsterdam.

Moore, M.W., Brumm, A., 2007. Stone artifacts and hominins in island Southeast Asia: new insights from Flores, Eastern Indonesia. Journal of Human Evolution 52, 85–102.

Moore, M.W., Sutikna, T., Morwood, M.J., Brumm, A. in press. Continuities in stone flaking technology at Liang Bua, Flores, Indonesia. Journal of Human Evolution.

Morwood, M.J., Brown, P., Jatmiko, Sutikna, T., Wahyu Saptomo, E., Westaway, K.E., Rokus Awe Due, Roberts, R.G., Maeda, T., Wasisto, S., Djubiantono, T., 2005. Further evidence for small-bodied hominins from the Late Pleistocene of Flores, Indonesia. Nature 437, 1012–1017.

Morwood, M. J., Soejono, R.P., Roberts, R.G., Sutikna, T., Turney, C.S.M., Westaway, K.E., Rink, W.J., Zhao, J.-x., van den Bergh, G.D., Rokus Awe Due, Hobbs, D.R., Moore, M.W., Bird, M.I., Fifield, L.K., 2004. Archaeology and age of a new hominin from Flores in eastern Indonesia. Nature 431, 1087–1091.

Niven, J.E., 2007. Brains, islands and evolution: breaking all the rules. Trends in Ecology and Evolution 22, 57–59.

Odell, G.H., 2004. Lithic Analysis. Manuals in Archaeological Method, Theory, and Technique. Kluwer Academic, New York.

O'Sullivan, P.B., Morwood, M., Hobbs, D., Aziz, F., Suminto, Situmorang, M., Raza, A., Maas, R., 2001. Archaeological implications of the geology and chronology of the Soa Basin, Flores, Indonesia. Geology 29, 607–610.

Richards, G.D., 2006. Genetic, physiologic and ecogeographic factors contributing to variation in Homo sapiens: *Homo floresiensis* reconsidered. Journal of Evolutionary Biology 19, 1744–1767.

Semaw, S., Rogers, M.J., Quade, J., Renne, P.R., Butler, R.F., Dominguez-Rodrigo, M., Stout, D., Hart, W.S., Pickering, T., Simpson, S.W., 2003. 2.6-million-year-old stone tools and associated bones from OGS-6 and OGS-7, Gona, Afar, Ethiopia. Journal of Human Evolution 45, 169–77.

Taçon, P.S.C., 2005. Modern human behaviour and the implications of small-bodied hominin finds from the Late Pleistocene of Flores, Indonesia. Before Farming 4, 1–6.

Tomášková, S., 2005. What is a burin? Typology, technology, and interregional comparison. Journal of Archaeological Method and Theory 12, 79–115.

Westaway, K.E., 2006. Reconstructing the Quaternary landscape evolution and climate history of western Flores: an environmental and chronological context for an archaeological site. Ph.D. Dissertation, University of Wollongong.

Westaway, K.E., Morwood, M.J., Roberts, R.G., Zhao, J-x., Sutikna, T., Saptomo, E.W., Rink, W.J., in press. Establishing the time of initial human occupation of Liang Bua, western Flores, Indonesia. Quaternary Geochronology.

White, J.P., 1968. Fabricators, outils écaillés or scalar cores? Mankind 6, 658–666.

White, J.P., 1977. Crude, colourless and unenterprising? Prehistorians and their views on the Stone Age of Sunda and Sahul. In: Allen, J., Golson, J., Jones, R. (Eds.), Sunda and Sahul: Prehistoric Studies in Southeast Asia, Melanesia and Australia. Academic Press, London, pp. 13–30.

Wilke, P.J., Schroth, A.B., 1989. Lithic raw material prospects in the Mohave Desert. Journal of California and Great Basin Anthropology 11, 146–174.

Wong, K., 2005. The littlest human. Scientific American 292, 56–65.

7. Methodological Considerations in the Study of Oldowan Raw Material Selectivity: Insights from A. L. 894 (Hadar, Ethiopia)

Talia Goldman-Neuman*
The Institute of Archaeology
The Hebrew University of Jerusalem
Mt. Scopus, Jerusalem 91905
Israel
tali.neuman@mail.huji.ac.il

Erella Hovers
The Institute of Archaeology
The Hebrew University of Jerusalem
Mt. Scopus, Jerusalem 91905
Israel
hovers@mscc.huji.ac.il

Keywords: Raw material selectivity, Oldowan, Hadar, mobility, Plio-Pleistocene

Abstract

A study of raw material selection patterns at a Late Pliocene Oldowan site (A. L. 894, Hadar, Ethiopia) raised methodological questions inherent specifically to the study of the very early lithic assemblages. These in turn have a significant effect on our understanding of patterns of raw material selection and use as well as inferences about hominin cognitive abilities and ecological adaptations. Here we identify some of these methodological problems and propose potential ways of addressing them.

7.1 Introduction

A growing number of *in situ* occurrences in eastern Africa, mainly those from Gona (Semaw 2000; Semaw et al. 2003), Lokalalei (Roche et al. 1999) Kanjera (Plummer et al. 1999) and Hadar (Kimbel et al. 1996; Hovers et al. 2002), discovered in the course of the last two decades, have extended the temporal depth of uncontested tool-making by humans into the Late Pliocene. Contributions in this volume as well as many others published in the course of the last decade attest to the amount of information gained from these early sites through increasingly focused and detailed analyses of paleoenvironments, technological and taphonomic processes, leading to new insights into the inter-play between these three main agents of creating the archaeological record. Within this context, raw material procurement strategies have received much attention in recent research. These are approached by two separate (albeit sometimes complementary) analyses of transport patterns and of selectivity.

Raw material composition is an important source of assemblage-level variation throughout various prehistoric periods, as an indication of ranging distances of hominin groups and technological organization of their lithics-related activities (Stiles et al. 1974; Geneste 1985; Bamforth 1986; Potts 1988; Roebroeks et al., 1988; Dibble 1991; Féblot-Augustins 1993, 1997; Merrick et al. 1994; Stiles 1998; Kuhn 2004; Minichillo 2006; Negash and Shackley 2006; Negash et al. 2006). Patterns of raw material transport in the Oldowan until very recently have been inferred through technological analyses (some of

* Address for correspondence: tali.neuman@mail.huji.ac.il

E. Hovers and D.R. Braun (eds.), *Interdisciplinary Approaches to the Oldowan*, DOI: 10.1007/978-1-4020-9059-2,

the Olduvai data being a notable exception; see below). Those have been recently combined and/or complemented by sourcing studies. A shared pattern that emerges from this body of research is that Oldowan hominins have regularly procured and transported lithic raw materials over short distances (usually less than 5 km from the site), most often as cobbles from channel beds (Isaac et al. 1976; Merrick and Merrick 1976; Toth 1985, 1997; Schick 1987; Kibunjia et al. 1992; Isaac and Harris 1997; Kyara 1999; Semaw 2000; de la Torre et al. 2003; Plummer 2004; Harmand 2005; Negash et al. 2006). Somewhat longer transport distances have been documented in Olduvai Gorge, where phonolite, gneiss and quartzite may have been transported over a distance of 8–13 km from their sources during Bed I times (Hay 1976), and some lava artifacts in sites on the western Bed I lake margin might have been brought from a source 15–20 km away (Blumenschine et al. 2003). A relatively large portion of the lithics found in the Kanjera South localities are made on raw materials not available in the local drainages that may have been brought from a distance of over 20 km (Plummer 2004; Braun et al. 2008). Finally, the occurrence of flakes made on chert, without indications for on-site reduction of this raw material, in the 2.6 Ma assemblages of OG-7 speaks to spatial fragmentation of lithic reduction sequences and transport of selected elements across the landscape (Semaw et al. 2003; although this issue is not emphasized by Stout et al. 2005).

Understanding raw material selectivity in early lithic assemblages is not straightforward. Here we identify selectivity through the non-random discrepancies between archaeological assemblages and geological sources in the appearance of raw material characteristics. Selectivity may be inferred in a relatively direct manner when long distance raw material transport can be shown. Regardless of the type of occupation (either a "central place" or simply a place on the landscape to which hominins returned for various activities), the effort involved in transporting artifacts or blocks of specific, non-local raw material over some distance is usually perceived as indication that this raw material was selected for. The evidence is more equivocal when raw materials were moved over short distances, as is the case for many of the Oldowan assemblages. In such cases identification of intentional selection hinges on inferring from the data more nuanced, less readily recognized selection criteria. These may be responsive to the particular technologies used in the production of the lithic artifacts.

A review of the available literature indicates that various authors have different ways of documenting and quantifying raw material presence in archaeological assemblages, which in turn influences their appreciation of raw material selectivity in these assemblages. In many cases rock type frequencies in the Oldowan assemblages are a reflection of local availability. Such is the case with the use of basalts in Koobi Fora (Toth 1985; Schick 1987; Isaac and Harris 1997) and quartz in Omo's Shungura F sites (Merrick and Merrick1976). The shift to the dominant use of chert in Lower Bed II in Olduvai, likely motivated by the hominins' appreciation of the superior qualities of this raw material (Stiles et al. 1974; Stiles 1998), nonetheless reflects the temporary visibility and availability of chert exposures when the waters of paleo-lake Olduvai retreated (Hay 1976). Kimura (2002) showed that the trend reversed when the re-expanding lake again covered the chert sources.

In other cases rock types were clearly selected purposefully. Trachyte was the preferred raw material in the Gona sites 2.6–2.5 Ma, and its frequencies in the assemblages are substantially higher than in the geological sources (Semaw 2000; Semaw et al. 2003; Stout et al. 2005). Similarly, quartz was preferentially selected at the ~2.0 Ma locality Fj-1a at Fejej (de Lumley et al. 2004), and phonoloite was used preferentially in Lokalalei 2C (Delagens and Roche 2005; Harmand 2005). In other instances, including some where rock types were opportunistically used, subtler selection criteria have been identified. Schick and Toth (1993, and references therein) suggest that toolmakers in Koobi Fora systematically avoided vesicular and weathered basalt cobbles, whereas in Olduvai Bed I times (1.89–1.71 Ma) volcanic cobbles from the local channels were selected for their sizes and quality (Hay 1976, Schick 1987). Minimally reduced cobbles of low quality (at sites DK and FLK) possibly reflect testing of the raw material and its rejection by the tool-makers (Ludwig and Harris 1998; de la Torre and Mora 2005). In Lokalalei 2C as well as in A. L. 894 the existence of appropriate flaking angles that enabled initial knapping appears to have been an important selection criterion (Hovers 2001; Goldman 2004; Delagnes and Roche 2005).

Because systematic quantitative studies of selectivity are a recent research strategy where the Oldowan is concerned (Plummer 2004), it is not surprising that there have not yet emerged a coherent, accepted methodology and analytical procedures for it. Given that selectivity of raw material figures prominently in scenarios of planning abilities, capacities for technological organization, and for understanding their interaction with the environment, the absence of a unified body of methodologies to study selectivity in early assemblages is a cause for some concern. Science indeed thrives on plurality and diversity, but in this case, while all selectivity studies ask similar questions, we need to consider whether they address them in *comparable* ways.

It is these issues that we raise through the analysis of raw material at the Late Pliocene site A. L. 894 (Hadar, Ethiopia). Specifically, we turn to questions of identifying raw material sources and to the effects of various sampling procedures on understanding the raw material selectivity of early hominins. While such questions are inherent to raw material studies of any assemblages, they are particularly pertinent in the framework of studying the very early assemblages of lithics because of their broader implications with regard to technological proficiency, decision-making and cognitive abilities.

Clearly these issues are tied to questions about lithic technology. Here we deliberately severe these links, aiming to focus only on certain, specific aspects of the technological processes.

7.2 A. L. 894, Hadar, Ethiopia

Located in the Makaamitalu basin in the Afar region (Figure 7.1), the site is stratigraphically positioned in the Busidima Formation. Overlain by 3 m of variable alluvial silt-clay sediments, capped in turn by the 2.36±.06 Ma BKT-3 tephra (Campisano 2007), the locality is the oldest known archaeological occurrence in Ethiopia outside the Gona area and one of several eastern African sites with dates clustering around 2.4–2.3 Ma.

In the course of three field seasons (2000–2002) an area of 21.5 m^2 was excavated. A cluster of artifacts and fauna was discovered extending over an area of 20 m^2 in a 30 cm thick horizon. Within this area, a highly dense concentration over ca. 10 m^2 was surrounded by areas of much lower densities. All the lithic artifacts (ca. 5,000) were deposited anthropically. While some were surface finds, refits with pieces from the *in situ* horizon attest to the integrity of the assemblage and to the mild nature of through-site post-depositional disturbances (Hovers et al. 2002; Hovers 2003).

The assemblage consists of large and small flakes, flake fragments and angular fragments; only few flaked pieces (i.e. cores) were found, while retouched artifacts are extremely rare (Hovers 2003). All the artifacts were knapped from volcanic materials (mostly rhyolites, trachytes and basalts), which occurred as river cobbles in the river bed adjacent to the site, now represented by the KH-7 conglomerate (Kimbel et al. 1996; Yemane 1997; Campisano 2007). Flakes in the assemblage are characterized by low scar counts, the occurrence of cortex on the dorsal face, and untreated (cortical) or lightly treated (plain) striking platforms.

7.3 Characterizing Raw Material Sources

Characterizing and identifying raw materials is an issue that depends on both the lithology involved (i.e the degree of distinctiveness of the raw materials) and the analytical procedures used. Clearly, chemical and petrographic characterizations of both artifacts and sources are extremely helpful and may reveal patterns that are not immediately recognizable by low-tech observations. However, this is not always a feasible research option when researching the early assemblages (Goldman 2004; see also Harmand 2005; Stout et al. 2005 for West Turkana and Gona sites, respectively). Most importantly, the formal classification of rock lithology is by itself insufficient to inform us of ancient selection behavior. Low-tech approaches, focusing on visual inspections of variables related to raw material quality and texture (angularity, grain-size, rock homogeneity, phenocrysts size) may in fact be a closer analogue of the practices of the ancient tool-makers that relied only on visual assessment and cobble-testing.

Unless the raw materials exploited at a given locality are conspicuously exotic to a site's immediate surroundings (e.g. the use of quartz in Olduvai Bed I sites [Hay 1976], chert in Gona [Semaw et al. 2003], multiple raw materials in Kanjera [Braun et al. 2008]), it is assumed in most studies that cobbles were collected from the closest river bed. These are usually identified in current settings as sheet conglomerate layers or as buried channels in local stratigraphic sequences. In the majority of cases this is based on the assumption that hominins adopted least effort solutions (Isaac 1996 (1981)) to the problem of obtaining lithic raw material.

In the Makaamitalu Basin there are two distinct conglomerates that record the nature of cobbles in the

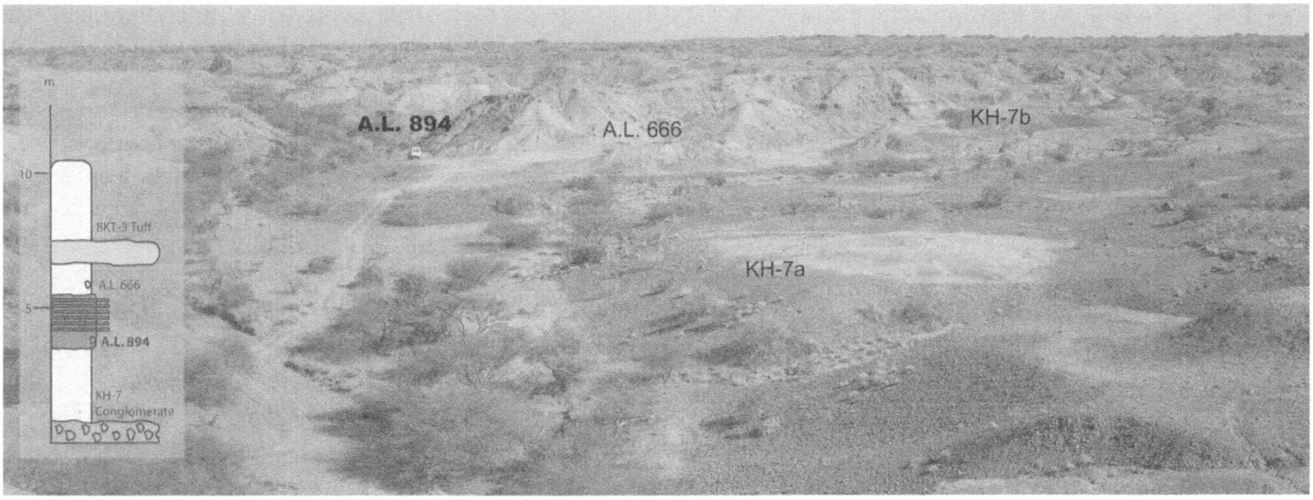

FIGURE 7.1 Location of A.L. 894, A.L. 666, and the conglomerate sampling locations KH-7a and KH-7b. Photograph taken from the exposure of kh-7+, looking to the north-east. Inset: schematic stratigraphic section of the Maka'amitalu basin (re-drawn from the original by C.J. Campisano). See text for details on distances and directions.

paleo-channels. The KH-7 conglomerate, stratigraphically lower than the site, represents an extensive paleo-channel that is recognized in numerous exposures in the Basin. A null hypothesis of least effort (Isaac 1996 (1981)) implicates the KH-7 conglomerate as the source of raw materials found at the site because of its proximity to the archaeological locality. The stratigraphic relationship to the archaeological site of kh-7+, a discontinuous conglomerate, could not be securely defined[1]. This conglomerate, located some 800 m south-west of A. L. 894, is less extensively exposed in the area.

The presence of more than a single conglomerate sheet required that we sample and study both conglomerates. Because it is not possible to identify the exact section from which raw materials were obtained, we might erroneously accept the null hypothesis that KH-7 was the source of raw material (a type II error) if the cobbles in both conglomerates were sufficiently similar so as not to affect selection criteria applied by the early hominins. Since the distances involved are rather small, such an error would have little influence on our understanding of the selection strategy of the early hominins. If, on the other hand, the two conglomerates were found to be different, the dual sampling would help characterize them in terms of lithology, metric attributes and shapes, and provide tools for a more rigorous test of the null hypothesis that KH-7 was exploited for raw material.

Similar problems arose with regards the internal variability of the KH-7 conglomerate, and similar analytical procedures were applied. This large sheet conglomerate clearly does not derive from a single flood event. The paleo-channel represented by this conglomerate may have shifted course and carried through time lithic materials from various sources, thus sampling and embedding different raw materials in different conglomerate horizons. Alternatively, flow energy may have changed through time and the size and level of attrition of cobbles from various raw materials can vary from one spot in the channel to the other pending the overall geometric configuration of the landscape. The paleogeography of the region is poorly known (Quade et al. 2004; Stout et al. 2005; Campisano 2007), so that these and other factors remain largely unmonitored. This lack of information, in turn, requires that variability in cobble properties within a single conglomerate be examined and accounted for.

7.3.1 Sampling Procedures

The two conglomerates were sampled during the 2001 and 2002 field seasons. The massive sheet conglomerate of KH-7 was sampled at two distinct locations in order to assess its internal variability in terms of cobble size, angularity and lithology. Two samples were collected (hereafter named KH-7a and KH-7b; see Figure 7.1 for the location of samples in relation to the site). Both sample points are about 400 m from the site. The KH-7b sample was collected about 350 m northeast of KH-7a. While the two samples represent different phases of the same stream (KH-7b is stratigraphically higher than KH-7a), both represent the same conglomerate and seem to originate from the same volcanic source (Feibel, pers. comm). An additional sample was taken from the second conglomerate, informally referred to here as kh-7+, located some 400 m to the south from the KH-7a sampling point. Cobbles were collected from three gridded square meters, each laid out randomly in a different conglomeratic horizon. The first 100 cobbles larger than ca. 7 cm (fist size), were collected and analyzed. A similar sampling procedure was employed in the Gona and Lokalalei sites (Semaw 1997; Semaw et al. 2003; Harmand 2005; Stout et al. 2005).

Differences between the two KH-7 samples were obvious already during field sampling. The KH-7a cobbles were embedded in loose sediments with obvious presence of salt minerals. The cobbles were weathered and fragmentary, and excavation into the conglomerate reached a depth of 30 cm before the sample of 100 whole cobbles larger than 7 cm was obtained. At the KH-7b sampling spot the conglomerate was lightly cemented and large, whole cobbles were abundant in the few top cm. At the kh-7+ sample spot the cobbles were found in a loose matrix. Their density was much lower than that seen at KH-7 sampling spots and a second square had to be excavated to reach the 100 cobble sample.

The differences in conglomerate appearance, cobble density and cobble breakage within KH-7 can be explained by local processes. The immediately observed differences in cobble size are possibly related to stream dynamics. Differences between the two sub-samples from KH-7 may then indicate separate deposition events or different locations along the same stream.

Both the geological and the archaeological samples (see below on sampling procedures of the archaeological sample) were analyzed as to lithology, grain size, rock homogeneity (vesicularity, banding[2] and phenocrysts presence), cobble size and shape. These parameters reflect the rock's inner properties and are proxies of its knapping qualities (Crabtree 1967; Cotterell and Kamminga 1979; Inizan et al. 1992, 1999; Andrefsky 1994, 2000; Brantingham et al. 2000).

7.3.2 Analysis

Cobble shape. Given the variability observed in the samples, seven morphotypes were defined, ranging from rounded to angular shapes (Figure 7.2; Table 7.1). Sphericity and roundness reflect geological processes affecting the cobbles (and thus serve to heuristically characterize the local conglomerates

[1] While it does not offer a decisive solution to the question of stratigraphic relations between kh-7+ and the site, the recent detailed study by Campisano (2007) suggests that kh-7+ may be younger than the site. It thus provides independent support to our conclusion (below) that this was not a likely source of the raw material for A. L. 894.

[2] Preferential fracture plane in the rock.

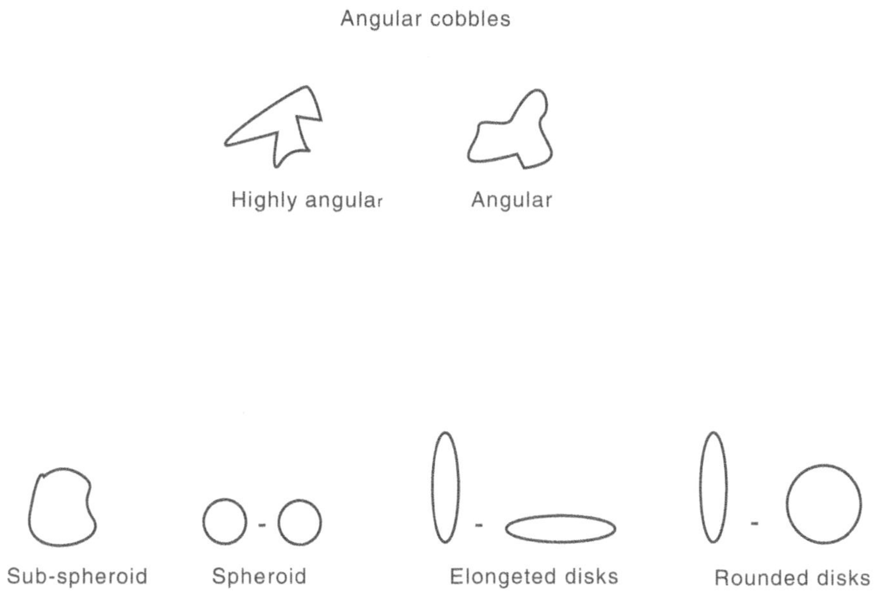

FIGURE 7.2. Schematic classification of cobble shapes in the geological samples.

in relation to each other). At the same time cobble shape may be an important selection criterion, given that certain shapes provide natural ridges that facilitate initial removals and therefore more conducive for lithic production.

In the KH-7a and KH-7b samples angular and highly angular shapes constitute 56.31% and 60%, respectively, while in kh-7+ these groups constitute 84% of the sample. Two of the rounded types are altogether missing from kh-7+ and the other rounded types are poorly represented. Despite the observed differences in preservation between the two subsamples of KH-7, the distribution of cobble shapes is similar (Table 7.1). These results indicate that conglomerate kh-7+ was embedded by a channel of a different nature than KH-7. Cobbles were either transferred from a different source or had been subjected to lower hydraulic energy than the one responsible for the deposition of KH-7. The second option is supported by the low density of cobbles in kh-7+, its appearance as a thin horizon and the higher frequencies of angular cobbles.

TABLE 7.1. The frequencies of cobble shapes in the various conglomerates*

Shape	KH-7A		KH-7B		KH-7+	
	N	%	*N*	%	*N*	%
Highly angular	21	20.39	37	35.24	46	46
Angular	37	35.92	26	24.76	38	38
Subspheroids	12	11.65	11	10.48	7	7
Spheroids	2	1.94	9	8.57	1	1
Bread loaf	17	16.50	11	10.48	-	-
Elongated disks	4	3.88	3	2.86	-	-
Rounded disks	10	9.71	8	7.62	8	8
Total	103	100.00	105	100.00	100	100.00

*X_2 tests were conducted on unified groups: angular (including highly angular and angular cobbles), spherical (including spheroids and sub-spheroid cobbles and bread loaf-shaped shaped cobbles) and disks (including elongated and rounded disks). The differences between KH-7 sub samples and kh-7+ were statistically significant (kh-7+ and KH-7a: $X_2 = 19.92$, df = 2, $p = 0.00$; kh-7+ and KH-7b: $X^2 = 16.93$, df = 2, $p = 0.00$; KH-7a and KH-7b: $X^2 = 0.55$, df = 2, $pl = 0.76$).

Metric measurements. Maximum length, maximum width and thickness were measured for each cobble (Table 7.2). kh-7+ cobbles are slightly bigger than those in the KH-7a and KH-7b samples (Figure 7.3). This is consistent with the suggestion that kh-7+ cobbles had undergone less abrasion as discussed above. Additionally, the similarity in the size tendencies of cobbles from the KH-7 subsamples, located at distances of 400 m from one another, suggests that they represent the size range available to hominins in the Makaamitalu Basin.

Lithological "types". Field identifications were conducted (KH-7a and KH-7b) or supervised (kh-7+) by C. Feibel. All cobbles were weathered with comparatively high alteration of the phenocrysts. The cortex in all rock types was abraded.

Identification of volcanic rocks is based on their mineralogical composition (Le Maitre 1989). A hand lens (X10) was used to identify the visible minerals (phenocrysts) in the rock samples, which were then classified based on a protocol for geological field classification of volcanic rocks by a QAPF diagram (Prinz et al. 1978; Le Maitre 1989). In principle this means that, if a rock is not porphyritic (i.e., has visible crystals in it) it could not be classified. Most of the rocks from

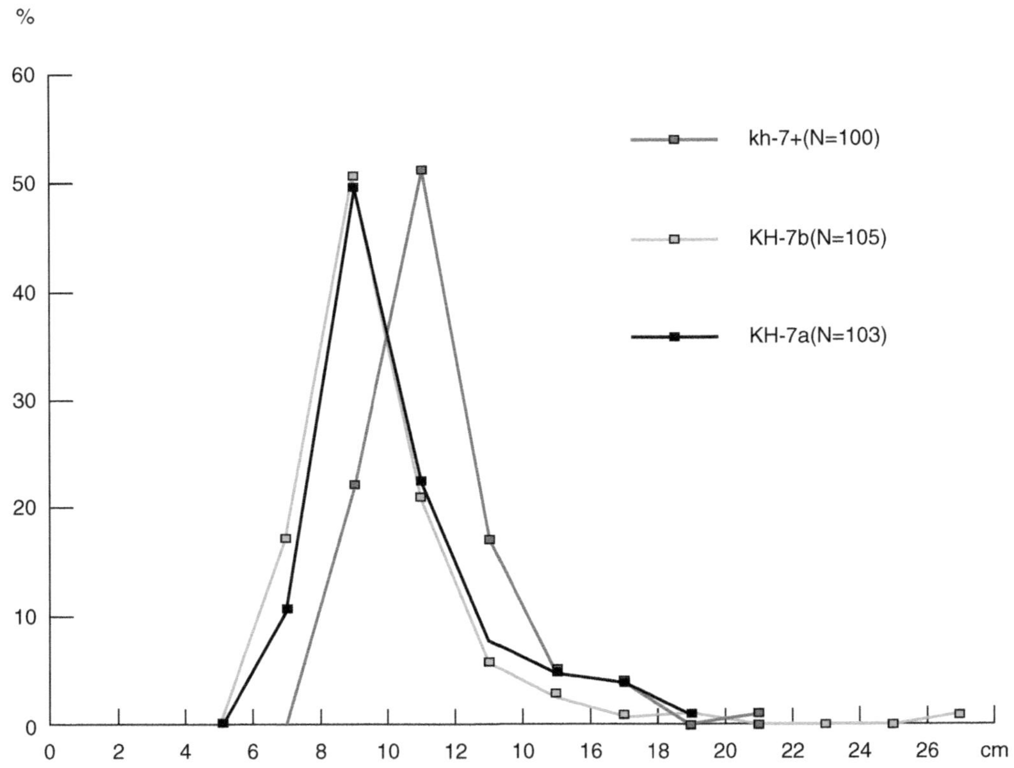

FIGURE 7.3. Length distribution of cobbles in the three conglomerate samples. The X-axis presents length categories with 2 cm increments. X^2 test was invalid due to empty cells.

the Makaamitalu conglomerates were porphyritic and could be classified in this manner.[3]

Table 7.3 presents the distribution of rock types thus recognized. Rhyolite is the dominant rock type in all three conglomerates, most notably in kh-7+. The second common rock type differs among samples: in the KH-7 subsamples basalt appears in high frequencies while trachyte comes second in kh-7+. However, rhyolite is practically the only rock type represented in this sample. Finally, small and very weathered obsidian cobbles occurred in small quantities *only* in KH-7a.

The lithological compositions of KH-7a and KH-7b differ significantly (see Table 7.3), the main difference being in the relative frequencies of trachyte in each subsample. This variability within KH-7 emphasizes the fact that the exact sections from which raw materials were obtained cannot be pinpointed. At the same time, significant differences between KH-7 and kh-7+ support the distinction between the two potential sources of raw material and facilitate a probabilistic identification of one of them as the source of lithics for A. L. 894.

Matrix. Four properties of the matrix, which attest to the mechanical properties of the raw material (and thus its quality for knapping), were examined and ranked: grain size, phenocryst coverage, vesicularity and banding. Grain size is a trait of the matrix and is independent of the amount and size of the phenocrysts and of the presence (or lack thereof) of vesicles and bands. All the geological samples are dominated by fine-grained rocks (Figure 7.4a).

Cobbles with vesicles appear in similar frequencies in KH-7a and KH-7b, in which they are somewhat more abundant than in kh-7+ (Figure 7.4b). Frequencies of cobbles with banding differ among the three samples. kh-7+ cobbles are the most banded whereas KH-7a cobbles are the least banded.

TABLE 7.2. Descriptive statistics of cobble sizes (in cm) in the different conglomerates

Measurements		Mean	Std. dev.	Std. error.	Count	Min.	Max.
Length	kh-7+	10.36	2.11	.21	100	7.40	20.70
	KH-7a	9.22	2.39	.24	103	6.10	17.50
	KH-7b	8.74	2.64	.26	105	5.70	25.00
Width	kh-7+	7.47	1.50	.15	100	4.00	12.20
	KH-7a	6.57	2.04	.20	103	4.10	14.00
	KH-7b	6.46	1.74	.17	105	3.00	12.80
Thickness	kh-7+	4.89	1.36	.14	100	2.20	9.50
	KH-7a	4.35	1.56	.15	103	1.80	10.30
	KH-7b	4.27	1.31	.13	105	2.10	10.30

[3] Highly weathered specimens are excluded from the analysis.

TABLE 7.3. Rock type frequencies in the three conglomerate samples*

	kh-7+		KH-7a		KH-7b	
Rock type	*N*	%	*N*	%	*N*	%
Rhyclite	76	76.77	52	50.48	56	53.33
Basalt	9	9.09	43	41.75	31	29.52
Trachyte	10	10.10	3	2.91	14	13.33
Obsidian	-	-	2	1.94	-	-
Varia**	3	3.03	-	-	-	-
Weathered	1	1.01	3	2.91	4	3.81
Total	99	100.00	103	100.00	105	100.00

* Obsidian, weathered and Varia were excluded from the X^2 test. The differences between KH-7 sub samples and kh-7+ were statistically significant (KH-7a and KH-7b: $X^2 = 9.17$, df = 2, $p = 0.01$; kh-7+ and KH-7a: $X^2 = 30.46$, df = 2, $p = 0.00$; kh-7+ and KH-7b: $X^2 = 15.63$, df = 2, $p = 0.00$).

** Varia category in 7+ includes breccia, anortite and ignimbrite.

In sum, the analysis of conglomerates in the study area revealed that the two occurrences can in fact be distinguished on the basis of cobble properties. While internal variability in the distributions of cobble properties does occur within KH-7, it is negligible in comparison to the differences between this conglomerate and kh-7+. The main attribute in which the two KH-7 subsamples varied was the distribution of rock types. The presence of small quantities of obsidian cobbles in KH-7a as well as in the archaeological sample suggests that KH-7 is more likely the source of raw material. It is this property, together with the differences in the distribution of practically all other attributes of kh-7+ cobbles, that separates the two conglomerates.

The comparative analysis of the two potential raw material sources does not permit rejection of the null hypothesis that KH-7 was the source of lithic raw material. This, however, is not the same as taking for granted that the nearest potential source was indeed the raw material source simply because early hominins operated by default under principles of least effort when procuring raw material (Isaac 1996, (1981). Even when raw materials are not exotic and appear to be highly similar in the site's vicinity, the detailed analysis enables a more probable identification of a main source of lithic material for flaking, which could nonetheless be part of a larger "source area" (as discussed, for example, by Stout et al. 2005).

7.4 Matching Sources and Assemblage

The identification of patterns of raw material selection is based as a rule on comparisons between sources and assemblages. Such comparisons are sometimes based on the presence in potential lithic sources of a specific and well-defined raw material used for knapping. Such is the case of obsidian in A. L. 894. This material exists only in one of the three geological samples and thus points to the probable use of this conglomerate horizon by the site's occupants. To identify *selectivity*, however, some quantitative comparison should be made in order to understand if materials had been chosen from an available source in a non-random manner. If so, we may consider the qualities that rendered them attractive to the early hominins.

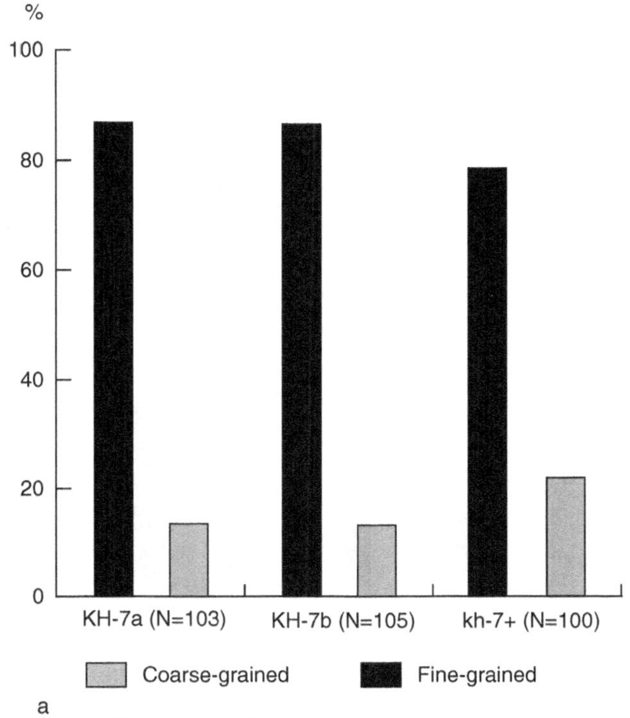

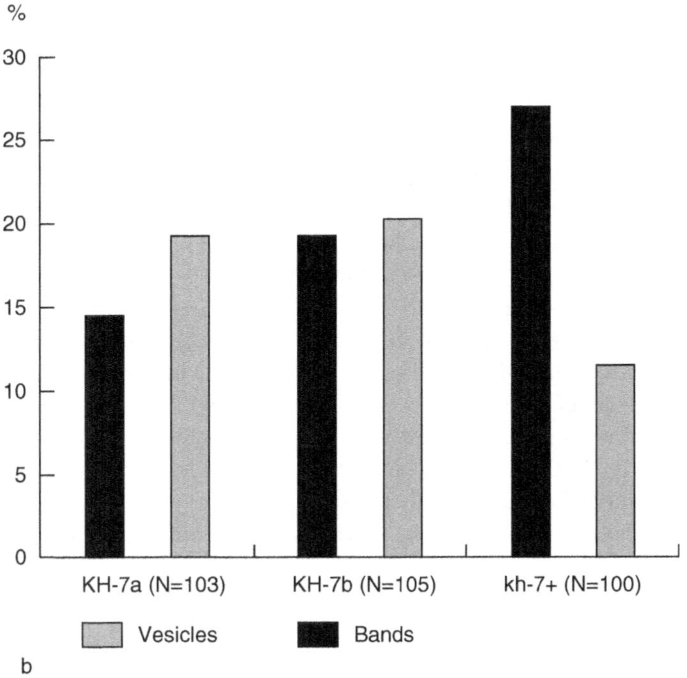

FIGURE 7.4. Frequencies of matrix types in the conglomerate samples: a. grain-size frequencies. b. impurity frequencies.

Quantification, it turns out, is not a straightforward process, because of the fundamental differences between the two types of samples involved. One constitutes natural cobbles whereas the other encompasses worked items that were derived from similar cobbles but had undergone significant changes. Artifacts in archaeological assemblages are flaked or detached (i.e., cores or flakes) and do not bear necessarily resemblance either in shape, cortical cover or dimensions to the original cobbles. This places much of the burden of evidence on parameters that do not change through the reduction process, namely rock lithology. However, archaeological stone artifacts are sometime complete but quite often broken due to knapping accidents or post-depositional processes (e.g. Hovers 2003). This raises the problem of a realistic number of detached pieces in an assemblage ("MNI", Hiscock 2002) and how this number can be used to reveal patterns of raw material selection. If comparisons are carried out by artifact numbers, it is necessary to understand the breakage patterns of the various rock types. Brittle/fragile rock types may be represented by more pieces than the more durable rocks. They would be over-represented in archaeological contexts and be erroneously interpreted as the ones favored by the early stone knappers.

Some workers attempted to circumvent this difficulty by including only complete flakes in the archaeological samples compared with geological sources (Semaw 2000). It should be clear from the ongoing discussion that while such a procedure eliminates one type of error (i.e. reduces the risk of over-estimating the number of artifacts of any rock type) it does not solve the biases that may arise if rocks break differently. Such a procedure would lead to elimination of many of the artifacts from the study of early Oldowan assemblages. At A. L. 894 preliminary observations indicated that complete flakes were less than half the assemblage. Hence exclusion of broken artifacts was not a satisfactory course of option in this case.

7.4.1 Procedures of Sampling the Archaeological Assemblage

Due to logistical constraints we were unable to include all the excavated lithics in the raw material study. Our sampling procedure consisted of the analysis of the complete collection excavated during the first (2000) season. Of the materials excavated during the 2001 and 2002 season every fifth artifact (20%) was analyzed[4], regardless of its size, completeness and typological identification. In order to evaluate the effect of breakage on our understanding of raw material selectivity, the material from the 2000 season was used to establish the basic patterns. In the following paragraphs only the latter collection is discussed (unless stated otherwise).

Three breakage states were recognized in A. L. 894. "Complete" and "broken" relate to the breakage state of detached and flaked pieces, whereas "fragmented" describes those pieces where knapping could not be identified with confidence, and which might have resulted from natural processes at A. L. 894 (Hovers 2003). These breakage states result from thresholds of rock fragility (i.e. a rock's tendency to break) or from its brittleness (i.e. the tendency to produce high quantities of debris when force is applied, as would happen during knapping). In the lack of engineering measures or experimental knapping, we do not attempt here to establish a direct correlation between breakage states and the specific fracture mechanics of the different rock types. Still it is clear that recognizing which rocks tend to break or crumble more often than others was an important step in the knapper's evaluation of the rock suitability for flaking.

7.4.2 Understanding Breakage Patterns — Artifact Frequencies and Weight Distributions

Breakage states. Figure 7.5 shows the frequencies of rock types according to breakage states. The *overall pattern* of distribution of rock lithology within the three breakage states is widely similar. Rhyolites, being the most common raw material on site, are the most common also within each breakage category (Figure 7.5). We crossed breakage states and rock types only within the 2000 collection. Superficially, it would seem on the basis of these data (Table 7.4, Figure 7.5) that complete artifacts may suffice for reconstructing raw material selectivity. However, it is the inclusion of incomplete pieces

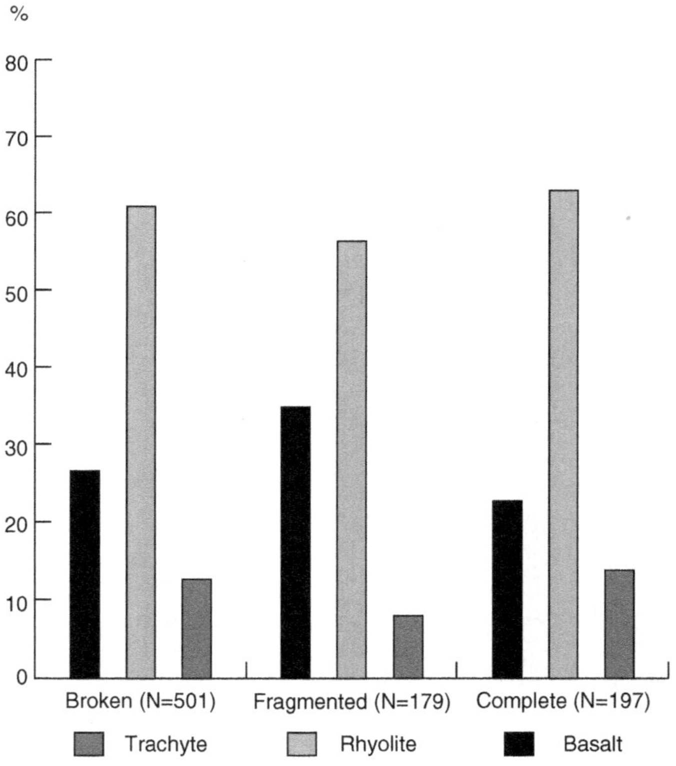

FIGURE 7.5. Frequencies of rock types according to breakage states within the 2000 collection.

[4] We compared the results of the fully analyzed 2000 collection with the sampled ones from the 2001-2 seasons and found the sampled collections to be representative.

TABLE 7.4. Rock type frequencies of the complete artifacts compared to the total number* of artifacts in the 2000 collection.

Rock type	Complete artifacts		All artifacts	
	N	%	*N*	%
Rhyolites	125	63.45	241	60.66
Basalts	45	22.84	532	27.48
Trachytes	27	13.71	104	11.86
Total	197	100.00	877	100.00

* regardless of breakage state
* X^2 test on the two samples indicate that differences between them are insignificant (X^2 =1.95, df=2, p=0.37).

in the study which reveals more nuanced selection patterns, as will be discussed below.

Breakage by lithology. The broad patterns of breakage states within each rock type are similar (Figure 7.6). However, this similarity accentuates the fact that the mild differences in breakage states among raw materials are not random, so that fragmented basalts are over-represented compared to the frequency of this raw material type in the geological sample (Figure 7.7; see discussion below). Basalts are either more fragile (having more fragmented pieces) or are more brittle (with higher fragmentation while knapping) than the other raw materials in the assemblage. A technological analysis of the A. L. 894 lithics does not indicate that basalts were knapped more intensively than other raw materials (Hovers, work in progress), which undermines the notion that basalt cobbles broke more extensively due to knapping procedures. Thus basalts are the least represented rock type in the two breakage states associated with knapping (in the sense defined above). With the highest frequencies of fragments (possibly derived from geogenic processes) basalts may indeed have been broken taphonomically more often compared to other raw materials.[5] Since breakage state designations were not available at the time of study for all the analyzed artifacts, we incorporated the distribution data of breakage states with additional analyses of weight and size distributions. We first established that there were no significant differences in the specific weights of the three rock types, which assured that the comparison between weight and raw frequencies are comparable. Comparison between the distribution of rock types by weight and by raw numbers of pieces (Table 7.5) assists in understanding potential biases in the sample that are due to nonrandom breakage.

Table 7.5 shows the distribution of artifact numbers and weight in the analyzed sample (2000–2002; see above). The biggest discrepancy between the two data sets is for trachytes, whose numbers are under-represented relative to weight. This discrepancy is attributed to the large size of trachyte flaked and detached pieces and to their relative lack of breakage (Figure 7.5, 7.6). Basalts exhibit the opposite trends, being smaller in size than both trachytes and rhyolites (Figure 7.8) and more numerous compared to their weight, due to their higher fragmentation (Figure 7.5, 7.6; Table 7.5). All this suggests that the relative frequencies of basalts are inflated. When frequencies of rock types are compared between the geological source and archaeological assemblage, they should be more realistically approximated by taking into account the differences in breakage states of the various rock types in order to avoid comparing apples and oranges (as suggested in Figure 7.7).

[5] Not all basalts share the same mechanical properties; some small, high quality basalt cobbles were utilized.

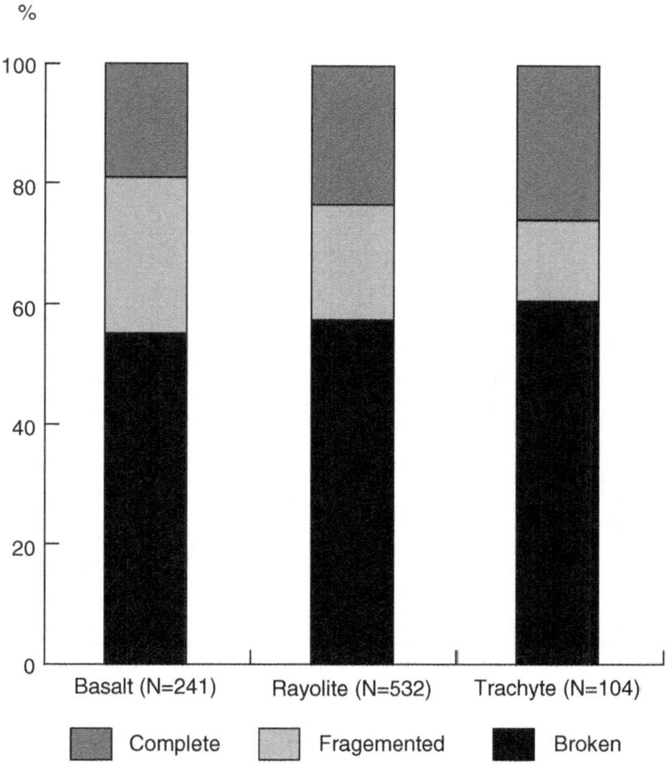

FIGURE 7.6. Breakage states according to rock types within the 2000 collection. The differences between the three breakage states are statistically significant (X^2=9.48, df=2, p=0.05). The difference between the complete and incomplete (=broken and fragmented) specimens is statistically insignificant (X^2=3.05, df=2, p=0.21).

7.5 Discussion

The onset of tool production represents a process of intensification or a strategy of risk reduction on the part of early hominins. It arguably provided technological aids with which to more effectively exploit their ecological niches. As raw material acquisition is the first step in the operational chain of stone tool-making, selectivity and the specific criteria involved in procurement inform researchers about cognitive processes,

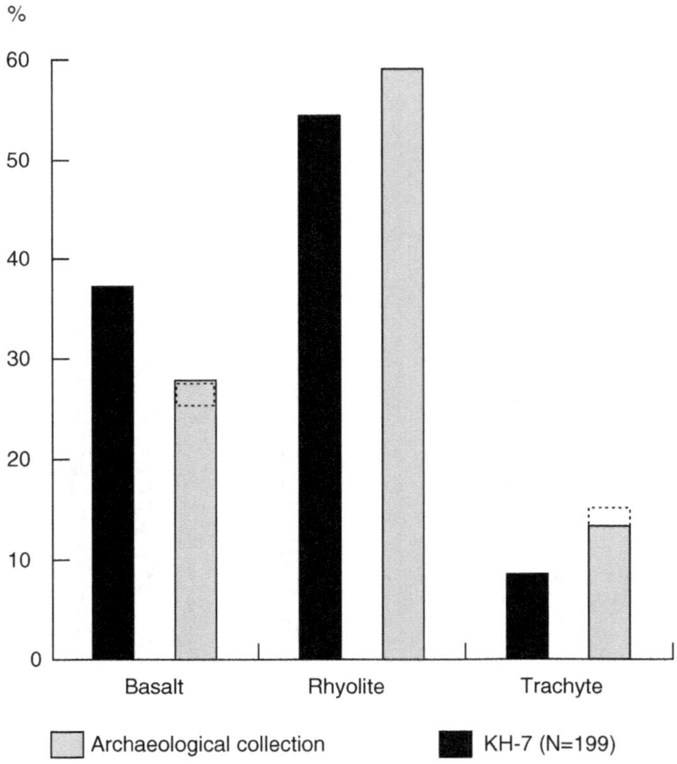

FIGURE 7.7 Comparison of rock types between the archaeological and geological samples. Dashed line represents speculated "calibrated" frequencies based on analysis of the breakage states of each rock type.

for instance the execution of tasks according to preconceived mental templates (or lack thereof), i.e. pre-planning. Why and how raw materials were selected for tool production have become increasingly important questions in attempts to understand hominins' adaptive strategies.

Studies dealing with these questions have shared a number of analytical and methodological commonalities, relating either implicitly or explicitly to transport distances and raw material qualities as indirect and direct proxies (respectively) of selectivity. When exotic raw materials are absent, it is often posited that early hominin problem-solving was guided by least effort principles, such that raw materials for lithic artifacts had been sought in the nearest source available at the time of site occupation. Discrepancies in rock qualities between assemblages and sources, on the other hand, are thought to

TABLE 7.5. Comparison of rock type frequencies and weight distribution according to rock type in the archaeological sample (all the analyzed artifacts)

	Weight		Rock type frequencies	
	Gram	*%*	*N*	*%*
Basalts	2,875	20.48	407	28.70
Rhyolites	8,170	58.20	878	61.92
Trachytes	2,992	21.31	133	9.38
Total	14,037	99.99	1,418	100.00

reflect selectivity on the part of early hominins. Searching for quality selection involves examination (by either visible properties or chemical and mechanical laboratory studies) of rock type quality, looking at rock characteristics.

While raw material studies are anchored in these relatively consensual premises, they are not standardized. Comparisons are often quantified to some degree, but details of sampling protocols, analytical approaches and manner of quantification are sometimes ambiguous. As a result, it is not clear to what degree our reconstructions of ancient behavioral patterns are based on *comparable* data sets. This lack of standardization in research protocols informed the procedures adopted for the study of patterns of raw material selection at A. L. 894. A number of methodological challenges were identified that should be addressed formally and explicitly. Those are particularly pertinent in contexts where raw material selectivity does not reflect the use of exotic raw material. As discussed in the introduction to this paper, this is the case with the majority of the Late Pliocene archaeological sites.

The first issue is related to comparison between the archaeological and geological samples. While this comparison is a common means to evaluate the degree of selectivity, its strength is compromised if not subject to testing. In the case of A. L. 894 a comparative study of two potential sources enabled us to focus on one of two examined exposures as a main source of raw material, in close proximity to the archaeological site (and possibly part of a larger "source area"). Internal variability with regards rock types, but not in terms of metrics and matrix[6], was observed within this source. Recognizing this variability helped in calibrating the variation seen in the archaeological sample against the spatial and temporal variability within the geological source, and to assess whether it is contained within the range observed in the conglomerate.

Another issue that appears unclear in many studies of raw material procurement and selection are the units of comparison. Assessment of the similarity and differences of raw material source and the archaeological assemblage is inherently difficult. It is essentially a comparison between artifacts that emulate a starting point of the reduction process (i.e. the complete cobbles from the natural exposures) with items that represent variable points along (possibly) several reduction sequences (i.e. the flaked and detached pieces in the archaeological assemblage). The mechanisms of transformation from one state to the other involve both fracture mechanics of the rocks and technological practices of the knappers. Comparison between source and assemblage needs to take into account the changes in size and shape of the products compared to the original cobble, as well as the inflated frequencies of each rock type in the assemblage that result from knapping a single cobble into a number of products. Almost all raw material studies rely to variable degrees on some manner

[6] Obviously, this is not necessarily the case elsewhere, and intra-source variability can be recognized in other properties of the raw material (e.g., cobble sizes, differences in rock properties).

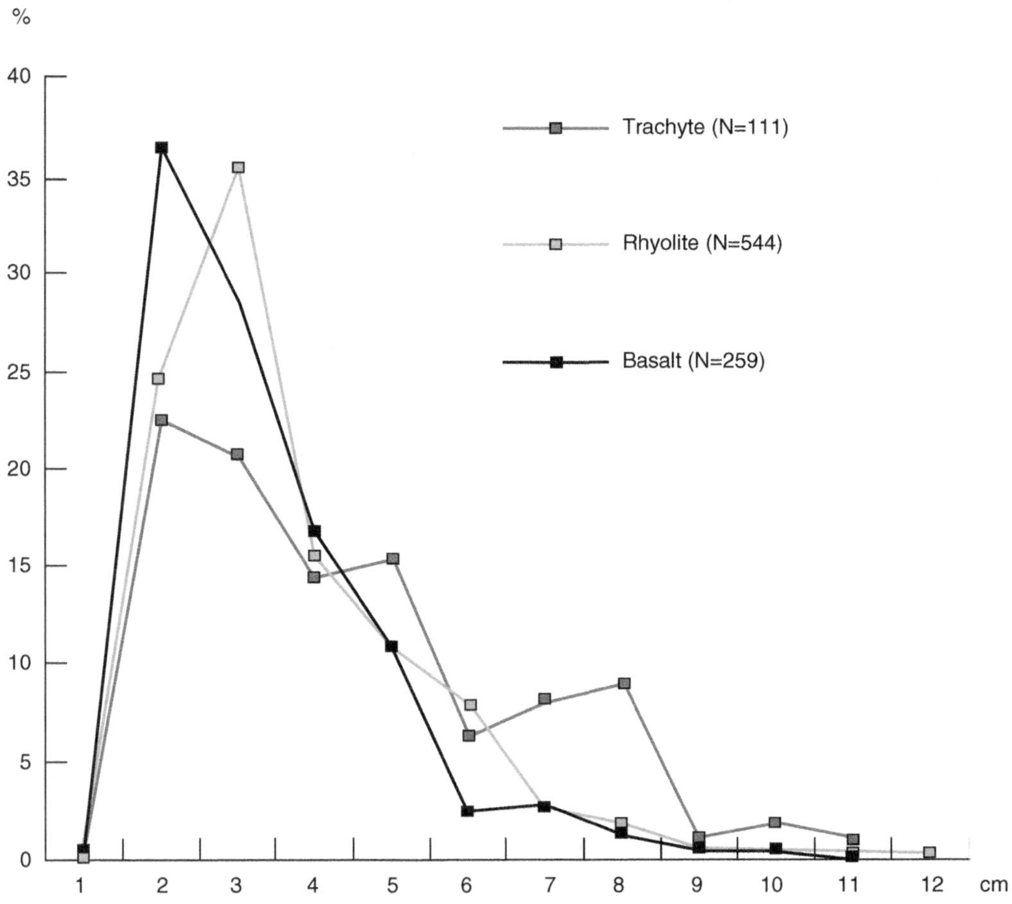

FIGURE 7.8 Length frequencies by rock lithology of the artifacts collected in the 2000 season.

of quantification of rock properties (or their frequencies) in both samples. Bridging the distance between a package of raw material and its products is thus better understood if multiple quantitative as well as qualitative measures are used in tandem.

Since technological analyses of the A. L. 894 assemblage did not reveal differences in knapping practices between the various raw materials (preliminary results from work in progress), we focused on the effects of fracture mechanics as the potential explanation of the discrepancy in frequencies of various rock types within the assemblage and in comparison to the geological source. Combined with taphonomic observations, we inferred that basalts were over-represented due to their tendency to break more often and could address the question of selectivity in a more informed manner. Rock lithology does not appear to have been a major selection criterion used by the site's occupants (Figure 7.7). Rather, they appear to have been partial to rock quality (i.e. fine-grained, relatively homogeneous and with few or small phenocrysts; Goldman 2004.).[7] Recognizing the existence of these subtle yet significant selection criteria would have been difficult, if not impossible, on the basis of rock type frequencies by themselves, or by analyzing only complete flakes.

The latter point is noteworthy in the general framework of raw material studies, but it may be of special significance in the narrower context of studying the earliest lithic assemblages. In the majority of cases, such assemblages contain high frequencies of incomplete artifacts, often accounting for more than half of the assemblage (e.g. Hovers 2003; Figure 7.4). Excluding this information from studies on assemblages of small sizes may too often lead to unwarranted biases in our understanding of raw material selectivity on the parts of early hominins.

Another raw material characteristic that has been used to make inferences about hominin life ways is the distance

[7] Another selection criterion that emerged from the analysis (not discussed here) is cobble angularity (Hovers 2001; Goldman 2004). Whenever determination of initial cobble shape is possible, the flaked pieces in the archaeological assemblage suggest that angled cobbles, where at least one acute angle enabled initiation of the flaking process, were preferentially selected. While cobbles in kh-7+ are both larger and more angular, (Table 7.1, Figure 7.3) rock qualities differ from those seen in the archaeological assemblage. This on the other hand emphasizes the specific combination of traits that were selected for.

traveled to raw material sources. Féblot- Augustins (1997) documented 14 African Pliocene sites for which distances to raw material sources were either not specified or were reported as less than 1 km away from the site. This pattern, which is consistent with hypotheses predicting that hominins were tethered to raw material sources in the early stages of stone tool technology (e.g. Harris and Capaldo 1993; Rogers 1994), has persisted through the intensive research of the last decade, with a growing number of raw material reports from an increasing number of sites dates >2.0 Ma showing very short transport distances. A notable change is discerned in the period 1.9–1.3 Ma, when distances as long as 13 km have been documented for 26 Early Pleistocene sites. Following this, Marwick (2003) notes that while chimps move lithic and wood objects over short distances and exploit ca. 17% of their home ranges, hominins in sites dated 1.9–1.3 Ma have exploited and moved raw material over nearly 100% of their postulated home ranges. This difference between Early Pleistocene hominins and chimps indicates that the former invested effort and energy in raw material procurement. While investment in raw material transport may have been confined to limited home ranges defined by hominins' physical build (Steele 1996; see also Jones 1994), it would have been guided also by specific selection criteria. Suitability for knapping can be expected to have figured importantly in this process, as would be the case with suitability for intended tasks (Braun et al. 2008; Noll 2000). If such criteria can be identified, they inform us on the cognitive abilities (e.g. planning depth) of the early tool-makers while engaged in ecological problem-solving. Indeed, Marwick (2003) relates this difference in pattern of raw material transport between Early Pleistocene hominins and chimps to the higher planning capacities of the former (see Roebroeks et al. 1988).

This in turn begs the question whether Late Pliocene hominins settled for whatever was available in their immediate vicinity, which might explain the short transport distances noted in the majority of sites. Where varied raw materials were available, this was not the case (e.g. at the Lokalalei, Gona and Hadar sites; Semaw 2000; Goldman 2004; Stout et al. 2005; Delagnes and Roche 2005; Harmand 2005, 2009). Although search and transport constraints were minimal in these sites, careful selection criteria were applied during raw material acquisition. The accumulated data from A. L. 894 and many other Oldowan sites shows that there is no inherent correlation between large transport distance and raw material selectivity. This holds true both for the Late Pliocene as well as for Upper Pleistocene occurrences (e.g., Minichillo 2006 on the Howiesons Poort in Klasies River).

At A. L. 894 no clear preference for a single rock type was observed. At first glance hominins appear to have opportunistically used rock types in frequencies that resembled those in the geological source. In the lack of material science (e.g. Noll 2000) or experimental studies (e.g. Jones 1994; Toth 1997; Madsen and Goren-Inbar 2004; Harmand 2005), analyses based on detailed observations and on multiple units of comparison made possible the identification and evaluation of the selectivity underlying the raw material composition in this early assemblage. This selectivity can now be incorporated into explanatory scenarios of lithic technology as both an adaptive tool and a reflection of hominin behaviors.

Acknowledgments. Field and laboratory work at A. L. 894 and on finds from the site was supported by NSF grant BCS-0080378 (to W. H. Kimbel), NGS grant #7352-02 and grants from the L. S. B. Leakey Foundation (2002, 2005) to EH. We thank the Ethiopian Authority for Research and Conservation of Cultural Heritage and the Afar Regional Government for their administrative help, and the staff of the National Museum of Ethiopia in Addis Ababa, where our laboratory work was conducted for many months, for their assistance. Special thanks go to Mamitu Yilma, Menkir Bitew and Alemu Ademasu for their help when working in the museum. We thank Gerry Eck, Bill Kimbel, Caley Orr, Yoel Rak, Amy Rector, Karen Shcollmeyer, and especially Craig Feibel and Chris Campisano for their support and help in the field. We are also grateful to Ilan Sharon for his help in statistical matters and to Gonen Sharon, Naama Goren-Inbar and David Braun, as well as two anonymous reviewers, for their comments on various drafts of this paper. The graduate students of the prehistory lab at the Institute of Archaeology of the Hebrew University were a tough critical audience to bounce ideas off. Last but not least, the Addis crew and the Afar people of Eloha and the Hadar region have been great hosts and great friends to us, and we thank them for that.

References

Andrefsky, W., 1994. Raw material availability and the organization of technology. American Antiquity 59, 21–34.

Andrefsky, W., 2000. Lithics – Macroscopic Approaches to Analysis. Cambridge University press, Cambridge.

Bamforth, D. B., 1986. Technological efficiency and tool curation. American Antiquity 51, 38–50.

Blumenschine, R.J., Peters, C., Madao, F.T., Clarke, R.J., Deino, A.L., Hay, R.L., Swisher, C.C., Stanistreet, I.G., Ashley, G.M., McHenry, L.J., Sikes, N.E., van der Merwe, N.J., Tactikos, J.C., Cushing, A.E., Deocampo, D.M., Njau, J.K., Ebert, J.I., 2003. Late Pliocene *Homo* and hominid land use from Western Olduvai Gorge, Tanzania. Science 299, 1217–1221.

Brantingham, P.J., Olsen, J.W., Rech, J.A. Krivoshapkin, A.I., 2000. Raw material quality and prepared core technologies in Northeast Asia. Journal of Archaeological Science 27, 255–271.

Braun, D.R., Plummer, T., Ditchfield, P., Ferraro, J.V., Maina, D., Bishop, L.C., Potts, R. 2008. Oldowan behavior and raw material transport: perspectives from the Kanjera Formation. Journal of Archaeological Science 35, 2329–2345.

Campisano, C.J., 2007. Tephrostratigraphy and Hominin Paleoenvironments of the Hadar Formation, Afar Depression, Ethiopia. Ph. D. thesis, Rutgers University

Cotterell, B., Kamminga, J., 1979. The Mechanics of flaking. In: Hayden, B. (Ed.), Lithic Use-Wear Analysis. Academic Press, New York, pp. 97–113

Crabtree, D.E., 1967. Notes on experiments in flintknapping: 3. the flintknapper's raw material. Tebiwa 10, 8–25.

de la Torre, I., Mora, R., 2005. Technological Strategies in the Lower Pleistocene at Olduvai Beds I & II. ERAUL 112. Université de Liège, Liège.

de la Torre, I., Mora, R., Dominguez-Rodrigo, M., de Luque, L., Alcalá, L., 2003. The Oldowan industry of Peninj and its bearing on the reconstruction of the technological skills of Lower Pleistocene hominids. Journal of Human Evolution 44, 203–224.

Delanges, A., Roche, H., 2005. Late Pliocene hominid knapping skills: the case of Lokalalei 2C, West Turkana, Kenya. Journal of Human Evolution 48, 435–472.

de Lumley H., Beyene Y. (Eds.), 2004. Les sites préhistoriques de la région de Fejej, sud-Omo, Éthiopie, dans leur contexte stratigraphique et paléontlogique. ADPF association, Paris.

Dibble, H.L., 1991. Local raw material exploitation and its effects on Lower and Middle Paleolithic assemblage variability. In Montet-White, A., Holen, S. (Eds.), Raw Material Economies among Prehistoric Hunter-Gatherers. University of Kansas Publications in Anthropology. University of Kansas, Lawrence, pp. 33–46.

Féblot-Augustins, J., 1993. Mobility strategies in the late Middle Palaeolithic of central Europe and western Europe: elements of stability and variability. Journal of Anthropological Archaeology 12, 211–265.

Féblot-Augustins, J., 1997. La circulation des matières premières au Paléolithique. ERAUL 75. Univesité de Liège, Liège.

Geneste J.-M., 1985. Analyse lithique d'industries moustériennes du Périgord: une approche technologique des comportements des groupes humains au Paléolithique moyen. Ph. D dissertation, Université de Bordeaux.

Goldman, T. 2004. Raw material Selection Pattern in A. L. 894, Hadar, Ethiopia. MA thesis, the Hebrew University of Jerusalem.

Harmand, S., 2005. Matières premières lithiques et comportements techno-économiques des hominiés Plio-Pléistocènes du Turkana occidental, Kenya. Ph. D. thesis, Université de Paris X - Nanterre.

Harmand, S., 2009. Variability of raw material selectivity and techno-economic behavior in the early Oldowan: evidence from the Late Pliocene sites of Lokalalei. In: Hovers, E., Braun, D.R. (Eds.), Interdisciplinary Approaches to the Oldowan. Springer, Dordrecth, pp. 85–97.

Harris, J.W.K., Capaldo S.D., 1993. The earliest stone tools: their implications for an understanding of the activities and beahviour of late Pliocene hominids. In: Berthelet, A., Chavaillon, J. (Eds.), The Use of Tools by Human and Non-Human Primates. Clarendon Press, Oxford, pp. 197–224.

Hay, R.L., 1976. Geology of the Olduvai Gorge. University of California Press, Berkeley.

Hiscock, P., 2002. Quantifying the size of artefact assemblages. Journal of Archaeological Science 29, 251–258.

Hovers, E., 2001. Stone knapping in the Late Pliocene of Hadar, Ethiopia. Roux, V. Knapping Stone – a Uniquely Human Behavior? conference abstracts http://web.mae.u-paris10.fr/knappingstone/anglais/resume.htm.

Hovers, E., 2003. Treading carefully: site formation processes and Pliocene lithic technology. In: Martínez-Moreno J., Torcal, R. M., de la Torre, I. (Eds.), Oldowan: Rather More than Smashing Stones. An Introduction to "The Technology of First Humans" Workshop. Universitat Autonoma de Barcelona, Barcelona, pp. 145–163.

Hovers, E., Schollmeyer, K., Goldman, T., Eck, G., Reed, K., Johanson, D., Kimbel, W., 2002. Late Pliocene archaeological sites in Hadar, Ethiopia. Paper read in the meeting of the Paleoanthropology Society, Denver Colorado, March 2002. *Journal of Human Evolution*, A17.

Inizan, M.L., Reduron-Ballinger, M., Roche, H. Tixier, J. 1999. Technology and Terminology of Knapped Stone. CREP, Nanterre.

Inizan, M. L., Roche, H. Tixier, J. 1992. Technology of Knapped Stone. CREP, Meudon.

Isaac G.L., Harris J.W.K., 1997. The stone artefact assemblages: a comparative study. In: Isaac, G.L., Isaac, B. (Eds.), Koobi Fora Research Project. Vol. 5: Plio-Pleistocene Archaeology. Clarendon Press, Oxford, pp. 262–362.

Isaac, G.L., Harris J.W.K., Crader D., 1976. Arcaheological evidence from the Koobi Fora Formation. In: Coppens, Y., Howell, F.C., Isaac G.L., Leakey R.E.F. (Eds.), Earliest Man and Environments in the Lake Rudolf Basin. Chicago University Press, Chicago, pp. 533–551.

Isaac, G.L.,1996 [1981]. Small is informative. The application of the study of mini-sites and least-effort criteria in the interpretation of the Early Pleistocene archaeological record of Koobi Fora, Kenya. In: Isaac B. (Ed.), The Archaeology of Human Origins. Papers by Glynn Isaac. Cambridge University Press, Cambridge, pp. 259–268.

Jones, P.R., 1994. Results of experimental work in relation to the stone industries of Olduvai Gorge. In: Leaky, M.D., Roe, D.A. (Eds.), Olduvai Gorge, pp. 254–298.

Kibunjia, M., Roche, H., Brown, F.H., Leakey, R.E., 1992. Pliocene and Pleistocene archeological sites west of Lake Turkana, Kenya. Journal of Human Evolution 23, 431.

Kimbel, W.H., Walter, R.C., Johanson, D.C., Reed, K.E., Aronson, J.L., Assefa, Z., Marean, C.W., Eck, G.G., Bobe, R., Hovers, E., Rak, Y., Vondra, C., Yemane, T., York, D., Chen, Y., Evenson, N.M., Smith, P.E., 1996. Late Pliocene *Homo* and Oldowan tools from the Hadar Formation (Kada Hadar Member), Ethiopia. Journal of Human Evolution 31, 549–561.

Kimura, Y., 2002. Examining time trends in the Oldowan technology at Beds I and II, Olduvai Gorge. Journal of Human Evolution 43, 291–321.

Kuhn, S.L., 2004. Upper Paleolithic raw material economies at Uçagizli cave, Turkey. Journal of Anthropological Archaeology 23, 431–448.

Kyara, O., 1999. Lithic Raw Materials and Their Implications on Assemblage Variation and Hominid Behavior During Bed II, Olduvai Gorge, Tanzania. Ph. D. dissertation, Rutgers.

Le Maitre, R.W. (Ed.),1989. A Classification of Igneous Rocks and Glossary of Terms. Blackwell Scientific Publications, Oxford.

Madsen, B. Goren-Inbar, N., 2004. Acheulian giant core technology and beyond: an archaeological and experimental case study. Eurasian Prehistory 2, 3–52.

Marwick, B., 2003. Pleistocene exchange networks as evidence for the evolution of language. Cambridge Archaeological Journal 13, 67–81.

Merrick, H.V., Merrick, J.P.S., 1976. Archaeological occurrences of earlier Pleistocene age from the Shungura Formation. In: Coppens, Y., Howell, F.C., Isaac G. ,L., Leakey R.E.F. (Eds.), Earliest Man and Environments in the Lake Rudolf Basin. Chicago University Press, Chicago, pp. 574–584.

Merrick, H.V., Brown, F.H., Nash, W.P., 1994. Use and movement of obsidian in the Early and Middle Stone Ages of Kenya and northern Tanzania. MASCA Research Papers in Science and Archaeology 11 (supplement), 29–44.

Minichillo, T.J., 2006. Raw material use and behavioral modernity: Howiesons Poort lithic foraging strategies. Journal of Human Evolution 50, 359–364.

Negash, A., Shackley, M.S., 2006. Geochemical provenience of obsidian artefacts from the MSA site of Porc Epic, Ethiopia. Archaeometry 48, 1–12.

Negash, A., Shackley, M.S, Alene, M., 2006. Source provenance of obsidian artifacts from the Early Stone Age (ESA) site of Melka Konture, Ethiopia. Journal of Archaeological Science 33, 1647–1650.

Noll, M.P., 2000. Components of Acheulean Lithic Assemblage Variability at Olorgesailie, Kenya. Ph. D. Dissertation, University of Illinois.

Plummer, T.W., 2004. Flaked stones and old bones: biological and cultural evolution at the dawn of technology. Yearbook of Physical Anthropology 47, 118–164.

Plummer, T., Bishop, L.C., Ditchfield, P. Hicks, J., 1999. Research on late Pliocene Oldowan sites at Kanjera South, Kenya. Journal of Human Evolution 36, 151–170.

Potts, R., 1988. Early Hominid Activities at Olduvai. Aldine de Gruyter, New York.

Prinz. M., Harlow. G., Peters, J. (Eds.), 1978. Simon & Schuster's Guide to Rocks and Minerals. Simon & Schuster inc.

Quade, J., Levin, N.E., Semaw, S., Stout, D., Renne, P.R., Rogers, M.J., Simpson, S., 2004. Paleoenviroments of the earliest stone toolmakers, Gona, Ethiopia. Geological Survey of America Bulletin 116, 1529–1544.

Roche, H., Delagnes, A., Brugal, J.P., Feibel, C.S., Kibunjia, M., Mourre, V., Texier, P.J., 1999. Early hominid stone tool production and technical skill 2.34 Myr ago in West Turkana, Kenya. Nature 399, 57–60.

Roebroeks, W., Kolen, J., Resnik, F., 1988. Planning depth, anticipation and the organization of Middle Palaeolithic technology: the "archaic natives" meet eve's descendants. Hellinium XXXVIII,17–34.

Rogers M.J., Harris J.W.K., Feibel C.S., 1994. Changing patterns of land use by Plio-Pleistocene hominids in the Lake Turkana Basin. Journal of Human Evolution 27, 139–158.

Schick, K.D., 1987. Modeling the formation of Early Stone Age artifact concentrations. Journal of Human Evolution 16, 789–807.

Schick, K.D., Toth N., 1993. Making Silent Stones Speak: Human Evolution and the Dawn of Technology. Simon & Schuster, New York.

Semaw, S., 1997. Late Pliocene Archaeology of the Gona River Deposits, Afar, Ethiopia. Ph. D. dissertation, Rutgers.

Semaw, S., 2000. The world's oldest stone artifacts from Gona, Ethiopia: their implications for understanding stone technology and patterns of human evolution between 2.6-1.5 Million years ago. Journal of Archaeological Science 27, 1197–1214.

Semaw, S., Rogers, M.J., Quade, J., Renne, P.R., Bulter, R.F., Dominguez-Rodrigo, M., Stout, D., Hart, W.S., Pickering, T. Simpson, S.W., 2003. 2.6 million-year-old stone tools ans associated bones from OGS-6 ans OGS-7, Gona, Afar, Ethiopia. Journal of Human Evolution 45, 169–177.

Steele, J., 1996. On predicting hominid group sizes. In: Steele, J. Shennan, S. (Eds.), The Archaeology of Human Ancestry: Power, Sex and Tradition, pp. 230–52. London: Routledge.

Stiles, D., 1991. Early homind behaviour and culture tradition: raw material studies in Bed II, Olduvai Gorge. The African Archaeological Review 9, 1–19

Stiles, D., 1998. Raw material as evidence for human behaviour in the Lower Pleistocene: the Olduvai case. In: Petraglia, M., Korisetter, R. (Eds.), Early Human Behavior in the Global Context: The Rise and Diversity of the Lower Paleolithic Period. Routledge, New York, pp. 133–150.

Stiles, D., Hay, R.L., O'Neil, J.R., 1974. The MNK chert factory site, Olduvai Gorge, Tanzania. World Archaeology 5, 285–308.

Stout, D., Quade, J., Semaw, S., Rogers, M.J., Levin, N.E., 2005. Raw material selectivity of the earliest stone toolmakers at Gona, Afar, Ethiopia. Journal of Human Evolution 48, 365–380.

Toth, N., 1985. The Oldowan reassessed: a close look at early stone artifacts. Journal of Archaeological Science 12, 101–120.

Toth, N., 1997. The artefact assemblages in the light of experimental studies. In: Isaac, G.L., Isaac, B. (Eds.), Koobi Fora Research Project. Clarendon Press, Oxford, pp. 363–388.

Yemane, T., 1997. Stratigraphy and Sedimentology of the Hadar Formation, Afar, Ethiopia. Ann Arbor: UMI.

8. Variability in Raw Material Selectivity at the Late Pliocene sites of Lokalalei, West Turkana, Kenya

Sonia Harmand*
Laboratoire de Préhistoire et Technologie
Maison de l'Archéologie et de l'Ethnologie
Boîte 3, 21 allée de l'Université
92023 Nanterre Cedex
France
sonia.harmand@mae.u-paris10.fr

Keywords: Raw material selection, Resource availability, Volcanic petrography, Lithic technology, Assemblage variation

Abstract

The current techno-economic study of two well-preserved Late Pliocene lithic assemblages from the Nachukui Formation, in the West Turkana region in North Kenya, provides new evidence of planning and foresight in raw material procurement, and management from 2.34 Ma onwards, testified to by the selection of specific raw materials. One of the most noteworthy results ensuing from this study, carried out in combination with geological surveys, petrographic analyses and lithic assemblage analyses, is the existence of substantial differences in raw material provisioning and management between Late Pliocene assemblages, geographically and chronologically close. These differences are related to the degree of selectivity for raw material sizes and morphologies as well as to the way they were processed, rather than to variations in resource availability.

8.1 Introduction

Since the 1970s, research addressing ancient hominin behavior in relation to lithic procurement has been particularly focused on hominin ranging and foraging behavior during the Early Pleistocene. This research has been carried out mainly in two regions of eastern Africa, Olduvai Gorge in Tanzania, and Koobi Fora, east of Lake Turkana in Kenya (e.g., Leakey, 1971, 1975, 1994; Hay, 1976; Isaac, 1977; Isaac and Harris, 1978; Jones, 1979; Clark, 1980; Toth, 1982, 1987). Undertaken at a regional scale, raw material studies have furthered the construction of Early Pleistocene hominin activity models (e.g., "home base hypothesis", Leakey, 1971; "routed foraging model", Binford, 1980; "favored places model", Schick, 1987; Schick and Toth, 1993), and have shown evidence of stone transport and management during the Early Pleistocene (Toth 1985; Schick 1987; Isaac et al. 1997). From 1.9 Ma onwards, the preferential use of a particular raw material is generally interpreted as a result of local abundance rather than choice on the part of the toolmakers (e.g. Merrick and Merrick 1976; Toth 1985; Schick 1987; Isaac et al. 1997), although site provisioning could involved distances of several kilometers (up to 13 km at Olduvai: Hay 1976; overview in Féblot-Augustins 1997). However, recent lithic studies based on raw material sourcing and characterization from a series of sites from the West Turkana region, Kenya, have provided a much more complex picture of resource availability management by Oldowan hominins, with evidence for selection patterns although a local procurement in stream channels and bedrock outcrops (Harmand 2005, in press). These involve a certain level of anticipation of the effects of raw material properties and further contradict the assumption generally made for the Early Stone Age of an opportunistic gathering of rocks. For the time being, published evidence for raw material transport and selectivity at Late Pliocene remains limited (Plummer et al. 1999; Hovers et al. 2002; Stout et al. 2005; Goldman-Neu-

* Address for correspondence: sonia.harmand@mae.u-paris10.fr

E. Hovers and D.R. Braun (eds.), *Interdisciplinary Approaches to the Oldowan*,
DOI: 10.1007/978-1-4020-9059-2,

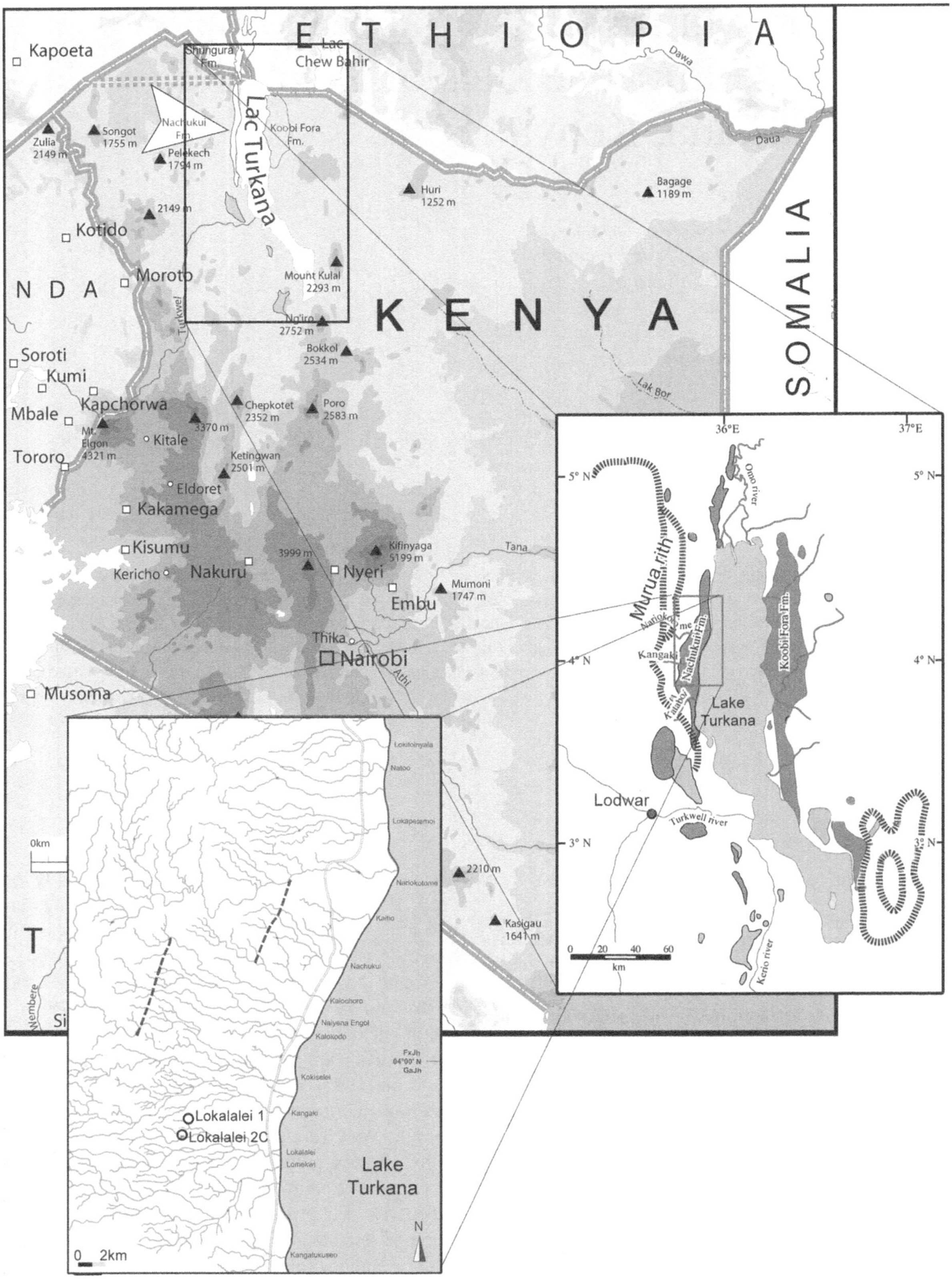

FIGURE 8.1. Lake Turkana Basin, Nachukui Formation, and location of the Lokalalei 2C, and Lokalalei 1 sites, modified after Roche et al. 2003.

man and Hovers 2009; Harmand in press). Recent investigations at the sites of Gona have provided evidence for selection patterns as early as 2.6 Ma (Semaw 2000; Stout et al. 2005), and evidence for selection patterns was documented also at the somewhat later site A.L. 894 in Hadar (Goldman-Neuman and Hovers 2009). At Kanjera South (Late Pliocene sites from KS1 and KS2 Beds), a small portion of the raw materials could also indicate a degree of selectivity and provisioning patterns from non-local sources (Plummer et al. 1999; Braun et al. 2008). In this context, further detailed studies of raw material, considered from the viewpoint of their provenance and use, are thus needed for a better assessment of raw material procurement and exploitation behaviors at Late Pliocene.

This chapter focuses on the results of a comparative techno-economic study that addresses the lithic procurement and exploitation patterns brought into play in the two earliest known archaeological assemblages in Kenya (ca. 2.34 Ma), the Lokalalei sites west of Lake Turkana. Besides providing renewed data on the emergence of early technological developments, the Lokalalei assemblages offer the opportunity to compare raw material procurement activities at early Oldowan (Late Pliocene) archaeological localities found in temporal and geographical proximity, and to document potential differences in lithic provisioning and exploitation in ancient contexts.

8.2 Geographical, Geological and Chronological Contexts

The geographical area covered by our study is part of the Lake Turkana Basin, namely the Nachukui Formation. It stretches on the western shore of the lake to the Labur and Murua Rith escarpments (Figure 8.1).

This Formation is characterized by Plio-Pleistocene sedimentary deposits that reach a thickness of about 730 meters and are exposed in an area of about 700 km^2. The Nachukui Formation comprises eight members (from 4 to 0.7 Ma), using widespread volcanic tuffs as bed markers (Harris et al. 1988; Feibel et al. 1989, 1991). This sequence is one of the longest and more complete in East Africa, and has yielded very rich and well-preserved archaeological sites of great antiquity, occupied by hominins between 2.34 and 0.70 Ma (Kibunjia et al. 1992; Roche and Kibunjia 1994; 1996; Roche et al. 1999, 2003). These sites were excavated by the West Turkana Archaeological Project, a joint project led by H. Roche (*Mission Préhistorique au Kenya*, France) and M. Kibunjia (National Museums of Kenya).

The Lokalalei archaeological sites are located in the South of the Nachukui Formation, one km apart (Fig. 8.1). Excavated in 1991 (Lokalalei 1), 1996 and 1997 (Lokalalei 2C), they are the oldest sites known so far in Kenya and belong to the very few African Pliocene sites (Roche et al. 2003). Locally, Lokalalei 1 and Lokalalei 2C are correlated by a mollusc-packed sandstone, which occurs beneath both sites. This marker was used as the local boundary between the Lokalalei and Kalochoro Members. The Kokiselei and Ekalalei Tuffs lying below the Lokalalei 1 and Lokalalei 2C sites are correlated with Tuffs E and F1 of the Shungura Formation, respectively. The Kokiselei Tuff has a stratigraphically scaled age of 2.4±0.05 Ma, while the Ekalalei Tuff is slightly younger than 2.34±0.04 Ma. An age of 2.34±0.05 Ma is thus estimated for the Lokalalei archaeological sites. The stratigraphical position of Lokalalei 2C, while slightly higher in the section than Lokalalei 1, is judged compatible with a chronological attribution within the same time interval (see Delagnes and Roche 2005; Tiercelin et al. in prep.).

8.3 Analytical Approach

The raw material provisioning and exploitation patterns at the Lokalalei sites are inferred from the techno-economic analysis of the lithic assemblages (e.g. Geneste 1989, 1991), which aimed more particularly to highlight the relationship between raw materials and tool production processes. This approach is linked to the notion of *chaîne opératoire* (e.g. Leroi-Gourhan 1964, 1971), whereby artifacts are analyzed as the result of a process, that is to say the strategies of reduction and the technical skills involved in tool-production. The implemented techno-economic analysis involved several stages of investigation. It required the systematic sourcing, mapping and sampling of raw materials by field surveys so as to determine their geographical provenience and to assess the opportunities for procurement according to the evolution of the geological and the post-depositional geomorphological environment and in relation to the geographical position of the sites. Petrographic analyses were carried out in order to evaluate the diversity of the raw materials available, and to assess the relative abundance of each type of rock in relation to their petrographic, structural and granular patterns. The natural morphologies and sizes of the collected boulders and cobbles were recorded to appraise the range of shapes and sizes available in the potential sources. Rock mechanics tests through knapping experiments were conducted in the field on rocks similar to those found in the archaeolgoical assembalges. These aimed to estimate the importance of the constraints imposed by the different properties of the various raw materials found at sites (here referred to as rock qualities). These knapping experiments were conducted according to the knapping techniques (direct hard hammer percussion) and the strategies of core reduction documented at Lokalalei (no preparation of the cores, natural striking surfaces and angles used). For each rock types (~50 cobbles), an average of ten flakes were removed from the initial cobbles using different lava cobble hammerstones (Harmand 2005). The flakes produced through the knapping tests were used for plant and goat processing to obtain precision on edge efficiency and durability. Table 8.1 presents the results of the raw material characterization.

TABLE 8.1 Characteristics and properties of the volcanic rocks found at sources and at sites at Lokalalei, inferred from classical petrological classification and knapping experiments

Rock type	Texture	Grain size	Colour	Mineral composition	Quality
Phonolite	Equigranular. Aphyric.	Medium grained (1 to 5 mm)	Often light black or grey.	Mainly composedof sanidine, amphibole and clinopyroxene. No quartz.	High quality rock, compact and homogeneous. Often rounded compact pebbles or cobbles. Parallel alignement of crystals. Advantage of breaking easily along foliation planes. Regular fracture and predictability of fracture orientation. Very suitable for obtaining durable and resistant sharp cutting edges.
	Equigranular. Aphyric.	Fine grained (crystals <1 mm).	Often light black or grey.	Mainly composed of san dine, amphibole and clinopyroxene. No quartz.	Often rounded pebbles or cobbles. Less control of the fracture. High frequency of fractured surfaces. Sharp cutting edges but fragile and no resistance. Low durability of the material.
	Inequigranular. Porphyritic.	Fine grained groudmass.	Often light black or grey.	Sanidine, amphibole and clinopyroxene. Sanidine and plagioclase phenocrysts.	Often rounded compact pebbles or cobbles. High frequency of medium crystals in the groundmass. Low predictability of fracture orientation.
Basalt	Equigranular. Aphyric.	Fine grained.	Often dark black.	Mainly composed of plagioclase feldsparand pyroxene.	Rounded compact pebbles or cobbles and often angular blocks. High density rock. Fracture resistant and difficult to break. Suitable for hammering.
	Inequigranular. Porphyritic.	Fine grained groudmass.	Often dark black.	Plagioclase feldspar, pyroxene. Olivine and pyroxene phenocrysts.	Rounded compact pebbles or cobbles and often angular blocks. High density rock. Fracture resistant and difficult to break. High frequency of large crystals in the groundmass. No predictability of fracture orientation.
Trachyte	Equigranular. Aphyric.	Fine to medium grained.	Usually pale in colour.	Mainly composed of alkali feldspar. Quartz in small amounts.	Often rounded compact pebbles or cobbles. Less compact and homogeneous rock. Tiny bubbles or vesicles. Irregular fracture and low fracture predictability. Friable edges. Low edge durability.
	Inequigranular. Porphyritic.	Finegrained groudmass.	Usually pale in colour. Banded or layered.	Alkali feldspar phenocrysts.	Often rounded compact pebbles or cobbles. High frequency of medium crystals in the groundmass. Irregular fracture and low predictability of fracture orientation. Friable edges. Low edge durability.
Rhyolite	Equigranular. Aphyric.	Very fine grained.	Often red.	Mainly composed of quartz and feldspar	Diaclasic angular blocks and fragments. High quality rock. Breaks with a conchoidal fracture like glass. Sharp cutting edges. Less homogeneous rock due to frequent internal fissures, vesicles and spherulites.
Syenite	Equigranular. Aphyric.	Coarse grained (>5 mm).	White or shades of grey.	Mainly composed of feldspars. Quartz in small amounts.	Diaclasic angular fragments. Poorly welded crystals. No fracture predictability. Tiny bubbles or vesicles. Friable edges. Low edge durability.

Ultimately, the relationship between raw materials and desired end products has been determined and analyzed based on the results of the technological study of the reduction sequences documented for the Lokalalei sites (Kibunjia 1994, 1998; Delagnes and Roche 2005; Harmand 2005)

8.4 Results

8.4.1 Geographical Provenience and Opportunities for Procurement During the Late Pliocene in the Nachukui Formation

Paleogeographic reconstructions of the Turkana Basin indicate that prior to 2.0 Ma, the Turkana Basin was dominated by a river system interpreted as the paleo-Omo, which flowed through the Basin from the north and exited east into the Indian Ocean (Figure 8.2) (Brown and Feibel 1988; Feibel et al. 1991; Rogers et al. 1994).

Lying in the southern area of exposure, the Lokalalei sites are enclosed within paleosols that formed in the alluvial plain of the Paleo-Omo fluviatile system, within the Kalochoro Member. The sandstones associated with these paleosols suggest that the sites were located in the proximal part of the alluvial plain where small East-flowing streams joined the main axial system (Figure 8.2) (Brown and Feibel 1988; Roche et al. 1999). Due to the development of these East-flowing streams, debris-flow outcrops were accumulated in the western margin of the Turkana basin.

Although it has not been possible to locate the exact point where the knappers collected their raw materials, specific debris-flow outcrops lying stratigraphically below the archaeological layers or within the same layers were identified and sampled according to the geomorphological framework previously established for the region (Brown and Feibel 1991; Rogers et al. 1994; Feibel 2001) and based on extensive survey of the surrounding sediments by the geologist (see Harmand 2005). These debris-flow outcrops are located at lateral distances a few meters away from the sites (150 m from Lokalalei 1 site; 30 m from Lokalalei 2C site) (see Harmand 2005). For each conglomerate, the raw materials were sampled according to a three dimensional pattern to assess the variability of rocks within a certain volume of sediments. The samples correspond to collecting areas of two meters in lateral extent and one meter deep to approximate the horizontal distribution of raw materials in the conglomerates. The thickness of the collecting areas is equivalent to the thickness of the conglomerates.

The systematic sourcing of the poorly sorted debris-flow outcrops available in the vicinity of the archaeological sites indicates the predominance of rounded pebbles or cobbles of igneous and extrusive volcanic rocks (Table 8.1), all originating from flows of lava from the Murua Rith and Labur Miocene volcanic ranges which border the Basin to the West (Walsh and Dodson 1969; Feibel 2001) (Figure 8.2). These rocks, carried by the network of rivers, were grouped into several categories, differing in terms of color, grain, texture and homogeneity (Table 8.1). Five major volcanic rock types were identified in the Lokalalei study area based on classical petrologic classification (MacKenzie and Adams 1996): phonolite, basalt, trachyte, rhyolite, syenite (Table 8.1), and a few siliceous rocks. The raw material attribution was carried by eye identification. One artefact per major raw material group was devoted to a thin section for petrographic attributions. Each of the rock types was divided according to groundmass features (micro - to - cryptocrystalline), grain sizes (fine - to - coarse-grained fabrics), and textures (aphyric - to - porphyritic), resulting in distinct knapping and functional properties and varying initial morphologies and sizes (Table 8.1). For instance, the medium-grained phonolite displays good physical features in terms of its flaking qualities, the aphyric fine-grained basalt is difficult to break and more resistant to hard-hammer percussion.

At Lokalalei 2C area, 207 cobbles, pebbles and boulders were randomly collected from the different sampled gravel bars, out of which a total of 71 clasts of a diameter >40 mm were studied. These dimensions correspond to the smallest pebble used at the archaeological site (Table 8.2). At Lokalalei 1 area, a total of 173 cobbles, pebbles and boulders were randomly collected from the different sampled gravel bars (Table 8.3).

Each of the gravel bars sampled at both sites areas (Lokalalei 2C and Lokalalei 1) is located a short distance from the archaeological sites (both located near sources of raw materials, Roche et al. 1999) and includes a relatively large variety of raw materials of adequate size to make stone-tools. The shared range of volcanic rocks in the bars consists of phonolites, basalts, trachytes and rhyolites. Among them, trachytes repeatedly constitute the majority of the rock sampled, followed by phonolites, basalts and rhyolites (Table 8.2 and Table 8.3, see below).

The characteristics of the raw materials (types, sizes and cortical surfaces) found at archaeological sites and used for artefacts match with the ones sampled at sources (see below). This suggests that there is no reason to suppose that site provisionning at Lokalalei involved distances beyond a radius of 0.5 km. Raw material procurement at Lokalalei consisted of collecting and carrying raw materials from the local debris flow outcrops available in the immediate vicinity of the sites. As a result, the cost of search, acquisition and transport of raw materials was minimal.

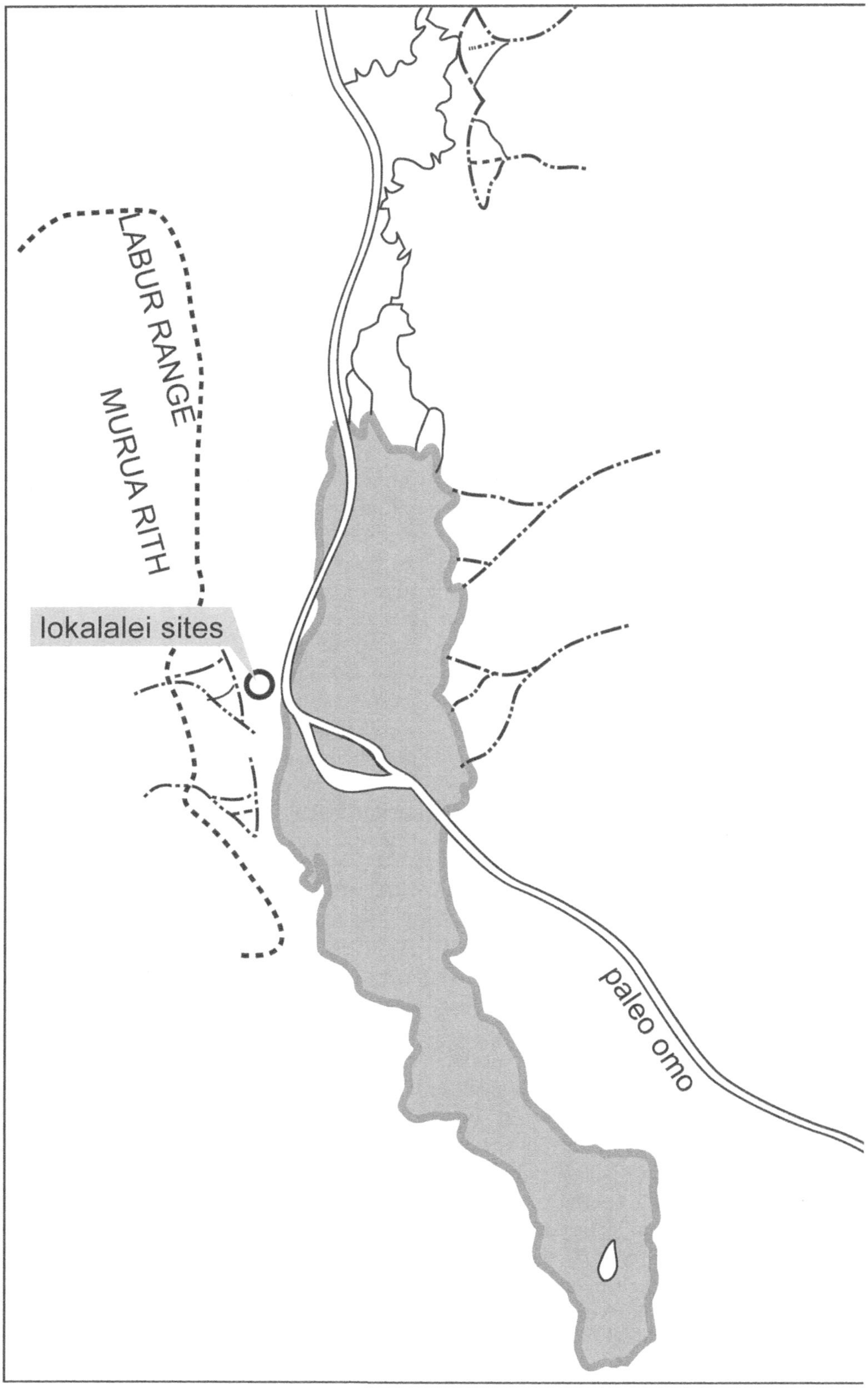

FIGURE 8.2. Paleogeographic setting of the Lokalalei sites at ca. 2.4 – 2.3 Ma, modified after Brown and Feibel 1988.

TABLE 8.2. Distribution of rock types in the lithic assemblage components at Lokalalei 2C and in the sampled outcrop

	Lokalalei 2C site						geological sample	
	débitage (cores, flakes, flake fragments)		hammerstones		unmodified or split cobbles		pebbles, cobbles, boulders*	
	N	%	*N*	%	*N*	%	*N*	%
Phonolites	1916	77.63	2	11.11	43	31.16	20	28.17
Basalts	350	14.18	3	16.67	20	14.49	17	23.95
Trachytes	175	7.09	13	72.22	66	47.83	26	36.62
Rhyolites	5	0.20	0	0.00	2	1.45	4	5.63
Syenites	1	0.04	0	0.00	0	0.00	0	0.00
Non det.	21	0.86	0	0.00	7	5.07	4	5.63
Total	2468	100.00	18	100.00	138	100.00	71	100.00

* from a total of 207 pebbles, cobbles and boulders collected, only the ones displaying a diameter >40 mm were studied. They correspond to the smallest pebbles used at the archaeological site (see Harmand 2005:136).

8.4.2 Raw Material Procurement and Exploitation Patterns at Lokalalei Sites

8.4.2.1 Lokalalei 2C

The randomly sampled outcrop in the Lokalalei 2C vicinity is dominated by cobbles and pebbles of a light brown aphyric trachyte (36.6%) (Figure 8.3).

Cobbles and pebbles of a dark gray aphyric phonolite account for 28% of the rocks sampled on the outcrops at Lokalalei 2C (Figure 8.3). Among the phonolites, cobbles and pebbles of a medium-grained type are predominant (24% of the raw materials). Cobbles and pebbles of a dark black porphyritic or aphyric basalts account for 24% (Figure 8.3) of the rocks sampled on the outcrop at Lokalalei 2C which comprises also smaller proportions of red rhyolites (5.6%, Figure 8.3).

At Lokalalei 2C, the excavation covered an area of 17 m^2 from which an abundant *in situ* lithic assemblage (n=2624) has been recovered, including cores, whole or broken flakes, a very few possibly retouched pieces, unmodified split cobbles and hammerstones (Table 8.2).

Although the archaeological deposit is partly truncated by erosion, the remaining part of the site shows evidence of good preservation, as indicated by a high ratio of very small elements and the freshness of the artifacts (Delagnes and Roche 2005). The high proportion of small elements (length <2 cm), the significant quantity of cores and flakes, and the presence of hammerstones are consistent with knapping activities carried out on site (Table 8.2), where an estimate of 190 cobbles or fragments of cobbles along with a few unmodified split cobbles were introduced from nearby sources (Harmand 2005). As documented by Delagnes and Roche (2005), the assemblage is remarkable in many respects, including the use of specific morphologies to conduct organized and highly productive *débitage*[1] sequences. These result in extensive production of relatively well-standardized flakes.

Regarding petrography, the knappers favored the medium-grained phonolite (52% of the on-site raw materials, Figure 8.3). This raw material displays a parallel mineral orientation that gives the rock a natural foliation. The rock has the mechanical advantage of breaking easily along the foliation plane when direct hard-hammer percussion is used, and therefore offers a measure of predictability in terms of fracture orientation (Table 8.1) (Harmand 2004, 2005). This type of raw material was thus most suitable in terms of flaking quality as well as cobble morphology for the production of large amounts of flakes. Medium-grained phonolite is also very suitable for obtaining potentially functional active edges (sharp cutting edges). At the source, the procurement patterns involved selecting phonolites in general over basalts and trachytes (Figure 8.3). The frequency of medium-grained phonolite is higher in the assemblage (~ 52% of the on-site raw materials of which 45 cores and a total of 745 flakes or flake fragments) than in the local conglomerates (~24% of the raw materials at source) (Figure 8.3). In comparison, the frequencies of trachytes and basalts are higher in the local conglomerates (respectively ~36.6% and ~24% of the raw materials at source) than in the assemblage (trachyte ~9.7% of the on-site raw materials of which only 8 cores and core fragments of trachyte; basalt ~14% of the on-site raw materials of which only 9 cores and core fragments not reduced extensively). The low frequencies of rhyolites and syenites in the assemblage (0.3% of the on-site raw materials) reflect their low frequencies in the local conglomerates (~5.6% of the raw materials at source).

Furthermore at Lokalalei 2C, medium-sized angular cobbles or fragments of cobbles (12 cm maximum dimension) of the medium-grained phonolite with naturally serviceable striking

[1] The word débitage is used in its original French meaning as a reflection of a specific reduction process, not as the residue of production (Inizan et al. 1995).

TABLE 8.3. Distribution of rock types in the lithic assemblage components at Lokalalei 1 and in the sampled outcrops

	Lokalalei 1 site						geological sample	
	débitage (cores, flakes, flake fragments)		hammerstones		unmodified or split cobbles		pebbles, cobbles, boulders (diameter > 40 mm)	
	N	*%*	*N*	*%*	*N*	*%*	*N*	*%*
Phonolites	238	68.19	2	28.57	25	69.44	69	39.88
Basalts	9	2.58	0	0.00	1	2.78	20	11.56
Trachytes	100	28.66	5	71.43	8	22.22	77	44.51
Rhyolites	0	0.00	0	0.00	0	0.00	2	1.16
Syenites	0	0.00	0	0.00	1	2.78	5	2.89
Siliceous	2	0.57	0	0.00	1	2.78	0	0.00
Total	349	100.00	7	100.00	36	100.00	173	100.00

surfaces and angles (n=95)[2] were preferentially selected for their suitability to be knapped without any preparation (Delagnes and Roche 2005). Flakes, cores and fragments are mainly made on these angular specimens of medium-grained phonolite (60% of the flakes and flake fragments, n=745). Selection at the sources was high, since such morphologies are poorly represented in the nearby sources, the medium-grained phonolite occuring mainly as rounded to subrounded cobbles, more rarely as angular cobbles (see Harmand 2005). The low representation of such specimens at the sources is probably one of the reasons for the deliberate breakage of large-sized (length >15 cm) to medium-sized (8 to 15 cm in length) rounded cobbles of medium-grained phonolite into several pieces to obtain suitable blanks for flaking, with serviceable striking platforms. This probably took place prior to transport and perhaps at the source, where the raw material was collected (Delagnes and Roche 2005; Harmand 2005). To maximize the reduction sequences, poorer quality cobbles of porphyritic phonolites and porphyritic basalts (Table 8.1) were exceptionally selected for their large dimensions (length >16 cm). Those were obviously partially flaked off-site prior to their transport (Delagnes and Roche 2005).

The initial selection of a specific morphology from a high-grade raw material (medium-grained phonolite), from which a large amount of flakes could be easily obtained, enabled the knappers to maximize the reduction sequences through a *débitage* system. As a result, the amount of flakes produced is relatively high (up to 80 for a single block) and they display sharp cutting edges, serviceable without any transformation by retouch (Delagnes and Roche 2005).

To a lesser extent, a more heterogeneous *débitage* by multidirectional removals was conducted on thick and globular cobbles mainly of medium-grained phonolite (6 cores) or on poorer quality cobbles or fragments of cobbles of porphyritic phonolites, aphyric trachytes or porphyritic basalts (8 cores, Figure 8.3). The presence of numerous phenocrystals of olivine and pyroxene within the porphyritic basalts significantly lessens the predictability of flake sizes and morphologies (Harmand 2005). In the same way, trachytes display low fracture predictability and generate low durability cutting-edges (Table 8.1). As a result, the cores display numerous knapping accidents from which only a small amount of small flakes were produced (Delagnes and Roche 2005).

In addition, a series of unmodified cobbles (n=54), mostly rounded cobbles (tested or not) of medium-grained trachyte or fine-grained basalt, were also brought to the site and probably stockpiled as resistant, massive, and difficult to break "manuports". Thirteen heavy and medium-sized rounded cobbles of a resistant medium-grained trachyte bear signs of percussion damage and are interpreted as hammerstones (Table 8.2) used for knapping most of the cores flaked at the site (Delagnes and Roche 2005). These hammerstones were selected from the cobbles most appropriate for percussion in terms of mass, size and shape, within the supply of raw materials brought to the site (Harmand 2004, 2005).

8.4.2.2 *Lokalalei 1*

The randomly sampled outcrops in the Lokalalei 1 area are also dominated by cobbles and pebbles of a light brown aphyric trachyte (44.5%) (Figure 8.4).

Pebbles and cobbles of a dark gray aphyric phonolite account for almost 40% of the rocks sampled (Figure 8.4). Among the phonolites, the medium-grained type is predominant (21% of the raw materials at source). Pebbles and cobbles of dark black porphyritic or aphyric basalts account for only 11.5% of the rocks sampled on the outcrops at Lokalalei 1 area (Figure 8.4). Various other rocks occur in smaller proportions: red rhyolites (1%), and a large-grained syenite (2.8%, Figure 8.4).

Compared to the Lokalalei 2C site, the Lokalalei 1 lithic assemblage is quite small (n=392). The excavation covered an

[2] Original shapes inferred from the refitting groups and documented by the study of the slightly modified cobbles (see Delagnes, Roche 2005).

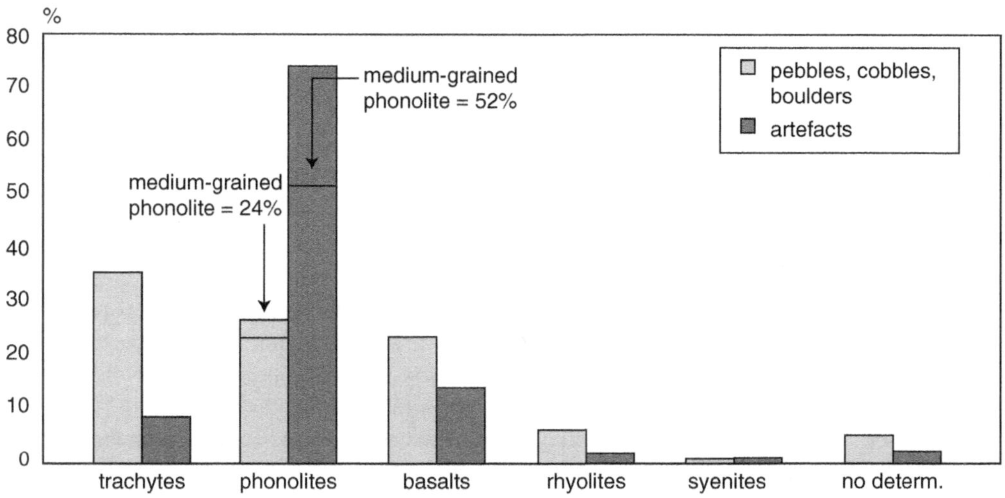

FIGURE 8.3. Distribution of rock types in the conglomerate samples (pebbles, cobbles, boulders) in the Lokalalei 2C area, and in the Lokalalei 2C assemblage (total of the artefacts). Rock types are organized from left to right in order of their declining relative frequency in the conglomerates. Note that the frequency of medium-grained phonolite is higher in the assemblage than in the local conglomerates. In comparison, the frequencies of trachytes and basalts are higher in the local conglomerates than in the assemblage. The low frequencies of rhyolites and syenites in the assemblage reflect their low frequencies in the local conglomerates.

area of 60 m^2 from which flakes and cores, unmodified split cobbles and hammerstones have been recovered (Table 8.3) (Kibunjia 1994, 1998; Roche et al. 2003; Harmand 2005). An estimated number of 110 boulders, cobbles or fragments of cobbles along with the few unmodified split cobbles were introduced into the site from a nearby source channel (Harmand 2005).

The selectivity towards raw material quality highlighted for the site of Lokalalei 2C is also documented for Lokalalei 1 site. At Lokalalei 1, the frequency of medium-grained phonolite is higher in the assemblage (~56.5% of the on-site raw materials) than in the local conglomerates (~21% of the raw materials at source) (Figure 8.4). The selectivity towards raw material quality at Lokalalei 1 can also be inferred from the avoidance of poorer quality rocks for the *débitage*: the frequencies of trachytes and basalts are higher in the local conglomerates (~44.5% and 12% respectively of the raw materials at source) than in the assemblage (respectively ~ 29% and 2.6% of the on-site raw materials). Furthermore, the over representation of the medium-grained phonolite in the archaeological assemblage is evident when looking at the cores: 43 cores of these particular rock versus only three cores or core fragments of trachyte and four cores of basalt not

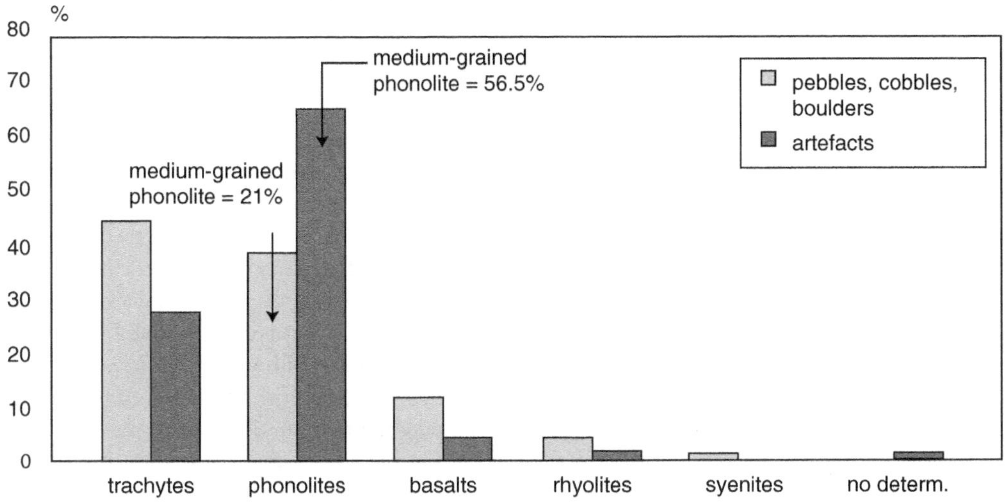

FIGURE 8.4. Distribution of rock types in the conglomerate samples (pebbles, cobbles, boulders) in the Lokalalei 1 area, and in the Lokalalei 1 assemblage (total of the artefacts). Rock types are organized from left to right in order of their declining relative frequency in the conglomerates. The frequency of medium-grained phonolite is higher in the assemblage than in the local conglomerates. The frequencies of trachytes and basalts are higher in the local conglomerates than in the assemblage. The low frequencies of rhyolites, syenites and siliceous in assemblage reflect their low frequencies in the local conglomerates.

reduced extensively. As noticed at Lokalalei 2C, the low frequencies of rhyolites, syenites and siliceous in the Lokalalei 1 assemblage (~ 1% of the on-site raw materials) reflect their low frequencies in the local conglomerates (~ 4% of the raw materials at source).

Nevertheless at Lokalalei 1, the selectivity of a raw material type has to be balanced when considering the evident opportunistic gathering of cobbles in terms of morphological types. The majority of the cobbles brought to the site are rounded, massive and globular in shape (medium to rather large thick, length >15 cm, width >10 cm). These cobbles are either of lower quality raw materials (aphyric phonolites and trachytes or porphyritic basalts) or of high-grade raw materials (medium-grained phonolite) but in any case, they display a scarcity of naturally serviceable striking surfaces (for 71% of the on-site cobbles). At Lokalalei 1, the initial selection of morphologies not suitable for carrying out *débitage* without any preparation led to low productivity reduction sequences. These consist of detaching small multidirectional flakes from two, three or more faces of the cores without any constant technical rules (as indicated by the direction and the number of negatives of removals on the cores). As a result, only a small amount of whole and serviceable flakes were produced: an estimate of 63 whole flakes of medium-grained phonolite removed for 43 cores; an estimate of 47 whole flakes for 18 cores made on other rocks (Harmand 2005).

The Lokalalei 1 assemblage also includes a relatively small amount of cores (19 cores) testifying to a more organized *débitage* conducted on small to medium-sized angular blocks (length > 12.5 cm) mainly on medium-grained phonolite, and displaying small portions of flat surfaces, naturally serviceable as striking surfaces. Longer reduction sequences were conducted on these angular blocks (from one or two surfaces) than on the rounded and globular ones. As a result, more than 10 flakes for one block were produced. Yet the exploitation sequences at Lokalalei 1 do not reach the highly productive sequences documented by Delagnes and Roche (2005) at nearby Lokalalei 2C. This has been related to a lower level of technical elaboration at Lokalalei 1, as indicated by the way raw materials were processed. At Lokalalei 1, the large and plane surfaces of cobbles were used as striking surfaces, whereas at Lokalalei 2C the large surfaces served as the flaking surfaces (see Delagnes and Roche 2005:468). Another reason cited for the differences between the sites is a lower level of manual dexterity, as testified to by cores whose negatives of removal indicate numerous knapping accidents, and repeated impact damage from failed percussions (see Kibunjia 1994; Delagnes and Roche 2005).

The few hammerstones in the assemblage (n=7) are mainly medium-size (maximum length: 12 cm), compact and rounded cobbles of aphyric trachytes, and aphyric phonolites in smaller proportions, appropriate for hammering hard rocks (Table 8.3). In addition, the lithic assemblage includes 36 relatively large-size (maximum length: 15 cm) rounded cobbles (tested or not) mainly from aphyric phonolite or trachyte, possibly stockpiled to be knapped and to serve as resistant and massive "manuports" and/or hammerstones (Table 8.3) (Harmand 2005).

8.5 Conclusion

This study conducted on two Late Pliocene assemblages and based on raw material sourcing and characterization has provided evidence for selection patterns as early as 2.34 Ma despite a local procurement in stream channels and/or bedrock outcrops. Early hominin provisioning behaviors at both Lokalalei sites show the antiquity of decision-making testify to by the selection of a specific raw material for its flaking quality and for the durability of the sharp cutting-edge generated and by the avoidance of lower grade raw materials. They suggest sensitivity to the quality of raw materials, involving a certain level of knowledge and anticipation of the effects of raw material qualities.

This differential selection of raw materials brought to light in the Late Pliocene assemblages of Lokalalei is in keeping with the new and unexpected results obtained at the Early Oldowan localities of Gona and Hadar in Ethiopia, and Kanjera in Kenya (Plummer et al. 1999; Semaw 2000; Semaw et al. 2003; Plummer 2004; Stout et al. 2005; Goldman-Neuman and Hovers 2009) and contradict the assumption generally made for the Early Stone Age of an opportunistic gathering of rocks.

Our techno-economic analysis has highlighted significant differences in raw material procurement and management in the Late Pliocene of the Nachukui Formation, mainly related to the shape of the raw material brought onto the sites. While evidence of selectivity at Lokalalei 1 remains restricted to raw material quality, the procurement and exploitation strategies at Lokalalei 2C underscore a higher degree of planning and foresight. Those are testified to by the careful selection of advantageous rocks (high-grade rocks and particular morphologies) for the purpose of carrying out long reduction processes according to the level of skill the Lokalalei 2C knappers have mastered.

Since the raw material availability has no impact on artefact manufacture (similar high grade raw materials and similar morphologies available) and cannot explain the variation across the Lokalalei archaeological assemblages, the factors of assemblage diversity at Lokalalei are plausibly due to: (1) distinct patterns of hominin activity and site occupation or (2) unequal abilities to implement core reduction strategies (i.e. the way the raw materials were processed).

The first hypothesis favors a diversity related to specific tasks and needs. The Lokalalei 1 lithic assemblage may be the result of immediate needs of sharp cutting-edges expediently produced (lithic assemblage quite small in number, cores not reduced extensively) and used on site during short periods of occupation by a small group of hominins. At Lokalalei 2C, since the extensive reduction of cores does not reflect a high cost in terms of procurement energy, the assemblage may be the result of structured activities possibly during a longer period of occupation, and requiring planning and anticipation for greater needs for artefacts (long elaborate reduction sequences on advantageous rocks, cores discarded in significantly reduced states, high amount of cutting edges).

The second hypothesis favors the presence of contrasting levels of decision-making processes regarding raw material selection, along with different levels of manual dexterity and technical skills among Late Pliocene tool-makers. The latter, possibly robust Australopithecine (present at around 2.5 Ma in the West Turkana region; Walker et al. 1986) or Early Homo (found in close proximity to the Lokalalei sites, Prat et al. 2005) can then be identified as the main source of variation between the Lokalalei assemblages. For the time being, the question to know whether technological diversity at Late Pliocene is due to distinct patterns of site occupation or to unequal abilities is still an open issue and only further research in the future will help to solve this issue.

Our results lend strong support to recent investigations of African Early Oldowan lithic productions, which reveal a much more complex panorama of the first technical systems and their related behaviors, suggestive of temporal and spatial variability as early as Late Pliocene (Kimbel et al. 1996; Roche 2000; Hovers et al. 2002; Martínez-Moreno et al. 2003; Roche et al. 2003; de la Torre et al. 2003; de la Torre 2004; Delagnes and Roche 2005; Harmand 2005; Braun et al. 2008). At present, evidence of diversity in techno-economic patterns for Early Oldowan productions remains limited. More studies based on detailed and comparable techno-economic analysis from a larger number of sites are thus needed for a technological assessment of Early Stone Age assemblages, and ultimately to generate a clearer picture of potentially different responses to resource availability by Early Oldowan hominins in terms of raw material selection and management.

Acknowledgements. The author would like to thank Hélène Roche for her comments on a draft of this chapter, Craig Feibel for the geological aspect of our work and for the field surveys when the author undertook the petro-archaeological study of the Lokalalei sites, Pierre-Jean Texier for the experimental aspect of our work and Alain Demant for his precious help in rock identification. The author would like also to thank Mzalendo Kibunjia for allowing her to look at the raw materials at the Lokalalei 1 lithic assemblage. The author is pleased to thank the four reviewers for their comments and suggestions which improved the content and style of this chapter. We thank the Office of the President of Kenya and the National Museums of Kenya for providing us the excavation and research permits. The author is grateful to the West Turkana Archaeological Project codirected by Hélène Roche (Mission Préhistorique au Kenya) and Mzalendo Kibunjia (National Museums of Kenya) and funded by the French Ministry of Foreign Affairs (sous-direction de la Recherche) and the CNRS/INSU (ECLIPSE Program). The author is grateful to the board of the "bourse Lavoisier" (French Ministry of Foreign Affairs, EGIDE, sous-direction de la Recherche) for supporting this research. Many thanks to Total Kenya Nairobi for logistical support.

References

Binford, L.R., 1980. Willow smoke and dog's tails: hunther-gatherer settlement systems and archaeological site formation. American Antiquity 45, 4–20.

Braun, D.R., Plummer, T., Ditchfield, P., Ferraro, J.V., Maina, D., Bishop, L.C., Potts, R., 2008. Oldowan behavior and raw material transport: perspectives from the Kanjera Formation. Journal of Archaeological Science 35, 2329–2345.

Brown, F.H., Feibel, C.S., 1988. "Robust" hominids and Plio-Pleistocene paleogeography of the Turkana Basin, Kenya and Ethiopia. In: Grine, F.E. (Ed.), Evolutionnary History of the "Robust" Australopithecines, Aldine de Gruyter, New York, pp. 325–341.

Brown, F.H., Feibel, C.S., 1991. Stratigraphy, depositional environments, and paleogeography of the Koobi Fora Formation. In: Harris, J.M. (Ed.), Koobi Fora Research Project, vol. 3, The fossil ungulates: geology, fossil artiodactyls, and palaeoenvironments, Clarendon Press, Oxford, pp. 1–103.

Clark, J.D., 1980. Raw material and African lithic technology. Man and Environment 4, 44–55.

de la Torre, I., Mora, R., Dominguez-Rodrigo, M., Luque, L., Alcala, L., 2003. The Oldowan industry of Peninj and its bearing on the reconstruction of the technological skills of Lower Pleistocene hominids. Journal of Human Evolution 44, 203–224.

de la Torre, I., 2004. Omo Revisited. Evaluating the technological skills of Pliocene Hominids. Current Anthropology 45, 439–465.

Delagnes, A., Roche, H., 2005. Late Pliocene hominid knapping skills: the case of Lokalalei 2C, West Turkana, Kenya. Journal of Human Evolution 48, 435–472.

Féblot-Augustins, J., 1997. La circulation des matières premières au Paléolithique. Synthèse des données. Perspectives comportementales, vol. 75, ERAUL, Liège, 275 p.

Feibel, C.S., 2001. Archaeological sediments in lake margin environments. In: Stein, J., and Farrand, W.R. (Eds.), Sediments in archaeological context, University of Utah Press, Salt Lake City, pp. 127–148.

Feibel, C.S., Brown, F.H., McDougall, I., 1989. Stratigraphic context of fossil hominids from the Omo group deposits: northern Turkana Basin, Kenya and Ethiopia. American Journal of Physical Anthropology 78, 595–622.

Feibel, C.S., Harris, J.M., and Brown, F.H., 1991. Paleoenvironmental context for the Late Neogene of the Turkana Basin. In: Harris J.M. (Ed.), Koobi Fora Research Project, vol. 3, The fossil ungulates: geology, fossil artiodactyls, and palaeoenvironments, Clarendon Press, Oxford, pp. 321–370.

Geneste, J.-M., 1989. Economie des ressources lithiques dans le Moustérien du sud-ouest de la France. In: Otte M. (Ed.), L'Homme de Néandertal: La subsistance vol. 6, ERAUL, Liège, pp. 75–97.

Geneste, J.-M., 1991. L'approvisionnement en matières premières dans les systèmes de production lithique: la dimension spatiale de la technologie. In: Mora, R., Terradas, X., Parpal, A., and Plana, C. (Eds.), Tecnologia y cadenas operativas liticas, Barcelona: Universitat Autonoma de Barcelona, Treballs d'Arqueologia I, pp. 1–36.

Goldman–Neuman, T., Hovers, E., 2009. Methodological issues in the study of Oldowan raw material selectivity: insights from A.L. 894 (Hadar, Ethiopia). In: Hovers, E. and Braun, D.R. (Eds.), Interdisciplinary Approaches to the Oldowan. Springer, Dordrecth, pp. 71–84.

Harmand, S., 2004. Raw materials and economic behaviour of Late Pliocene Hominids: the case of Lokalalei 2C and Lokalalei 1 sites, West Turkana (Kenya). Poster presented at the Paleoanthropology Society Annual Meeting, Montreal (Canada), March 30–31st.

Harmand, S., 2005. Matières premières lithiques et comportements techno-économiques des homininés plio-pléistocènes du Turkana occidental, Kenya. Thèse, Université Paris X-Nanterre, 317 p.

Harmand, S., in press. Raw materials and techno-economic behaviors at Oldowan and Acheulean sites in the West Turkana region, Kenya. In: Adams, B., Blades, B. (Eds.), Lithic Materials and Paleolithic Societies. Blackwell Publishers, Oxford.

Harris, J.M., Brown, F.H., Leakey, M.G., 1988. Stratigraphy and paleontology of pliocene and pleistocene localities west of lake Turkana, Kenya. Contribution to Science. 399, 1–128.

Hay, R.L., 1976. Geology of the Olduvai Gorge: a study of sedimentation in a semi-arid basin. University of California Press, Berkeley.

Hovers E. 2001. Stone knapping in the Late Pliocene of Hadar, Ethiopia. Roux, V., Knapping Stone – a Uniquely Human Behaivor. Conference abstracts (http://web.mae.u-paris10.fr/knappingstone/anglais/resume.htm).

Hovers, E., Schollmeyer, K., Goldman, T., Eck, G., Reed, K.E., Johanson, D.C., Kimbel, W.H., 2002. Late Pliocene archaeological sites in Hadar, Ethiopia. Journal of Human Evolution 42, 1–17.

Inizan, M.-L., Ballinger, M., Roche, H., Tixier, J., 1995. Technology and Terminology of Knapped Stone. In: Préhistoire de la Pierre Taillée, vol. 5, CREP, Nanterre.

Isaac, G.L., 1977. Olorgesailie: archaeological studies of a Middle Pleistocene lake basin in Kenya. University of Chicago Press, Chicago, 272 p.

Isaac, G.L., Harris, J.W.K., 1978. Archaeology. In: Leakey, M.G., and Leakey, R.E. (Eds.), Koobi Fora Research Project. The fossil hominids and an introduction to their context, 1968–1974. Clarendon Press, Oxford, pp. 64-85.

Isaac, G.L., Harris, J.W.K., Kroll, E., 1997. The stone artifact assemblages: a comparative study. In: Isaac, G.L. (Ed.), Koobi Fora Research Project, vol. 5: Plio-Pleistocene archaeology. Clarendon Press, Oxford, pp. 262–299.

Jones, P., 1979. Effects of raw material on biface manufacture. Science 204, 835–836.

Kibunjia, M., 1994. Pliocene archaeological occurrences in the Lake Turkana basin. Journal of Human Evolution 27, 159–171.

Kibunjia, M., 1998. Archaeological investigations of Lokalalei 1 (GaJh5): a Late Pliocene site west of lake Turkana, Kenya. Ph.D. dissertation, The State University of New-Jersey, Rutgers, USA.

Kibunjia, M., Roche, H., Brown, F.H., Leakey, R.E., 1992. Pliocene and Pleistocene archaeological sites west of Lake Turkana, Kenya. Journal of Human Evolution 23, 431–438.

Kimbel, W.H., Walter, R.C., Johnason, D.C., Reed, K.E., Aronson, J.L., Assefa, Z., Marean, C.W., Eck, G.G., Bobe, R., Hovers, E., Rak, Y., Vondra, C., Yemane, T., York, D., Chen, Y., Evensen, N.M., Smith, P.E., 1996. Late Pliocene Homo and Oldowan tools from the Hadar Formation (Kada Hadar Member), Ethiopia. Journal of Human Evolution 31, 549–561.

Leakey, M.D., 1971. Olduvai Gorge: Excavations in Beds I and II, 1960–1963, vol. III. Cambridge University Press, Cambridge, 172 p.

Leakey, M.D., 1975. Cultural patterns in the Olduvai sequence. In: Butzer, K.W., and Isaac, G.L. (Eds.), After the Australopithecines, Mouton, pp. 477–493.

Leakey, M.D., 1994. Olduvai Gorge: Excavations in Beds III, IV and the Masek Beds, 1968–1971, vol. V. Cambridge University Press, Cambridge, 327 p.

Leroi-Gourhan, A., 1964. Le geste et la parole. Technique et langage. Albin Michel, Paris.

Leroi-Gourhan, A., 1971. Evolution et technique. L'homme et la matière. Albin Michel, Paris.

MacKenzie, W.S., Adams, A.E., 1996. Atlas d'initiation à la petrographie. Masson, Paris.

Martínez-Moreno, J., Mora Torcal, R., de la Torre, I., 2003. Oldowan, rather more than smashing stones: An introduction to The Technology of First Humans. In: Moreno, J.M., Torcal, R.M., and Sainz, I.T. (Eds.), Oldowan: Rather more than smashing stones, Centre d'Estudis del Patrimoni Arqueologic de la Prehistoria, Bellaterra, pp. 11–35.

Merrick, H.V., Merrick, J.P.S., 1976. Recent archaeological occurrences of earlier Pleistocene age from the Shungura Formation. In: Coppens, Y., Howell, F.C., Isaac, G.L., and Leakey, R.E. (Eds.), Earliest man and environments in the Lake Rudolf Basin. Chicago University Press, Chicago, pp. 574–584.

Plummer, T., 2004. Flaked stones and old bones: biological and cultural evolution at the dawn of technology. Yearbook of physical anthropology, Wiley Interscience 47, 118–164.

Plummer, T., Bishop, L. C., Ditchfield, P., Hicks, J., 1999. Research on Late Pliocene Oldowan sites at Kanjera South, Kenya. Journal of Human Evolution 36, 151–170.

Prat, S., Brugal, J.-Ph., Tiercelin, J.-J., Barrat, J.-A., Bohn, M., Delagnes, A., Harmand, S., Kimeu, K., Kibunjia, M., Texier, P.-J., Roche, H., 2005. First occurrence of early *Homo* in the Nachukui Formation (West Turkana, Kenya) at 2.3-2.4 Myr. Journal of Human Evolution 49, 230–240.

Roche, H., 2000. Variability of Pliocene lithic productions in East Africa. Acta Archaeologica Sinica, vol. 19, 98–103.

Roche, H., Kibunjia, M., 1994. Les sites archéologiques plio-pléistocènes de la Formation de Nachukui, West Turkana, Kenya. Comptes Rendus de l'Académie des Sciences 318, série II, 1145–1151.

Roche H., Kibunjia, M., 1996 Contribution of the West Turkana Plio-Pleistocene sites to the archaeology of the lower Omo/Turkana Basin. Kaupia 6, 27–30.

Roche, H., Delagnes, A., Brugal, J.-Ph., Feibel, C.S., Kibunjia, M., Mourre, V., Texier, P.-J., 1999. Early hominid stone tool production and knapping skill 2.34 Myr ago in West Turkana. Nature 399, 57–60.

Roche, H., Brugal, J.-Ph., Delagnes, A., Feibel, C.S., Harmand, S., Kibunjia, M., Prat, S., and Texier, P.-J., 2003 Les sites archéologiques plio-pléistocènes de la Formation de Nachukui (Ouest Turkana, Kenya): bilan préliminaire 1997–2000. Comptes Rendus Palévol 2, 663–731.

Rogers, M.J., Harris, J.W.K., Feibel, C.S., 1994. Changing patterns of land use by Plio-Pleistocene hominids in the lake Turkana basis. Journal of Human Evolution 27, 139–158.

Schick, K., 1987. Modelling the Formation of Early Sone Age artifact concentrations. Journal of Human Evolution 16, 789–808.
Schick, K., Toth, N., 1993. Making Silent Stones Speak. Human Evolution and the Dawn of Technology. A Touchstone Book, Simon and Schuster, New York.
Semaw, S., 2000. The world's oldest stone artifacts from Gona, Ethiopia: their implications for understanding stone technology and patterns of human evolution between 2.6-1.5 million years ago. Journal of Archaeological Science 27, 1197–1214.
Semaw, S., Rogers, M.J., Quade, J., Renne, P.R., Butler, R.F., Dominguez-Rodrigo, M., Stout, D., Hart, W.S., Pickering T., Simpson, S.W., 2003. 2.6-Million-year-old stone tools and associated bones from OGS-6 and OGS-7, Gona, Afar, Ethiopia. Journal of Human Evolution 45, 169–177.
Stout, D., Quade, J., Semaw, S., Rogers, M., Levin, N., 2005. Raw material selectivity of the earliest stone toolmakers at Gona, Afar, Ethiopia. Journal of Human Evolution 48, 365–380.
Tiercelin, J.-J., Schuster, M., Roche, H., Thuo, P., Brugal, J.-Ph, Trentesaux, A., Barrat, J.-A., Bohn, M. Prat, S., Harmand, S., Davtian, G., Texier, P.-J., in prep. Stratigraphy and paleoenvironments of the 2.3-2.4 Myr Lokalalei archaeological sites, Nachukui Formation, West Turkana, Northern Kenya Rift.
Toth, N., 1982. The stone technologies of early hominids at Koobi Fora, Kenya: an experimental approach. Ph.D. dissertation, University of California, Berkeley.
Toth, N., 1985. The oldowan reassessed: a close look at early stone artifacts. Journal of Archaeological Science 12, 101–120.
Toth, N., 1987. Behavioral inferences from early stone artifact assemblages: an experimental model. Journal of Human Evolution 16, 763–787.
Walker, A., Leakey, R., Harris, J., 1986. 2.5 Myr Australopithecus boisei from west of Lake Turkana, Kenya. Nature 322, 517–522.
Walsh, J., Dodson, R.G., 1969. Geology of Northern Turkana. Republic of Kenya: Ministry of Natural Ressources, Geological Survey of Kenya, 42 p.

9. Oldowan Technology and Raw Material Variability at Kanjera South

David R. Braun*
NSF-IRFP
Department of Archaeology
University of Cape Town
Rondebosch 7701
South Africa
david.braun@uct.ac.za

Thomas W. Plummer
Department of Anthropology
Queens College, CUNY and NYCEP
65-30 Kissena Blvd
Flushing, NY, 11367
USA
thomas.plummer@qc.cuny.edu

Peter W. Ditchfield
Research Laboratory for Archaeology
6 Keble Road Oxford, OX1 3QT
United Kingdom
peter.ditchfield@archaeology-research.oxford.ac.uk

Laura C. Bishop
School of Biological and Earth Sciences,
Liverpool John Moores University
Byrom Street Liverpool, L3 3AF
United Kingdom
L.C.Bishop@ljmu.ac.uk

Joseph V. Ferraro
Department of Anthropology
Baylor University
One Bear Place
Waco, TX, 76798
USA
joseph_ferraro@baylor.edu

* Address for correspondence: david.braun@uct.ac.za

E. Hovers and D.R. Braun (eds.), *Interdisciplinary Approaches to the Oldowan*,
DOI: 10.1007/978-1-4020-9059-2,

Keywords: Stone tools, Oldowan, Raw Materials, Technology

Abstract

Advances in the study of Oldowan research have suggested that the earliest tool-makers had the technological capabilities usually suggested in later time periods. Work in West Turkana and Gona research areas suggests that Pliocene hominins had a concise understanding of stone fracture mechanics and had a clear conception of how to reduce cores in a manner that maintained flaking surfaces. Here we investigate if these same patterns existed at the Pliocene site of Kanjera South in Western Kenya. Technological analyses suggest that although many of the technological capabilities described for other Oldowan sites are present in the Kanjera South assemblage, specific aspects of the context of the site (raw material variability) produced a different expression of these behaviors. The most obvious difference between the Kanjera South site and other Oldowan sites is that as reduction continues several different reduction patterns can be seen. This suggests that a reduction sequence or core reduction mode is not an immutable formula and can change depending on its context.

9.1 Introduction

The analysis of Oldowan technology has largely been focused on archaeological localities excavated in the East African Rift Valley. The majority of our knowledge of the earliest technology derives from archaeological sites from the Afar region (Kimbel et al. 1996; Semaw et al. 1997; Hovers et al. 2002; Hovers 2003), the Turkana Basin (Isaac 1972; Isaac and Harris 1997; Roche et al. 1999; Delagnes and Roche 2005), and Olduvai Gorge (Leakey 1971; de la Torre and Mora 2005). The Oldowan archaeological locality of Kanjera South is therefore initially important because of its geographic location. Placed between the two major rift valleys in East Africa, in the Kavirondo rift system, the Kanjera South Formation is host to the only site, other than Senga 5 (Harris et al. 1987) and Nyabusosi (Texier 1997), outside the East African rift valley. However, several aspects of the stone artifact collection from Kanjera South make it a vital part of the discussion of resource use and technological decisions in the Pliocene. The collection of artifacts from Kanjera South is one of the largest collections of stone artifacts from a Pliocene context (n=4474). Further it is associated with one of the largest excavated Pliocene faunal collections. The high diversity of raw materials in the local area makes the collections from Kanjera South an ideal sample to understand the interaction between technological decisions and raw material quality and availability. This study represents the initial description of the technology of Oldowan artifacts in relation to raw material sources in the Kanjera South Fm. The analysis of the stone artifact collections from the Kanjera South Fm. requires a comprehensive analysis of raw material sources in the South Nyanza District of western Kenya, where the site is located. Therefore this paper will outline different technological strategies within the Kanjera South Fm. assemblages relative to raw material differences. The technological analysis here complements previous work (Braun 2006) but does not include the detailed metric analysis. Rather this analysis focuses on the core production modes (Roche 2000) present in the Kanjera South Formation in order to fit it into a broader picture of technological modes in the Oldowan.

9.2 Homa Peninsula: Site Context and Raw Material Availability

The southern member of the Kanjera Formation is a relatively small group of deposits (~800 square meters) situated on the southern shore of the Kavirondo Gulf of Lake Victoria. It has six beds, from oldest to youngest KS-1 to KS-6 (Behrensmeyer et al. 1995; Ditchfield et al. ms).

The site is anomalous among Pliocene Oldowan archaeological sites because of its large assemblage of lithic material, diversity of raw materials, large zooarcheological assemblage, and C_4 pedogenic carbonate signal indicating an open habitat setting (Ditchfield et al. 1999; Plummer et al. 1999, 2009; Plummer 2004). Sedimentological analyses suggest the site was situated on the shores of a Paleo-Kavirondo Lake (Ditchfield et al. 1999). The Oldowan archeological occurrences are largely restricted to Beds KS-1 to KS-3. A combination of chronometric methods (paleomagnetism, biostratigraphy) indicates the archeological occurrences predate the base of the Olduvai Subchron at 1.95 Ma, so a date of ~2.0 Ma is used here (Ditchfield et al. 1999; Plummer 2004). Artifacts and archeological fauna were buried on an alluvial plain by KS-1 to KS-3 fine pebbly sands. Water flow was directed northwards towards a paleo-lake and was generally diffuse and of low energy. The lake transgressed from North to South through time, completely covering the archeological occurrences during the deposition of KS-4 clays. Taphonomic and zooarcheological analyses indicate that the lithic and faunal assemblages through the KS-1 to KS-3 sequence formed predominantly through hominin activity (Ditchfield et al. ms.; Ferraro et al. ms.). Sediments to the East and West of the Kanjera locality also support the evidence for lacustrine flats in this area from the Pliocene through the Pleistocene.

An understanding of stone-tool technology at Kanjera South must incorporate a comprehensive understanding of the context of raw material sources. The variety of raw materials incorporated into the technological repertoire of these hominins is a product of the extremely heterogeneous geology of southwestern Kenya. Fortunately, this variation allows for testing of the effect of different aspects of raw material quality and availability on technological organization (Braun 2006). Constructing a provenance framework of raw material sources must be tuned to the research goals of the project. Artifacts need not be linked directly to specific rock outcrops but a connection should be made between artifacts and groups of rock outcrops that have a similar lithology and can be obtained in similar locations (secondary sources) across an ancient landscape. An extensive raw material sourcing program associated with this study has isolated 9 major raw material sources each with various subsources that have been identified geochemically using ED-XRF techniques (Braun 2006; Braun et al. 2008). Reiterating the process of isolating these raw material sources and determining the varying levels of availability for these materials is beyond the scope of this paper. Studies of secondary sources has shown that those raw materials that had primary sources that were relatively large distances from the site (> 20 km) where only found in paleo-conglomerates that were greater than 10 km from the site. In contrast the relatively closer primary sources where available in paleo-drainages that were between 1 and 2 km from Kanjera South. A more detailed explanation of raw material availability can be found in Braun et al. (2008), however a brief description of primary sources will reflect the relative difference in availability of certain raw materials.

9.3 Geology of South Nyanza and the Kisii Highlands

The basement rock geology of the Kisii Highlands and the South Nyanza areas of western Kenya are very complex. A brief review of different regions and their underlying geology is necessary to contextualize the samples used in the artifact provenance study. Features referred to in this review can be found on the map in Figure 9.1.

9.3.1 The Homa Mountain Carbonatite Center

The Homa Mountain at the far western end of the Kavirondo Gulf is the product of carbonatite volcanism. The most abundant rock types in this region are extrusive volcanics from the Nyanzian System. These rocks were largely rhyolites and Dacites that were emplaced over a large area on what was once a flat plain prior to the doming of the Homa Mountain carbonatite edifice (LeBas 1977). In this relatively unaltered state (considering their Archaean age) these rocks can be found in a broad arc that has a 9 km radius around the Homa Mountain carbonatite center. These rocks are fine-grained and sometimes glassy, with numerous phenocrysts. Rhyolites can be distinguished by their porphyritic nature set in a finer-grained matrix of quartz and orthoclase (Saggerson 1952; McCall 1958). In areas closer to the Homa Mountain carbonatite

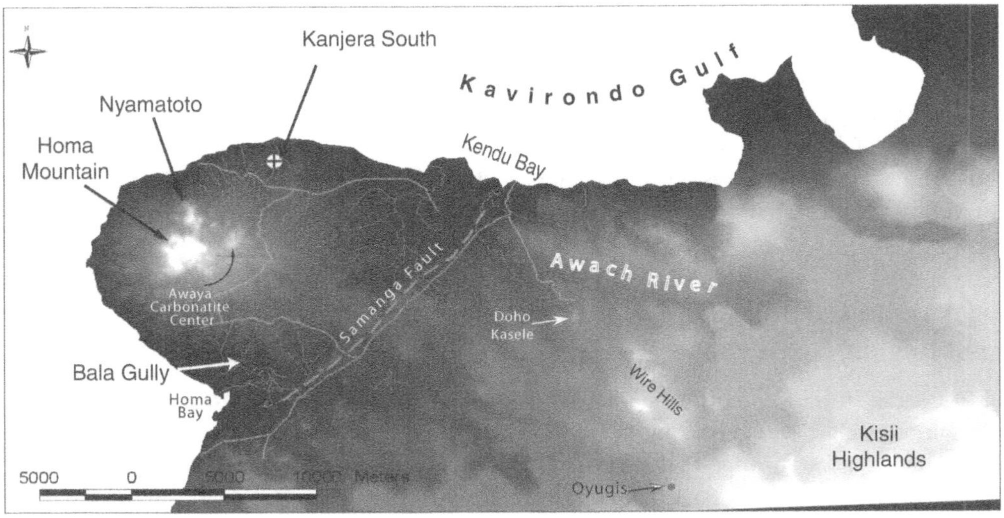

FIGURE 9.1. Map of the study region with areas mentioned in the text highlighted.

center these rocks underwent an intense metasomatism process. These metasomatized rocks are termed fenites and are largely brecciated and shattered with secondary minerals forming along the resulting joint surfaces. As this process becomes more intense the largely brownish-black rock can be wholly replaced by potassium feldspars working out from the veins and resulting in extensive penetration of the rock by amphibole and aegirine (LeBas 1977). Those rocks that received the least metasomatism are associated with the Awaya Hill center. Deep gully systems on either side of this large hill contain boulders of this fenitized Nyanzian rock.

At about 13 Ma the entire region around the present day Homa Mountain center began to dome up creating a series of steep sided hills. These hills are made primarily of various types of carbonatites mantled by fenitized Nyanzian rocks. The carbonatite rocks around the hills in this region are largely made of leucites (usually nepheline). By far the most abundant carbonatite rocks are Ijolites. These plutonic rocks are made from mostly aegerine-augite (40%) and nepheline (58%). Micro-ijolite, the finer grained variety, is somewhat common in the foothills of Awaya and along the Bala gully system (Saggerson 1952). As a general rule all volcanic rocks in and around the Homa Mountain center are extremely silica poor and alkali rich.

Around 2.5 Ma carbonatite volcanism on the Homa Peninsula reached a peak and then began to subside (LeBas 1977). This is recorded in the subsidence of the Homa Mountain center which today appears as a large ring structure with a depressed center. At about this time several phonolite plugs were emplaced in various areas around the Homa Mountain Complex. All of these phonolitic plugs have various levels of alkali feldspar. Depending on the percentage of alkali minerals, the rocks are termed nephelinites, phonolites or nephelinitic phonolites. These rocks are abundant throughout the Homa Mountain region and are likely tied to the phonolitic plugs of the Samanga Fault region.

The outer radius of the Homa Mountain carbonatite center is covered in a thick mantle of Plio-Pleistocene sediments. On the northern margin of Homa Mountain these sediments form the lacustrine and fluvial beds of the Rawi, Abundu, Kasibos, and Kanjera Formations (Ditchfield et al. 1999). These lake beds have several horizons of intercalated Limestones. The abundance of carbonate available in the system in this area probably makes for very favorable conditions for the formation of Limestone.

9.3.2 Plutonic and Extrusive Rocks of the Kendu Fault Region

Parallel to the road that passes from Kendu Bay on the southern shores of the Kavirondo Gulf to Homa Bay is a large and ancient fault. The fault marks an important lithological division because all of the rocks to the west of this fault are associated with the Homa Mountain and the country rocks affected (fenetized) by it. This fault is called the Samanga Fault (Saggerson 1952) named after the large phonolite plug that rises 70 meters out of the surrounding plain along this fault. This plug varies from pure nephelinite at the apex of the hill to pure phonolite at its base (LeBas 1977). Several more of these basic extrusive volcanic centers dot the length of the Samanga fault. To the Southwest of the Samanga fault near Homa Bay a series of dark Nyanzian acid igneous rocks have largely escaped secondary alteration and form large blocks of fine grained rhyolitic lava in this region. Further to the East a large area approximately 9 km wide (east-west) and 15 km long (north-south) is covered by a large granitic mass known as the Oyugis Granite. The granite cuts through Nyanzian rocks near Doho Kasele and the resulting rock is more microgranitic. Interactions between the granite and Nyanzian rocks throughout the area have resulted in various degrees of fine-grained and coarse-grained granites.

Three other rock types in this region are of note because of their appearance in the archaeological materials at Kanjera South. **1)** At the tops of some hills south of Kendu is a series of banded ironstone cherts from the Nyanzian system. These are very fine-grained rocks that are often brecciated and shattered but sometimes can form very thick bands of fine grained chert that is likely of colloidal origin (Shackleton 1946; Saggerson 1952). **2)** In the region near Doho Kasele along the southern banks of the Awach river system are large outcrops of Nyanzian Basalts. These rocks are very fine-grained and are blue to black in color. The complete lack of phenocrysts in this rock makes it extremely hard to distinguish from other basic lavas in the Homa Peninsula area in hand sample. **3)** Along the eastern side of the Oyugis-Kendu road are a series of very prominent hills referred to as the Wire Hills. They are comprised of a light grey to red rhyolitic rock.

9.3.3 Bukoban System of the Kisii Highlands

To the southeast of the Homa Peninsula is an upland region made of a series of steep ridges that are the product of the Bukoban system sometimes known as the Kisii series (Huddleston 1951). The base of the Bukoban system is composed of porphyritic basalts that are not widely exposed. They are dark green in color with large (15 mm) pale green feldspar phenocrysts. The middle of the Bukoban series is dominated by the upwards of 200 feet high scarps of banded quartzites. These rocks are very fine-grained grey to dark red and are made up of almost entirely silica. Microscopic analysis by Huddleston (1951) suggested these rocks were derived from very mature quartz beach deposits because of the complete lack of other minerals. The microcrystalline quartz grains are set in a siliceous fine-grained cement. Immediately overlying the quartzites in the Kisii highlands are a series of fine grained felsic volcanic rocks. In hand sample the rock can only be described as "fine grained felsic igneous" because of the lack of visible phenocrysts. Chemical studies have shown the rock to range from dacitic to rhyolitic composition.

Based on the previous description the nine major raw material groups that can be identified and used as analytical groups for investigating raw material effects on core-reduction modes are: Bukoban Basalt (BBA); Bukoban Felsite (BFE); Bukoban Quartzite (BQU); Fenetized Nyanzian Rhyolites/Dacites (FNY); Homa Peninsula Limestone (HLI); Homa Mountain Phonolite (HPH); Nyanzian Banded Ironstone Chert (NYC); Nyanzian Rhyolite (NYR); Oyugis Granite (OYG).

9.4 Technological Analysis

Various studies of Early Stone Age industries have suggested that raw material variability significantly affects the technological decisions. Some studies have shown that certain raw materials are consistently found in similar artifact forms. Some have suggested that different raw materials were selected for different tasks (Stiles 1991; Martínez 1998; White 1998; Kimura 1999, 2002). There is a possibility that the link between raw material groups and artifact categories suggested by some authors may actually be a product of varying intensities of artifact reduction (McPherron 1994, 2000). Here we investigate the types of core reduction modes applied to different raw material categories at Kanjera South specifically to see if different core reduction modes were associated with raw materials because of their specific physical properties or the availability of these materials. Here we will display core reduction modes through the aid of diacritic diagrams (Dauvois 1976; Inizan et al. 1999; de la Torre et al. 2003; Delagnes and Roche 2005). More traditional descriptions can be found in Braun (2006). The analysis of technological modes in the absence of multiple refit sequences is clearly restricted to inferences of core reduction patterns. Such analytical techniques have been employed successfully in the context of technological systems that are clearly more organized and standardized than Oldowan industries (e.g. Levallois; Boëda 1995). However, here we are interested in the investigation of modalities of core forms as described by technological attributes. We will limit this study to core forms because they represent a clearer picture of core reduction patterns. However, complimentary patterns can be found in debitage component of these assemblages. Although we describe the association of technological traits with specific raw material groups it should be noted that there is extensive variation within and across these groups. Indeed many forms appear to grade into each other through the course of reduction. This analysis is attempting to focus on description of these modes of core form rather than isolate discrete types. A review of different technological measures for each raw material can be found in Table 9.1.

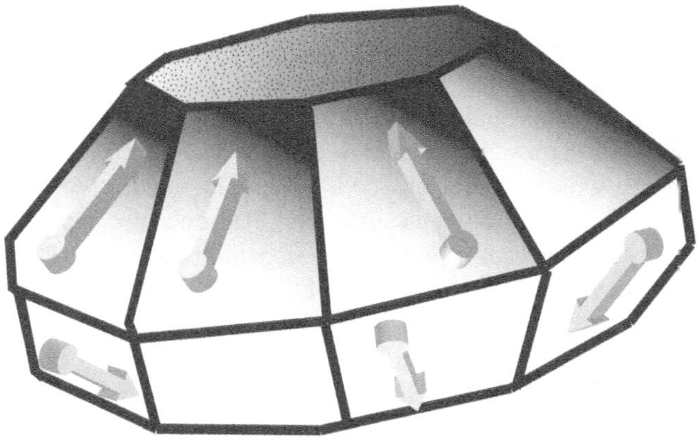

FIGURE 9.2. Core reduction characteristic of the Bukoban Basalt material

9.4.1 Bukoban Basalt (N=7)

The original form of this raw matetrial group is rounded river cobbles. Core reduction modes on this raw material seem to

TABLE 9.1. Review of technological patterns for each raw material

Raw Material	N	Original Form	Median Flake Scar Count	Median Core Mass (g)	Median Exploitation Surface Angle (degrees)
Bukoban Basalt	7	Rounded River Cobbles	8	85.4	88
Bukoban Felsite	18	Rounded River Cobbles	6.2	100.5	90
Bukoban Quartzite	23	Rounded River Cobbles	3	42.18	96
Fenetized Nyanzian Rhyolite/Dacite	125	Tabular Blocks	2	65.35	83
Homa Limsetone	17	Various	4	74.2	85.5
Homa Phonolite	53	Various/Tabular	4	61.85	85
Nyanzian Banded Ironstone Chert	7	Tabular/Rounded	4	57.8	90
Nyanzian Rhyolite	25	Rounded River Cobbles	7	51.2	90
Oyugis Granite	19	Rounded Blocks	8	137.1	59

be directed by the original rounded shape. Almost all core forms have employed a centripetal flaking pattern with cores usually divided into two hemispheres. These hemispheres appear to have equal amounts of volume on either side of the horizontal axis. There are some cores which seem to represent the standard "chopper" core forms. The majority of cores have two convex debitage surfaces. There does not appear to be any evidence of a hierarchical division between surfaces. Both surfaces appear to have been flaked at similar angles with most cores exhibiting relatively obtuse platform angles (median 88°). Removals are usually alternating and not alternate (Inizan et al. 1999), suggesting there is no attempt to prepare platforms for a second generation of removals. Cortex can be found on both surfaces suggesting exploitation surfaces were not divided into debitage and preparation surfaces (e.g. Figure. 9.2). Flaking was usually limited to two or three flaking surfaces on cores. However, the number of flake-scars per core was actually quite high relative to the size of the cores suggesting near exhaustive reduction of these cobbles (median flake-scars: 8; median mass: 85.48 g).

9.4.2 Bukoban Felsite (N=18)

This raw material sometimes appears as a porphyritic variety although this was very rarely used, and only two cores were recovered made on this variety. This material appears to be made almost exclusively on rounded river cobbles although the somewhat reduced nature of this material makes if difficult to determine the original form on many cores. This raw material has two major core reduction modes. A third group of cores have no real organization and appear to be the fragments of cores that were secondarily utilized to extract a few final flakes from near exhausted cores. **1)** The first core-reduction mode is similar to one that is found in Bukoban Quartzite reduction. Cores have two hemispheres with different levels of convexity. The lower surface usually retains a small amount of cortex and is markedly more convex that the upper surface. Further the upper surface has longer flake removals and more flakes per exploitation surface. However, the upper surface in these Bukoban Felsite cores follows an orthogonal removal pattern with platforms rarely reused (Figure. 9.3). Instead, the distal edge of the first generation of flake-scars is usually the surface that is exploited for the second generation of flake-scars. These cores are sometimes split in half in an attempt to expose new exploitation surfaces when platform angles become too obtuse. **2)** The second major core reduction mode is bifacial centripetal with a bipyramidal cross section. These cores have two clear hemispheres with each exploitation surface used as a platform for removals on the opposite surface. The center of mass of the cores appears to stay constant in these cores despite extensive reduction (median flake scar count: 6.2; median core mass: 100.5 g). Removals usually exploit previous scars as new platforms so that small portions of cortex are preserved on the base and top of the core. Although the generalized plan for these core-reduction modes is two separate exploitation surfaces with removals radiating in opposite directions, sometimes acute angles that were perpendicular to the axis of flaking on either the upper or lower surface were exploited during episodes of core-rotation (Figure. 9.4).

9.4.3 Bukoban Quartzite (N=23)

For almost all of the cores that retain cortex in this raw material it is clear that the initial raw material packages were rounded river cobbles. This raw material includes three major core-reduction modes. **1)** The most abundant one is the re-use of flakes as "flake cores." The platforms on these flake-cores tend to be relatively large (median platform area: 376 mm^2; mean platform area of quartzite flakes: 130 mm^2) even though the resulting cores are actually quite small suggesting significant reduction of these specimens (Dibble 1995). Distinguishing

FIGURE 9.3. Core reduction mode usually associated with Bukoban Felsite cores.

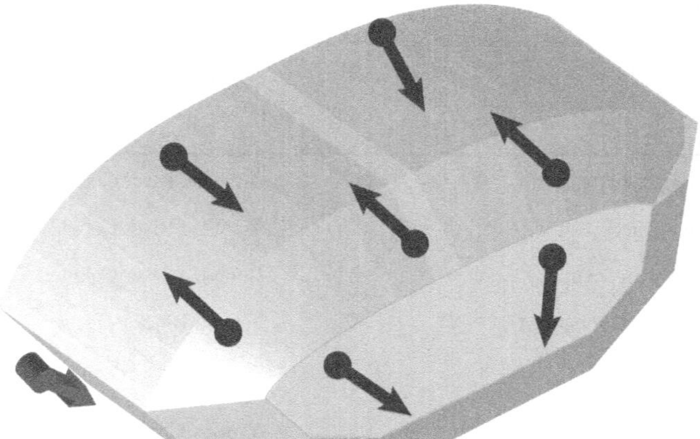

FIGURE 9.4. Core reduction mode usually associated with Bukoban Felsite cores.

between negative flake-scars that were removed prior to the production of the flake, which was subsequently used as a flake-core, and removals made after the production of the flake is difficult. However, initial removals are usually made from the internal (ventral) surface of the flake (Figure 9.5). The second generation of removals is usually removed from the external (dorsal) surface, and negative bulbs of percussion usually distinguish these removals from previous flake-scars. The internal surface of the flake-cores is usually centripetally flaked. The slight convexity of the ventral surface is usually exploited until this surface is rendered planar. Subsequent removals on the dorsal surface of the flake-core attempt to renew the angle between the dorsal and ventral surface, however tertiary removals on the ventral surface are rare. The distal extremity of the flake-core is almost always truncated to produce secondary platforms. The original platform of the flake-core is rarely secondarily flaked (which is why these pieces can be recognized as flake-cores). **2)** The second major core reduction mode is a bifacial reduction mode which usually divides the core into two hemispheres. These hemispheres do not show any pronounced hierarchical separation in terms of the nature of flake removals. However, there does seem to be a marked difference in the convexity of these surfaces. Removals on the upper surface are large and the surface tends to be less convex. Removals on this upper surface tend to be unipolar or bipolar with flake-scars usually intersecting at the center of the mass of the core (Figure. 9.6). Each series of removals is rather short with flaking surfaces having a median value of 2.2 removals per surface. The length of these debitage sequences was clearly restricted by the original size of the cores which was quite small (mean longest axis 45.01 mm). Secondary or tertiary series of removals on this surface can be orthogonal. The lower surface tends to have short flake-scars which are abrupt. The angle between these two surfaces is usually quite steep (median angle: 88°). This lower surface can quite often have a small portion of cortex. **3)** The third core reduction mode may or may not represent an advanced stage of the second core reduction mode. This mode of core reduction includes rotation of the core after an initial series of removals, and utilization of new angles that are produced via core rejuvenation flakes. These are also represented in the detached piece assemblage (18.7% of the whole flakes). In an extreme case after the first round of core-rotation depleted all available angles, bipolar reduction commenced along the longest axis of the core. This particular core resulted in 11 removals on a core that was only 25.2 grams. In this core reduction mode the center of gravity of the core is maintained despite core-rotation and often removals from the previous organization of the core are completely removed.

9.4.4 Fenetized Nyanzian Rhyolites/Dacites (N=125)

This raw material is characterized by a number of joints and cracks that lead to the material shattering during the course of reduction. Subsequently many of the exploitation surfaces show very short reduction sequences. Some cores even document reduction that was halted by the intersection of an exploitation surface and an internal joint surface. There are approximately 125 cores made on this material. However, many of these pieces are only casually flaked. In fact 23% of the cores made form this raw material have only one exploitation surface. Of these cores 62% of them have only two flake-scars in a unidirectional pattern. It is very difficult to characterize a real reduction mode amongst this raw material. The majority of cores have only a few removals despite adequate size of material available for reduction (median flake-scar count: 2; median core-mass: 85.3 g). In fact the most frequent reduction mode appears to be the use of natural angles produced by the joints in the rock that are then flaked unidirectionally until

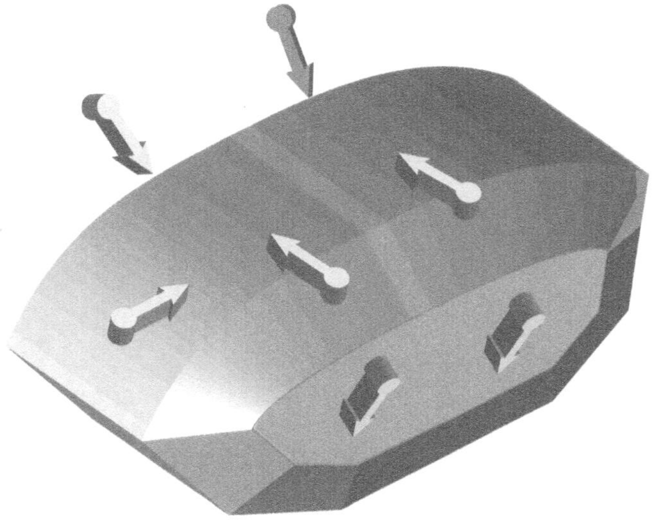

FIGURE 9.5. Core Reduction mode usually associated with Bukoban Quartzite and Nyanzian Rhyolite cores.

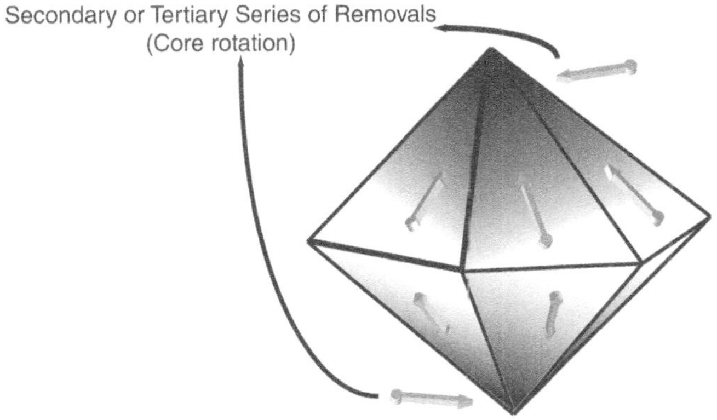

FIGURE 9.6. Core reduction mode usually associated with Bukoban Quartzite cores.

flaking is interrupted by the intersection of a flake-scar with another joint surface. In fact 74.3% of the cores made on this raw material were discarded because flaking was interrupted by internal flaws in the rock. Although unidirectional flaking dominates most debitage surfaces in this raw material, some cores exhibit orthogonal and opposed reduction methods. In a small number of cores a bipolar strategy was employed. However, the tendency of this material to shatter probably precluded this technique during reduction. Although the Fenetized Nyanzian rock was easily the most abundant rock near the Kanjera South site the numbers of surfaces exploited on these cores suggest that this raw material played only a minor role in the technology at Kanjera South. Although the material is quite hard, the internal fractures make it relatively impossible to continue with a particular flaking strategy. Hence it appears that fenetized Nyanzian rocks were used casually and only in times when a limited number of flakes were required.

9.4.5 Homa Limestone (N=17)

This material is quite often found in very large blocks or rounded cobbles near the excavation site (median core mass: 74.2 g). The material varies extensively with some pieces extensively silicified and others being quite chalky with fossil inclusions. Despite the large package size and the high availability of Homa Limestone, this material was rarely used for artifact manufacture. Cores do not seem to follow any standardized reduction mode with most flake-scars using natural surfaces or previous flake-scars as platforms. Cores exhibit a reduction pattern that causes the volumetric structure of the core to change dramatically through the course of reduction (Roche et al. 1999). Most cores represent cobbles or blocks that were split into two and then reduced further using the split cobble as a platform surface. In many cores, reduction did not extend much further than 2 to three removals. In fact this raw material had the lowest number of flake-scars per exploitation surface (median: 1.3). Core reduction modes show no preferred pattern with cores exhibiting all varieties of convex, concave and planar morphologies. Removals tend to capitalize on whatever acute angles happen to be on the core.

9.4.6 Homa Phonolite (N=53)

The phonolite plugs that ring the Awaya center make up most of the raw material in this category of cores. A large majority of the cores appear to be made on subangular blocks that were likely procured from the river systems that drain from the base of the Awaya carbonatite complex and empty into the Kavirondo Basin about .5–2 km east of the Kanjera site. This material has undergone extensive chemical weathering and it is often very difficult to ascertain the core reduction mode on many cores. The surfaces of these pieces tend to crumble and flake-scar surfaces sometimes become obscured. However on some cores the flake-scars are still very fresh. These cores are usually quite large (median core mass 61.8 g) but tend to have relatively few flake-scars per exploitation surface (median 1.6). There is great variation in the core reduction techniques and modes. Some core reduction modes are very casual with one or two flakes removed always in a unidirectional pattern. In these cores split cobbles or blocks are usually the dominant original form.

There are some cores which exhibit quite complex flaking systems with centripetal and bidirectional opposed flaking patterns. Cores flaked in this manner usually have a biconvex form with two distinct hemispheres that are bipyramidal

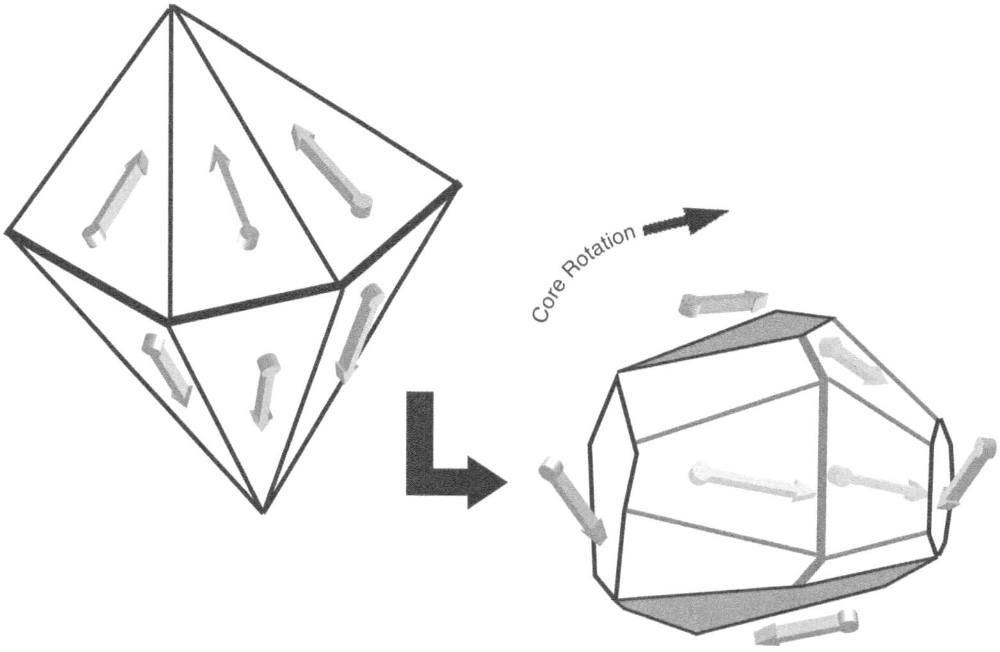

FIGURE 9.7. Core reduction mode found on various cores where core rotation and subsequent reduction obscures previous core orientation.

in their form (Figure. 9.4). Removals are usually alternate and the lateral edges of previous flake-scars on the opposite hemisphere are usually used as the platform for subsequent removals. Interestingly it appears that extensive reduction of the bipyrmaidal cores through core-rotation and secondary generations of removals allows for biconvex flaking system. In these cores the flake-scars from the previous core orientation are subsequently used as platforms with removals tangential to the original removal surface. The end result of this pattern is a core that appears multifacial but with a distinct planar surface on the top hemisphere (usually facilitated by one or two large removals) and a more convex hemisphere on the base. Although this pattern is somewhat complex it represent usually less than 8 flake-scars and a debitage series of no more than 3 or 4 flakes per exploitation surface (Figure. 9.7). This raw material appears to have been used rather casually and only sporadically was the material extensively reduced. Since this material was available near the site it is possible that this material was usually only casually flaked but when circumstances required increased flake production, these cores were utilized more extensively. This model could be tested by investigating complementary debitage patterns. If we are correct about the use of core-rotation and subsequent tangential removals we would expect flakes that display the initial bypyramidal pattern on their dorsal surface to be smaller than average. Further experimental research hopes to investigate these patterns.

9.4.7 Nyanzian Banded Ironstone Chert (N=7)

This raw material has a highly variable flaking quality. Some pieces of this material tend to be very glassy with little evidence of internal flaws. However, some fragments are banded with ironstone which rapidly oxidizes and causes planar fractures through the rock. Some of these lenses of ironstone can be as small as .5–1 mm and therefore are not easily seen at the beginning of a reduction sequence. However some core reduction sequences on this raw material are terminated by extensive step fractures. At least 50% of the attempts at core-reduction on this raw material were truncated by intersection of the flaking surface with an internal fracture plane in the rock. That said, when core reduction was allowed to continue it usually followed a pattern where a single exploitation surface was flaked until angles became too obtuse to allow further reduction. Then the distal extremity of the flake-scars of this first series of removals is subsequently utilized for a second generation of flakes. The result is a core with a series of flaking surfaces that are all perpendicular to each other yet without any meaningful connections between removal surfaces. These cores appear to have been flaked in a manner that attempted to exploit every possible acute angle on the core.

9.4.8 Nyanzian Rhyolite (N=25)

The rhyolite used for artifact manufacture is geochemically similar to the rhyolite found at the southern end of the Samanga Fault and is very fine grained with some phenocrysts but these rarely interrupt flaking. Much like the quartzite this raw material occurred almost exclusively as rounded river cobbles. This raw material follows many of the flaking patterns described earlier for Bukoban Quartzite. The most abundant core reduction mode is almost identical to the "flake-core" reduction mode described by Bukoban Quartzite. The convexity of the ventral surface is exploited using centripetal flaking patterns to reduce a large flake. The platform of the original flake is rarely exploited. These platforms are again quite large (median platform area 31.4 mm^2) suggesting that large flakes were subsequently used as cores. The other reduction that is quite common is the biconvex centripetal pattern. In this pattern the core has a lower surface that preserves a small portion of cortex and tends to be more convex. The upper surface is more planar and has larger removals, while the lower surface tends to have smaller and more abrupt removals (Figure. 9.6). Like the Bukoban Quartzite cores, the Nyanzian rhyolite cores are relatively small (median core mass: 51.2 g). Some rhyolite cores were made of more porphyritic materials that tend to fracture sporadically. These cores were reduced less extensively.

9.4.9 Oyugis Granite (N=19)

Of all of the raw materials exploited, granite has the largest grain size and particular aspects of this raw material dictated the core reduction modes. Fracture patterns in this rock are slightly different because of the large crystal size. Unlike most crypto-crstalline materials it is possible to remove a flake from a platform with a greater than 90 degree angle. This peculiar fracture pattern has been recorded for other coarse-grained materials (Sahnouni et al. 1997). The resultant pattern of flaking is similar to what has been described as "subspheroids" with several different exploitation surfaces and numerous flake-scars. These cores are the largest in the assemblage (median core mass 137.04 g). Almost all cores follow the same reduction mode where flaking is done multifacially on all surfaces of the core, usually using previous flake-scars as platforms. There are no relationships between core surfaces and only rarely are two opposed core hemispheres determined. Although many of these cores have numerous flake-scars these cores do not appear to have been reduced exhaustively. Most core forms had angles between exploitation surfaces that were still acute when they were discarded (median exploitation angle 58 degrees). Core reduction is usually centripetal and cortex is only rarely seen on small patches of the core (median cortex percentage 18.5%). It is most likely that this material was exploited for its very durable edges, even though continuous flaking of the material was difficult because of the coarse texture.

9.5 Discussion and Conclusion

The technology of the Kanjera South assemblage follows many of the patterns previously outlined for Oldowan technology. Although the study of the Kanjera South materials has not yet recovered extensive refitting sequences it is possible to determine some aspects of the flaking patterns and core-reduction modes. Unidirectional and multidirectional flaking patterns feature prominently as has been described for the West Turkana materials (Roche et al. 1999). Similarly, hominins at Kanjera South employed core reduction modes that enhanced the length of the flaking series (Delagnes and Roche 2005). Continuous removals of previous platforms and core-rotation to maintain acute angles allowed the production of several generations of flake removals (Figure. 9.7). Kanjera South cores also seemed to display a similar conservative pattern of keeping the volume of the core centered on a distinct center of gravity that carried over from one generation of removals to the next. Yet the consistent conservation of the convexity of a core surface described through refit sequences at Lokalalei 2C (Delagnes and Roche 2005) is only found in a few raw materials at Kanjera South (BFE, BBA, BQU and NYR). Similar to patterns seen in the core forms in the Omo collection, even relatively small cores still preserved quite complex flaking patterns (de la Torre 2004). In fact it is the smaller cores that reflect the most complex flaking patterns. Indeed the raw materials with the smallest core-volume (BQU and NYR; except for FNY which has numerous small core forms) have the high numbers of flake-scars per exploitation surface (BQU: 5-2 flake-scars per exploitation surface; NYR: 6-2 flake-scars per exploitation surface). Although many of the raw materials exhibit two separate flaking surfaces in cores at Kanjera South, there does not seem to be any hierarchical division between these surfaces as has been suggested for Oldowan assemblages later in time (de la Torre et al. 2003). In fact the development of two separate core reduction surfaces seems to be only a temporary aspect of core reduction. As core reduction becomes more exhaustive these divisions are ignored (Figure. 9.7). The presence of a core reduction mode that uses the convexity of dorsal surfaces of large flakes appears to be unique to Kanjera South.

Although it does appear that the technology of Kanjera South fits within the overall Oldowan technological system, a pattern expressed at Kanjera South suggests some slight differences. Studies of the earliest technology have begun to view these early industries as representative of a complex understanding of core reduction. In fact one of the most convincing pieces of evidence for this claim is the remarkable refits at the Lokalalei 2C site where there does appear to be defined set of rules that are followed in the course of core reduction (Roche et al. 1999; Delagnes and Roche 2005). These cores maintain a plano-convex cross section to allow for a long series of removals on one exploitation surface. However, in comparison to sites like Lokalalei 2C and Olduvai, Kanjera South cores are far smaller (Lokalalei 2C mean dimensions: 7.6cm x 5.4 cm x 4.2 cm; Kanjera South mean dimensions 5.6 cm x 4.3 cm x 2.8 cm; these are measures of the final size of these cores). The development of a long series of removals by creating and maintaining a single flaking surface over the length of the core reduction is not tenable on small cores. Instead the use of multiple core surfaces and subsequent removal of previous platforms to develop new exploitation surfaces characterizes the Kanjera South assemblage. Clearly difficulties with some raw materials at Kanjera South made extensive reduction sequences impossible. The frequent truncation of reduction sequences by step fractures caused by flaws in the rock is a testament to these difficulties. The Kanjera South assemblage does display an adaptability in technological systems. Core-reduction systems seem to co-vary with the availability and quality of different raw materials. In fact the most conspicuous feature of the Kanjera South assemblage is that raw materials that were procured from further distances (BQU; BFE; BBA; NYR, Braun 2006) were reduced more extensively than those that were available on site. The dominant core reduction modes in raw materials that were collected from distant sources utilizes multiple exploitation surfaces and maintains or produces convexities on multiple surfaces throughout the reduction of the core. In comparison raw materials that were procured on site or near the site (HLI, HPH, or FNY) show less complex reduction sequences.

This adds further problems to the development of an "Oldowan" technological scheme. The technological competency expressed in some raw materials at Kanjera South (e.g. BQU and NYR) are not reflected in the core reduction modes of other raw materials at Kanjera South (e.g. FNY or HLI). This diversity in reduction modes within one locality across different materials would make it very difficult to assign a singular description to "Oldowan" technological competency. When Oldowan hominins have access to large cobbles of adequate raw material they appear to express remarkable understanding of fracture mechanics and a volumetric understanding of knapping patterns. However, raw material variability is reflected in variable levels of core reduction and hominins seem to express these abilities relative to these constraints. If we are to understand the place of Oldowan technology in the evolution of human technological competency it will require a better understanding of the context of these behaviors. Especially when these behaviors are expressed in extremely different ecological contexts (e.g. Europe or China) it will be necessary to understand why hominins are producing stone artifacts. Considering the diversity of technological modes that can be found at one locality, where we may be able to assume constant technological skills and ecological context, it seems near-sighted to assume that direct comparisons between the morphology of artifacts in assemblages across vast amounts of space and time can be fruitful. If we are to really understand how the Oldowan affected the evolutionary trajectory of early *Homo* it seems necessary to understand these archaic assemblages in and out of Africa relative to the availability and quality of raw material as well as the ecological context of these finds.

Acknowledgements. We are grateful to the Office of the President of Kenya, and the National Museums of Kenya for permission to study the Kanjera fossils and artifacts. Further, we thank the National Museums of Kenya (especially Dr. Purity Kiura) for support and assistance. The Homa Peninsula field research is conducted through the cooperative agreement between the National Museums of Kenya and the Smithsonian Institution. Logistical support and funding was also provided by the Smithsonian's Human Origins Program. Funding from the L. S. B. Leakey Foundation, the National Geographic Society, the National Science Foundation, the Professional Staff Congress-City University of New York Research Award Program and the Wenner-Gren Foundation for Kanjera field and laboratory work is gratefully acknowledged. We would also like to thank Rick Potts and the Human Origins Program for support during all phases of the Kanjera research. DRB would like to thank the continuous support of Kathryn Underwood in helping with the completion of this project.

References

Braun, D.R., 2006. The Ecology of Oldowan Technology: Perspectives from Kanjera South and Koobi Fora. Ph.D., Rutgers University.

Braun, D.R., Plummer, T., Ditchfield, P., Ferraro, J.V., Maina, D., Bishop L.C., Potts, R., 2008. Oldowan behavior and raw material transport: perspectives from the Kanjera Formation. Journal of Archaeological Science 35, 2329–2345.

Boëda, E., 1995. Levallois: a volumetric construction, methods, a technique. In: Bar-Yosef, O., Dibble, H.L. (Eds.), Definition and Interpretation of Levallois Technology Prehistory Press, Madison, pp. 41–68.

Dauvois, M., 1976. Precis de dessin dynamique et structural des industries lithiques prehistoriques. Perigueux: Fanlac.

Delagnes, A., Roche, H., 2005. Late Pliocene hominid knapping skills: The case of Lokalalei 2C, West Turkana, Kenya. Journal of Human Evolution 48, 435–472.

de la Torre, I. 2004. Omo revisited. Evaluating the technological skills of Pliocene hominids. Current Anthropology 45, 439–465.

de la Torre, I., Mora, R., 2005. Unmodified lithic material at Olduvai Bed I: manuports or ecofacts? Journal of Archaeological Science 32, 273–285.

de la Torre, I., Mora, R., Dominguez-Rodrigo, M., de Luque, L., Alcala, L., 2003. The Oldowan industry of Peninj and its bearing on the reconstruction of the technological skills of Lower Pleistocene hominids. Journal of Human Evolution 44, 203–224.

Dibble, H.L., 1995. Middle Paleolithic scraper reduction: background, clarification, and review of the evidence to date. Journal of Archaeological Method and Theory 2, 299–368.

Ditchfield, P., Hicks, J., Plummer, T., Bishop, L.C., Potts, R., 1999. Current research on the Late Pliocene and Pleistocene deposits north of Homa Mountain, southwestern Kenya. Journal of Human Evolution 36, 123–150.

Hovers, E., 2003. Treading carefully: Site formation processes and Pliocene lithic technology. In: Martínez J., Mora R., de la Torre, I. (Eds.), Oldowan: Rather More than Smashing Stones. First Hominid Technology Workshop. Treballs d'Arqueologia. Barcelona: Centre d'Estudis del Patrimoni Arqueològic de la Prehistòria. Universitat Autònoma de Barcelona, Barcelona, pp. 145–164.

Hovers, E., Schollmeyer, K., Goldman, T., Eck, G.G., Reed, K.E., Johanson, D.C., Kimbel, W.H., 2002. Late Pliocene archaeological sites in Hadar, Ethiopia. Paper read in the meeting of the Paleoanthropology Society, Denver Colorado, March 2002. Journal of Human Evolution A17.

Huddleston, A., 1951. Geology of the Kisii District. Colony and Protectorate of Kenya, Mines and Geological Department 18.

Inizan, M.-L., Reduron-Ballinger, M. Roche, H., Tixier, J., 1999. Technology and Terminology of Knapped Stone. Vol. 5. Prehistoire de la Pierre Taillee. CREP, Nanterre.

Isaac, G.L., 1972. Some experiments in quantitative methods for characterizing assemblages of Acheulian artifacts. In: VIième Congrès Panafricain de Préhistoire et de l'Etude du Quaternaire, 1967. Imprimeries Réunies de Chambéry, Chambéry.

Isaac, G.L., Harris, J.W.K., 1997. The Stone Artifact Assemblages: A Comparative Study. In: Isaac, B., Isaac, G.L. (Eds.), Koobi Fora Research Project, vol. 5: Plio-Pleistocene Archaeology. Clarendon Press, Oxford, pp. 262–362.

Kimbel, W.H., Walter, R.C., Johanson, D.C., Reed, K.E., Aronson, J.L., Assefa, Z., Marean, C.W. Eck, G.G., Robe, R., Hovers, E., Rak, Y., Vondra, C., Yemane, T., York, D., Chen, Y. Evensen, N.M., Smith, P.E., 1996. Late Pliocene Homo and Oldowan tools from the Hadar Formation (Kada Hadar member), Ethiopia. Journal of Human Evolution 31, 549–561.

Kimura, Y., 1999. Tool-using strategies by early hominids at Bed II, Olduvai Gorge, Tanzania. Journal of Human Evolution 37, 807–831.

Kimura, Y., 2002. Examining time trends in the Oldowan technology at Beds I and II, Olduvai Gorge. Journal of Human Evolution 43, 291–321.

LeBas, M.J. 1977. Carbonatite-Nephelinite Volcanism: An African Case History. Cambridge: Cambridge University Press.

Martínez, M.M., 1998. Differential raw material use in the Middle Pleistocene of Spain: evidence from Sierra de Atapuerca, Torralba, Ambrona and Aridos. Cambridge Archaeological Journal 8, 15–28.

McCall, G.J.H., 1958. Geology of the Gwasi Area. Ministry of Commerce and Industry Geological Survey of Kenya 45.

McPherron, S.P., 1994. A Reduction Model for Variability in the Acheulian Biface Morphology. Doctoral, University of Pennsylvania.

McPherron, S.P., 2000. Handaxes as a measure of the mental capabilities of early hominids. Journal of Archaeological Science 27, 655–663.

Plummer, T., 2004. Flaked stones and old bones: biological and cultural evolution at the dawn of technology. American Journal of Physical Anthropology, 118–164.

Plummer, T., Bishop, L.C., Ditchfield, P., Hicks, J., 1999. Research on late Pliocene Oldowan sites at Kanjera South, Kenya. Journal of Human Evolution 36, 151–170.

Plummer, T.W., Bishop, L.C., Ditchfield, P.W., Ferraro, J.V., Kingston, J.D., Hertel, F., Braun, D.R., 2009. The environmental context of Oldowan hominin activities in Kanjera South, Kenya. In: Hovers, E. and Braun, D.R. (Eds.), Interdisciplinary Approaches to the Oldowan. Springer, Dordrecht, pp. 149–160.

Roche, H., Delagnes, A, Brugal, J.-P., Feibel, C.S., Kibunjia, M., Mourre, A., Texier, P.-J., 1999. Early hominid stone tool production and technical skill 2.34 Myr ago in west Turkana, Kenya. Nature 399, 57–60.

Roche, H., 2000. Variability of Pliocene Lithic Productions in East Africa. Acta Anthropological Sinica 19, 98–103.
Sahnouni, M., Schick, K.D., Toth, N., 1997. An experimental investigation into the nature of faceted limestone "spheroids" in the Early Palaeolithic. Journal of Archaeological Science 24, 701–713.
Saggerson, E.P., 1952. Geology of the Kisumu District. Republic of Kenya, Mines and Geological Department 21.
Semaw, S., 2000. The world's oldest stone artefacts from Gona, Ethiopia: their implications for understanding stone technology and patterns of human evolution between 2.6–1.5 million years ago. Journal of Archaeological Science 27, 1197–1214.
Shackleton, R.M., 1946. Geology of the Migori Gold Belt and Adjoining Areas. Colony and Protectorate of Kenya, Mining and Geological Department 10.
Stiles, D., 1991. Early hominid behaviour and culture tradition: raw material studies in Bed II, Olduvai Gorge. African Archaeological Review 9, 1–19.
Texier, P.-J., 1997. Ensemble lithique oldowayen sur quartz du site NY18 a Nyabusosi (province de Toro, Ouganda). Préhistoire Anthropologie Méditerraneenes 6, 227–237.
White, M.J,. 1998. On the significance of Acheulian biface variability in Southern Britain. Proceedings of the Prehistoric Society 64, 15–44.

10. Obsidian exploitation and utilization during the Oldowan at Melka Kunture (Ethiopia)

Marcello Piperno*
Dipartimento di Scienze Storiche
Archeologiche e Antropologiche dell'Antichità
Università di Roma "La Sapienza"
Via Palestro, 63 – 00185 Roma
Italy
m.piperno@tin.it

Carmine Collina
CEPAM UMR 6130 CNRS
Université Nice -Sophia Antipolis
250 rue Albert Einstein, bât. 1, 06560 Valbonne
France
And
Dipartimento di Scienze Storiche
Archeologiche e Antropologiche dell'Antichità
Università di Roma "La Sapienza"
Via Palestro, 63 – 00185 Roma
Italy
barnum_it@yahoo.fr

Rosalia Gallotti
Dipartimento di Scienze Storiche
Archeologiche e Antropologiche dell'Antichità
Università di Roma "La Sapienza"
Via Palestro, 63 – 00185 Roma
Italy
rosaliagallotti@yahoo.it

Jean-Paul Raynal
CNRS, UMR 5199 PACEA, IPGQ
Université de Bordeaux 1
Bâtiment B18, avenue des Facultés
33405 Talence
France
jpraynal@wanadoo.fr

Guy Kieffer
22 rue Paul Gaubin
63450 Le Crest
France
And
UMR 6042 CNRS

* Address for correspondence: m.piperno@tin.it

E. Hovers and D.R. Braun (eds.), *Interdisciplinary Approaches to the Oldowan,*
DOI: 10.1007/978-1-4020-9059-2,

Francois-Xavier le Bourdonnec
Institut de Recherche sur les Archéomatériaux
UMR 5060 CNRS-Université Bordeaux 3
Maison de l'Archéologie
Esplanade des Antilles
33607 Pessac
France
francois.lebourdonnec@etu.u-bordeaux3.fr

Gerard Poupeau
Institut de Recherche sur les Archéomatériaux
UMR 5060 CNRS-Université Bordeaux 3
Maison de l'Archéologie
Esplanade des Antilles
33607 Pessac
France
g.poupeau@u-bordeaux3.fr

Denis Geraads
UPR 2147 CNRS
44 rue de l'Amiral Mouchez
75014 Paris
France
dgeraads@ivry.cnrs.fr

Keywords: Oldowan, Melka Kunture, obsidian, exploitation strategies

Abstract

The Oldowan assemblages of Melka Kunture represent the earliest known example of obsidian utilization. The proximity of primary and secondary sources of Balchit obsidian, a high quality raw material easily available in large quantities, is a unique situation among East African Oldowan sites. Obsidian represents a large component of the lithic assemblages at Melka Kunture, not only during the Oldowan but during the Acheulian times as well. Other volcanic rocks are incorporated into the technological system at Melka Kunture such as basalts, ignimbrites, trachytes and trachybasalts, which present completely different characteristics for knapping.

10.1 Introduction

With a few exceptions, the exploitation of obsidian at Melka Kunture can be considered as a leitmotiv for more than 1.7 myr, because it represents the first utilization of this material during the Oldowan. Sites of various entities, including Oldowan sites (Karre I, Gombore I, Garba IV, Gombore Iγ), Acheulian and the Middle Stone Age sites (Garba XIIJ, Simbiro III, Gombore II, Garba III), with ages ranging between 1.7 and 0.2 Ma, show that obsidian was an important component of the lithic assemblages (Figure 10.1a). During the Late Stone Age and in recent times obsidian became the dominant raw material (Chavaillon et al. 1979; Chavaillon and Berthelet 2004; Chavaillon and Piperno 2004).

The obsidian-dominated Oldowan assemblages of Melka Kunture represent the earliest known example of systematic utilization of this raw material. At Balchit, 7 km north of Melka Kunture on the western border of the Main Ethiopian Rift, the primary source of obsidian is a dome-flow which belongs to the Pliocene rift margin silicic centers of the Wachacha Formation. However, Quaternary alluvial deposits constitute rich and numerous secondary sources in the area (Poupeau et al. 2004).

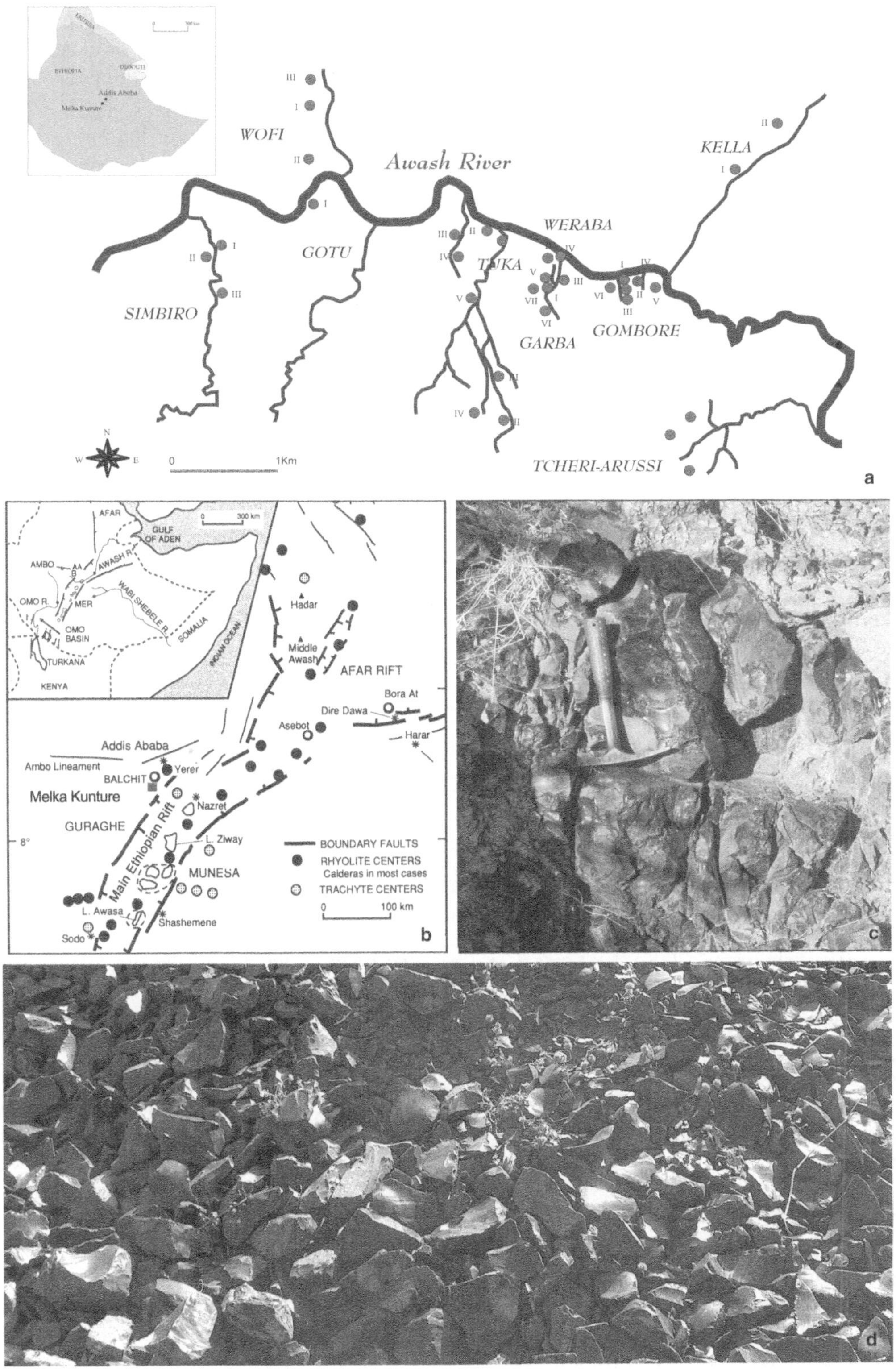

FIGURE 10.1. a: location of the Melka Kunture sites; b: location of the Balchit obsidian outcrops (general setting after WoldeGabriel et al. 1992); c: unweathered massive obsidian; d: view of obsidian debris at Balchit locality.

East Africa is one of the few geographic areas with abundant obsidian sources. Apart from the Ethiopian evidence, most of the sources are located in Kenya, close to the Lake Naivasha Basin and Mount Eburru. Potential, but apparently relatively minor sources of volcanic glass are present in the northern portions of Kenya, East of Lake Turkana and in the southern end of the Suguta Valley (Watkins 1981). The southern Kenyan Rift zone and northern Tanzania near Mount Kilimanjaro may also have been a significant source of obsidian (Merrick and Brown 1984).

The exploitation of obsidian is more or less continuous at Melka Kunture throughout the Acheulian. Otherwise it is known from only two other (Acheulian) sites — Kariandusi and Kilombe — around 0.7 Ma. However, even Kilombe only records a few pieces of worked obsidian, and at Kariandusi obsidian represents approximately 15% of the industry (Gowlett 1993; Gowlett and Crompton 1993). From the Middle Stone Age onwards, obsidian was frequently utilized in almost all East African sites (Merrick et al. 1994) and is generally dominant in Late Stone Age lithic assemblages in the region.

The physical properties of obsidian, in particular its mode of fracture, make it particularly suitable for the manufacture of many types of tools. In the case of Melka Kunture, the volcanic rocks utilized for knapping were different types of basalts, ignimbrites, trachytes and trachybasalts on one hand and obsidian on the other hand. These two groups of raw materials present completely different qualities for stone knapping. This unique situation raises a series of questions and could add important information to the current debate on the criteria that identify the Oldowan Industrial Complex (Texier 1995, 2005; de la Torre et al. 2003; Martínez-Moreno et al. 2003; de la Torre 2004; Delagne and Roche 2005; de la Torre and Mora 2005; Stout et al. 2005; Bishop et al. 2006).

10.2 Geological Background

The sites of Melka Kunture are located in a demi-graben depression that belongs to the Upper Awash Basin on the Ethiopian Plateau (Figure 10.1b). The Basin surface area is around 3,000 km^2 and it is delimited by Pliocene volcanoes. The main volcanic centers are Wachacha and Furi in the North, Boti and Agoiabi in the South. The eastern limit is marked by the main graben of the Ethiopian Rift belonging to the large East African Rift system (Mohr 1999). The Melka Kunture area is made up of valleys whose inner terraces resisted erosion. The visible thickness of these deposits is around 30 m, but the cumulative thickness of the various levels is about 100 m. A recent structural, tephrostratigraphic and lithostratigraphic approach provides new insights into the evolution of the environmental background for hominin activities in this area (Kieffer et al. 2002, 2004; Bardin et al. 2004; Raynal and Kieffer 2004).

Volcanism in the Melka Kunture region was characterized by multiple eruptions, correlated with the Mio-Plio-Pleistocene evolution of the Ethiopian Rift. Ancient episodes of this event are represented on the landscape as basaltic hills. With the exception of the products of the initial local eruptive manifestations, the secondary basaltic flows originated from volcanoes several tens of kilometres away from the site. The volcanoes were mostly explosive, as indicated by the differentiated nature of the magmas. Beginning about 4 or 5 Ma, these volcanoes underwent multiple eruptions. Some of the pyroclastic material arrived directly to localities that later became occupied by hominins. In particular, various ignimbrites, aerial ash and pumiceous fallouts from repeated phreato-magmatic eruptions fell on the sites. These types of eruptions have a high destructive strength. The periods of human occupation post-date these large eruptions that produced the wide sheets of welded ignimbrites. Nevertheless, very violent late eruptions have on several occasions completely erased any evidence of the presence of hominins along the course of the Upper Awash.

The valley of the Awash River has been a focus of hominin occupation since 4 to 5 Ma. The Awash regularly reestablished its course after each important volcanic episode and each time established a new basal level of erosion. The water flow of this river and its tributaries provided the sedimentary context of reworked volcanic materials that buried and preserved the archaeological sites within the Melka Kunture Formation. This sedimentation was a consequence of the reactivation of the border faults that provoked on several occasions the subsidence of the demi-graben. This process of burial was assisted by the input of pyroclastic materials during eruptions. The influx of sediment considerably increased the river's bed load. However, this cycle was largely controlled by the level of the Awash sill, at the exit of the Basin. This may have remained high for a long period of time, reducing the likelihood of local erosion. Its position upstream of the river gorges separated it until recently from the regressive erosion processes.

The alluvium of the right bank tributaries demonstrates an evolution in time related to the different geodynamic phases of the Melka Kunture fault and the associated volcanism. Those of the left bank recorded the different stages of dismantling of superficial formations covering the part of the valley between Melka Kunture and the Wachacha volcanic center.

10.3 The Volcanic Raw Materials

The lavas found near Melka Kunture are associated with the different volcanic episodes that occurred in the area over the last several million years. These lavas are abundant in the alluvial sediments of the Awash River and its tributaries in the vicinity of Melka Kunture. Moreover, the less fragile facies of the lavas cobbles are preserved in the different archaeological sites. We have identified (Kieffer et al. 2004): aphiric to porphyric basalts, microdoleritic basalts, trachybasalts, trachyandesites, trachytes, rhyolites including obsidian and various welded ignimbrites. We note here a magmatic bimodality, with melanocratic basalts on one hand and trachytes/rhyolites on the other. The intermediary facies seem less well represented, which is consistent with the regional magmatic phases (WoldeGabriel et al. 1992; Chernet et al. 1998). The different volcanic rocks have been introduced into the sites as cobbles, fragments or blocks. Rare small-sized flint and opal fragments are known, which are suitable for small tool shaping. These most probably correspond to precipitations of amorphous silica from hydrothermal circulations directly linked to volcanism. Among these different raw materials the use of obsidian is a distinctive feature of the oldowan assemblages of Melka Kunture. The closest primary obsidian sources are known at Balchit, seven kilometres north of the site.

The obsidian dome-flow of Balchit is spotted by extended flaking areas where cores, flakes, blades and debris have been accumulated on several thousands of square meters since prehistoric times. The first studies of the Balchit area and of the obsidian Late Stone Age assemblages were conducted in 1973 and in 1976 (Chavaillon 1976; Hivernel 1976; Hivernel-Guerre 1976; Soulier 1976). Since 1999, special attention was paid to obsidian artifacts and their primary and secondary sources. Analyses were performed on several obsidian samples from various sites including the outcrops of Balchit and reworked debris or pebbles and cobbles from different alluvial formations of the Awash River and its tributaries (Poupeau et al. 2004; Raynal et al. 2005).

The massif of Balchit belongs to the Pliocene Rift margin silicic centers of the Wachacha Formation, located on the western border of the Main Ethiopian Rift, in the Addis Ababa Rift Embayment (Figure 10.1b). Recently, the age of the massif has been established at 4.37±0.07 Ma by K-Ar measurements (Chernet et al. 1998). It is a flat dome-flow, outcropping over an area of about four square kilometres with a wide variety of eruptive facies. The formation is better exposed at the North-Northeast limit of the outcrop, in a gully a few metres deep. A well developed fluidal structure, almost vertical, could possibly indicate an extrusive flow or represent ramp structures in a flow. Perlites and greyish to white lithophysae are abundant in a sometimes perlitised finely banded lava; the lithophysae are either spherulithic growths of Feldspar or devitrified glass in which the original banding of the lava is still visible.

Amygdales up to 1 meter long of pure and massive obsidian are scattered among the lava dome-flow and preserved among the weathered rock (Figure 10.1c). The obsidian colour is dominantly black but locally blue, green, red and beige colours have been observed. It corresponds to an obsidian *sensu stricto*, and it is different from other volcanic glasses derived from quick cooling, such as the base of the nearby ignimbrites. The obsidian flow appears *in situ* only in peripheral banded and deformed facies found on the Jimjima Plateau and close to the village of Balchit. The entire structure could have been transformed by pumice formation and devitrification in a bright and fluidal lava, where subsisted amygdales veins and blocks of this obsidian. The unweathered lava that was selected for knapping is massive, uniformly black and very finely banded and breaks easily with conchoidal fracture, giving more or less translucent flakes with excellent cutting edges.

The obsidians from Melka Kunture are of calc-alkaline composition (Muir and Hivernel 1976). Four compositional types were recognized on the basis of concentrations derived from the ICP-MS analysis of 23 to 26 trace elements (Poupeau et al. 2004). Type A composition was isolated from two Balchit obsidians and six nearby samples from alluvial deposit of the Awash River. Three other samples from alluvial deposits have different compositions, labelled B, C and D. While the Type B composition differs from Type A group essentially by the content of rare earth elements (REE), Type C presents in general lower trace element content and Type D has a higher content of some trace elements, especially Y, Zr, Hf and the REE. More recently, Negash et al. (2006) analysed by XRF the composition of 10 Balchit obsidian samples. The contents obtained for Ti, Mn, Fe, Rb, Sr, Y, Zr and Nb are in excellent agreement with our Type A composition.

Thus to date, only one elemental composition was found for the 12 Balchit obsidians recently analysed by Poupeau et al. (2004) and Negash et al. (2006). The same composition was found for 10 artifacts from Gombore I and II and Garba IV (Negash et al. 2006) and two obsidians from alluvial deposits (Poupeau et al. 2003). These comparisons are significant, as an inter-laboratory comparison program in progress shows that these two groups of researchers obtain similar results on obsidians from various volcanic provinces (unpublished results). A comparison with the data by Muir and Hivernel (1976) would be more uncertain, although their results do not seem contradictory with a Type A composition.

Further field sampling and analyses will be necessary to understand if the B and C types of composition reflect only minor variations in Balchit obsidians or if they have to be referred to yet unidentified sources upstream or on the left bank Basin of the Awash River. The obsidian of Type D collected south of the Awash River in the Simbiro creek formation is a grey vitreous fluidal lava with a porphyric microstructure; quartz and feldspar crystals are oriented according to the fluidal structure. This material appears to have an ignimbrite facies and might be a bedsole rapidly cooled when in contact with the substratum. Similar ignimbrite facies were actually observed in this area.

Obsidian debris were widely distributed across the paleolandscape in secondary sources (Figure 10.1d) as products of erosion from the primary source. Large blocks, cobbles, and gravels are found in Quaternary alluviums and in minor river beds and form secondary sources which were available for prehistoric groups.

10.4 The Oldowan Sites

The Oldowan sites discovered at Melka Kunture are Karre I, Gombore I, Gombore Iγ and Garba IV, dated between 1.7 and 1.4 Ma (Schmitt et al. 1977; Westphal et al. 1979; Cressier 1980). The localities of Karre I and Gombore Iγ have been excavated over limited areas, respectively of around 10 m^2 and 12 m^2. The sites of Gombore I and Garba IV, characterized by several archaeological layers that have been extensively excavated and investigated in greater detail (Chavaillon 2004).

10.4.1 Methodology

In this paper we introduce the general features of the archaeological contexts related to the sites of Gombore I and Garba IV and give details of the preliminary results of a technological analysis of the lithic series of Layer E in Garba IV, which results from the research carried out in 2005. The lithic assemblages of Gombore IB2 and of Garba IVD are presented using Leakey's (1971) typological scheme, subsequently revised by Chavaillon et al. (2004). This scheme has been widely utilized during the eighties and nineties.

In Tables 10.1 and Tables 10.2 the term "shaping" refers to a "knapping operation carried out for the purpose of manufacturing a single artifact by sculpting the raw material in accordance with the desired form" (Inizan et al. 1999:138). The category "shaping products" includes the types classified by Chavaillon et al. (2004) as pebble tools (choppers, polyhedrons, spheroids, heavy end-scrapers, rabots, various pebble tools), handaxes and cleavers. The technological analyses in progress will try to verify if these artifacts were indeed the products of the activities of shaping or, alternately, the results (cores) of the activities of flaking. "Flaking" is used in this paper as "an intentional flaking of blocks of raw material, in order to obtain products that will either be subsequently shaped or retouched, or directly used without further modification" (Inizan et al. 1999:155).

The term "percussion material" refers to pebbles and cobbles with fairly numerous impact marks on one or several faces and to pebbles and cobbles with one, two, three or more fractures (Chavaillon 1979; Chavaillon et al. 2004).

The classification used for Garba IVE lithic assemblage is necessarily different in consideration of the different methodological approach; it considers the description of the structural criteria identifying the reduction process (de la Torre 2003; Delagne and Roche 2005). Until completion of the ongoing technological study, the lithic assemblages of Gombore IB2 and Garba IVD currently are not comparable with that of Garba IVE from a technological point of view.

10.4.2 Gombore I

This site, discovered in 1965 by J. Chavaillon, was excavated up until 1982, reaching an exposure of about 250 m^2. Various Oldowan layers were discovered. Layer B was divided into three sublayers: the most important (B2) is, at various locations, separated from the underlying layer B3 by a layer of volcanic ash. Two limited excavations have been carried out at this site in order to reach layers C and D. A left distal humerus of *Homo erectus* was discovered in 1976 in the southern sector of the excavation (Chavaillon and Coppens 1986). A total of 20,403 archaeological items have been discovered in B2, 10,411 (51%) of which are lithic artifacts, 1,832 (9%) faunal remains and 8,160 (40%) unmodified pebbles.

10.4.2.1 Stratigraphy and environment

The stratigraphy of the Gombore sequence has been recently reexamined (Raynal et al. 2004). The deposits at Gombore I belong to the lowest parts of the Melka Kunture Formation (Raynal et al. 2004), which consists mainly of the accumulation of pyroclastic air-fall and of pyroclastic material in secondary deposition within volcano-derived fluvial systems (Figure 10.2a). A test pit excavated at Gombore I shows the deposits below archaeological layer B of Chavaillon's (2004) excavations. From top to bottom these consist of:

- Silty coarse sands, very poorly sorted with a multi-modal grain-size curve.
- A tuff unit between archaeological layers B2 and B3 was sampled by J. Chavaillon during his excavations at Gombore I. This silty-sandy tuffaceous material is a rhyolitic air-fall, quasi-identical to the "Grazia tuff" at the bottom of the Garba IV series (Raynal, Kieffer 2004). The magnetic polarity of the tuff itself has not been established but units below and above show reverse polarity (Cressier 1980) corresponding to the Matuyama Reversal Chron.

We assume that earlier units form the bottom of the visible series, even if there is no observed connection between these deposits and those preserved and observed along the Gombore Gully sections.

The palynological analysis, based on several samples from Units IB and IC (Bonnefille 1976), clearly indicates thicket/scrub vegetation (bush) and the presence of a nearby forest. The percentage of *Juniperus* and *Podocarpus* pollens is high and these pollens are associated with those of high altitude thickets. The Gramineae represent 63% of the total pollens and the trees only 29%. Among the trees, *Juniperus* (80%) represents 21% of the total pollens while it accounts for only 1.5% nowadays. This indicates a humid climate, which cooled during the time of Gombore IB. A fragment of liana (*Cesalpinioxilon* sp.) was also recovered at the base of level B2.

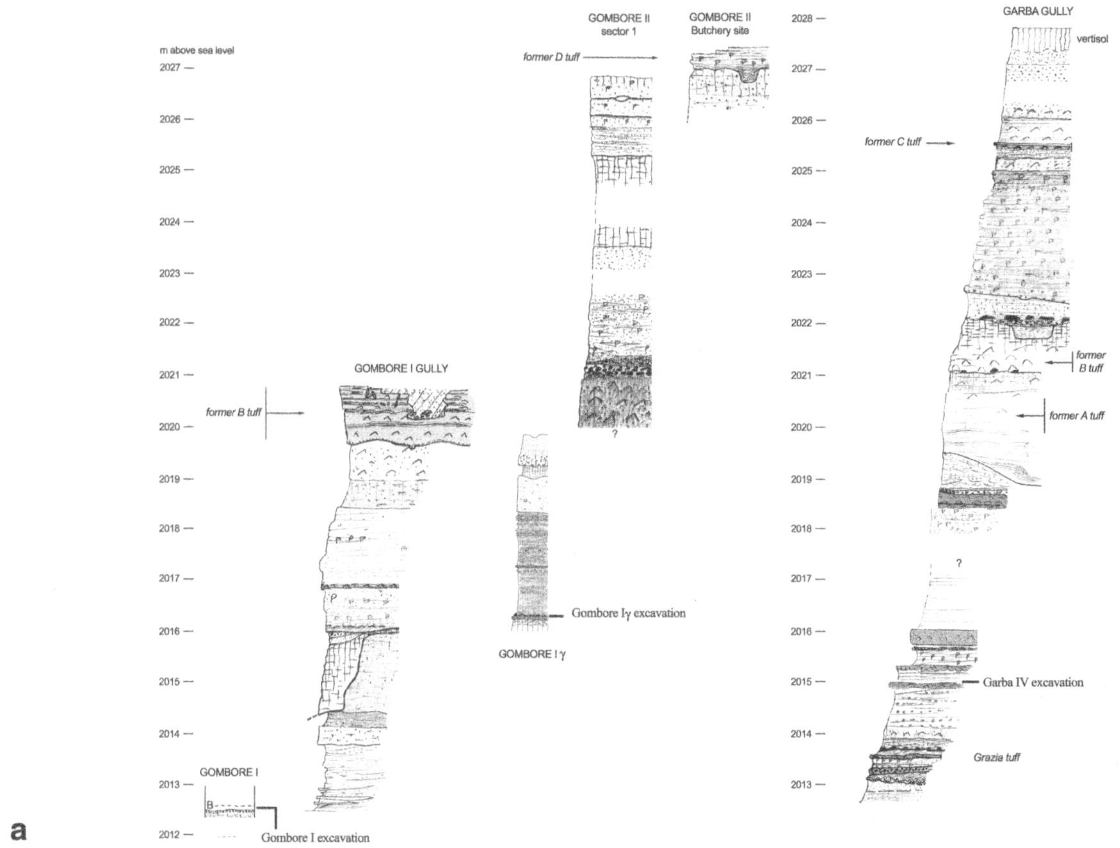

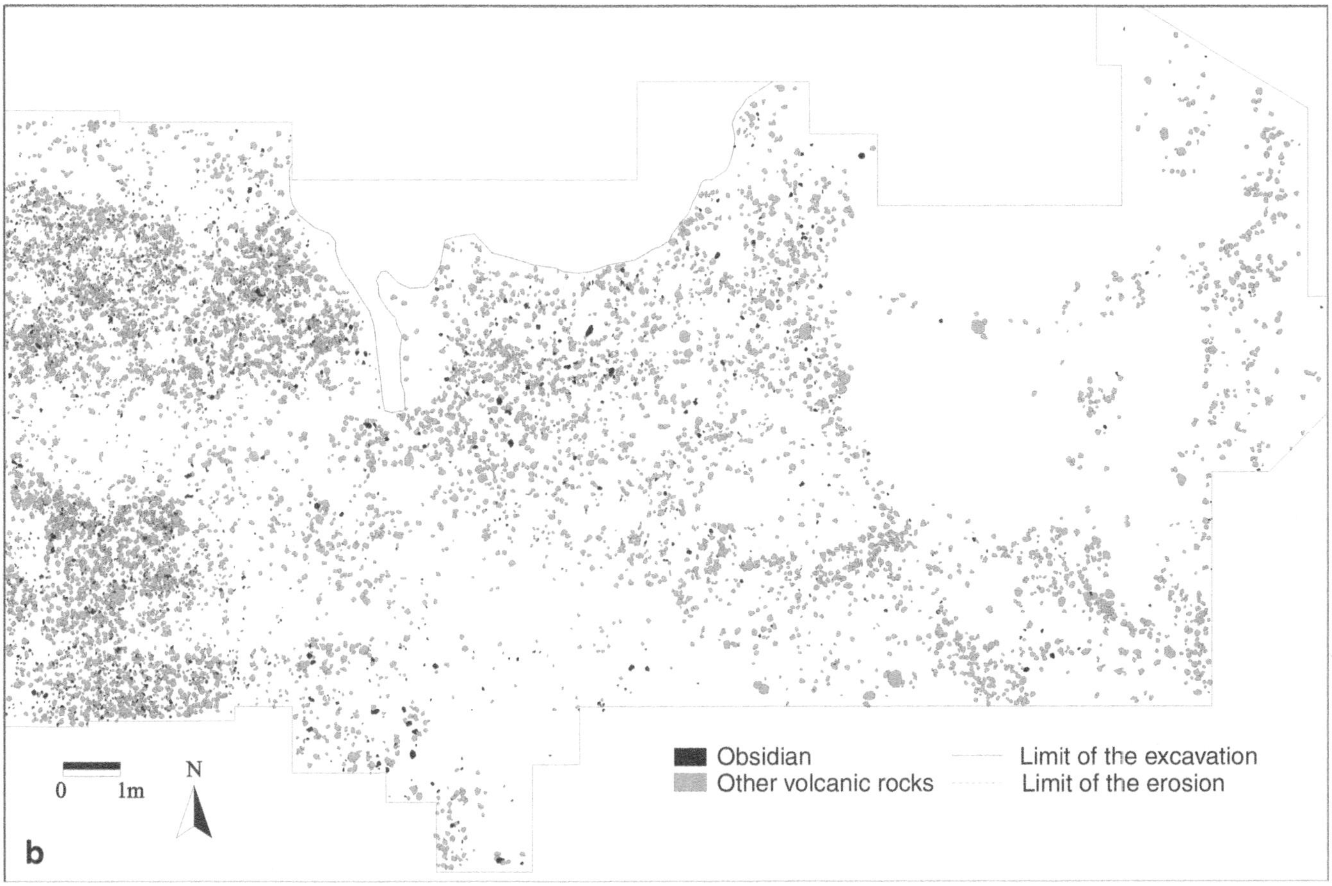

FIGURE 10.2 a: Gombore I series in the context of the Melka Kunture Formation; b: Gombore IB2, plan of the lithic artifacts

TABLE 10.1. Gombore IB2: lithic assemblage components, excluding percussion material.

	Obsidian			Others			Total	
Category	*N*	*% in relation to the total of the category*	*% in relation to the total of the obsidian artifacts*	*N*	*% in relation to the total of the category*	*% in relation to the total of artifacts on other volcanic rocks*	*N*	*%*
Shaping products	190	10.0	14.0	1719	90.0	57.9	1,909	44.1
Cores	120	48.0	8.8	130	52.0	4.4	250	5.8
Flakes	726	46.5	53.3	835	53.5	28.1	1,561	36
Retouched flakes	99	53.8	7.3	85	46.2	2.9	184	4.3
Tools on flake	172	48.5	12.6	183	51.5	6.2	355	8.2
Indeterminate fragments	54	77.1	4.0	16	22.9	0.5	70	1.6
Total	**1361**			**2968**			**4329**	

The fragmentary nature of the materials made it difficult to identify all faunal fragments, yet several specimens were identified to family or species level. The site nevertheless yielded well-preserved specimens of hippopotamus (*Hippopotamus amphibius*) and suids (*Metridiochoerus* and *Kolpochoerus*). There are also some remains of giraffe. Elephants (*Elephas recki*) and crocodiles are very rare. Bovids are abundant (*Connochaetes cf. gentryi* and *Damaliscus*) and equids are represented by *Hipparion*.

It is hard to draw precise biochronological estimates from this fragmentary fauna, but the *Connochaetes* is definitely not *C. olduvaiensis*, which appears at Olduvai Bed II, and more reminiscent of *C. gentryi* from earlier levels of Olduvai and the Turkana Basin (Geraads et al. 2004). Based on the magnetostratigraphy and the faunal evidence, we place the age of the site at 1.9–1.6 Ma.

10.4.2.2 Lithic production

The raw materials utilized for the lithic production (Figure 10.2b) consist of obsidian (19%) and other volcanic rocks previously described (81%).

A large part (58%) of the assemblage consists of percussion materials, namely cobbles and/or blocks of raw material with impact marks or fractures, whose anthropic origin is often difficult to identify. The most commonly used raw materials are the welded ignimbrites, various basalts and trachytes. obsidian is extremely rare. The composition of the remaining part of the lithic assemblage is shown in Table 10.1.

Excluding the percussion material, the obsidian constitutes 31.5% of the total lithic series and around 50% of each category, except for the shaping products (10%). The numbers of obsidian flakes and cores are more or less similar to those of other volcanic rocks. The production of obsidian flakes seems more or less equally related to shaping and flaking. Flakes of other volcanic rocks appear to be mainly the result of shaping. Fifty-one percent of all flakes are retouched and nearly half of them are on obsidian (Figure 10.3).

10.4.3 Garba IV

The site of Garba IV is located on the right bank of the Awash. The river has destroyed an unknown portion of the northern part of the site. The site, discovered by J. Chavaillon in 1972, was excavated from 1972 until 1982. The excavation and preliminary stratigraphic reconstruction allowed the identification of a sequence composed of five main archaeology-bearing stratigraphic units (C–G). The uppermost ones (C–D) were completely exposed, documented and removed.

In 1982 a 4 m^2 test trench was excavated below D in order to verify the thickness and nature of the underlying levels. Those were until then only seen in natural sections. An occupation level with fauna and lithic industry, corresponding to E, was reached about 60 cm below the base of D (Piperno and Bulgarelli 2004). In this level, a fragmented mandible of a 2/3 years old *Homo erectus* child was discovered (Condemi 2004).

A new series of excavations in 2005 explored E on a surface of 8 m^2 and F on a surface of 6 m^2, allowing the clarification of the lower part of the Garba IV stratigraphic sequence.

10.4.3.1 Lithostratigraphy

Three stratigraphic units (1–3 from the bottom to the top) were recognized within the sedimentary fluvial series (Figure 10.4a) with a thickness of about 3 m (Kieffer et al. 2002; Raynal and Kieffer 2004; Raynal et al. 2004).

Stratigraphic Unit 1 at the base of the sequence is a layer of greenish silty sands, typical sediment gravity flow deposit, only the top of which has been excavated.

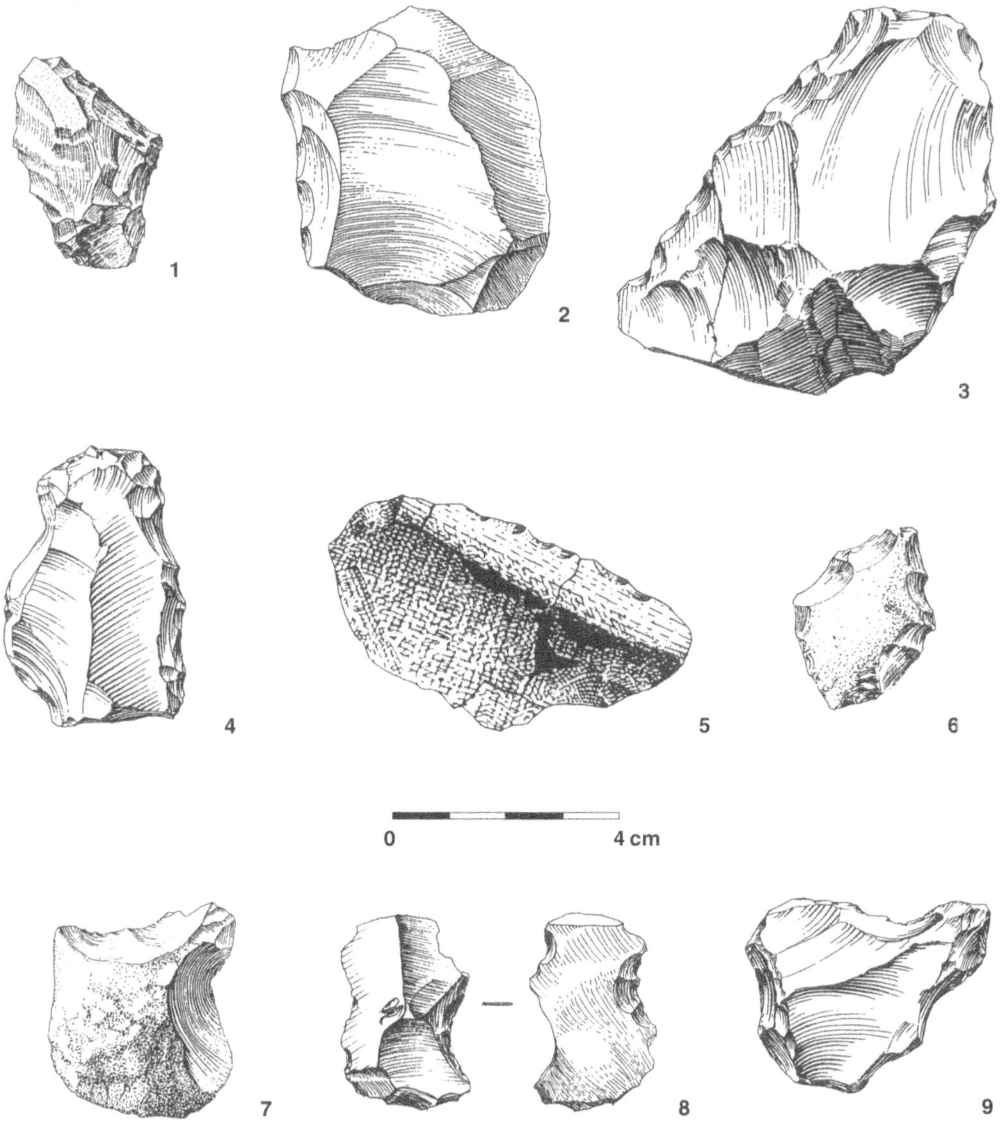

FIGURE 10.3. Gombore IB2. Obsidian. 1-3: straight simple side-scrapers; 4: convex simple side-scraper; 5: transversal convex side-scraper; 6: biconvex convergent side-scraper; 7-9: notches.

Stratigraphic Unit 2 is divided into 10 sub-units. From bottom to top it consists of:

1. Silty sands of a sediment gravity flow deposit containing the lower archaeostratigraphic unit G.
2. Light-gray ashy sands indicative of sediment-gravity to plane-bed flow deposit.
3. Silty sand layer of plane-bed flow deposit.
4. More or less coarse sands.
5. Grey pumiceous silty sand layer which includes archaeostratigraphic units F and E.
6. Gravel layer with obsidian granules.
7. Pumiceous sand layer with coarsely stratified Pumice, probably derived from a distant airfall ash.
8. White tuff of a distal direct airfall ash.
9. Fine sandy layer.
10. Green silty sands of sediment gravity flow deposit.

Stratigraphic Unit 3 is composed of eight sub-units. From bottom to top those are:

1. A clast supported massive gravel deposit that constitutes archaeostratigraphic unit D.
2. Upward refining (from coarse to fine) bedded sands.
3. Coarse massive sand containing archaeostratigraphic unit C.
4. Coarse sands and gravels with fine interbedded stratification, craddle and lenses, which indicate a lateral evolution of ephemeral shallow channels.
5. A cineritic layer of irregular thickness.
6. A layer of redeposited white tuff, muddy flow coulee type with surf structures.
7. A layer of obliquely stratified sands indicative of low flow regime.
8. A white sandy tuff (8).

Sub-units five to eight form a single reworked tuff unit.

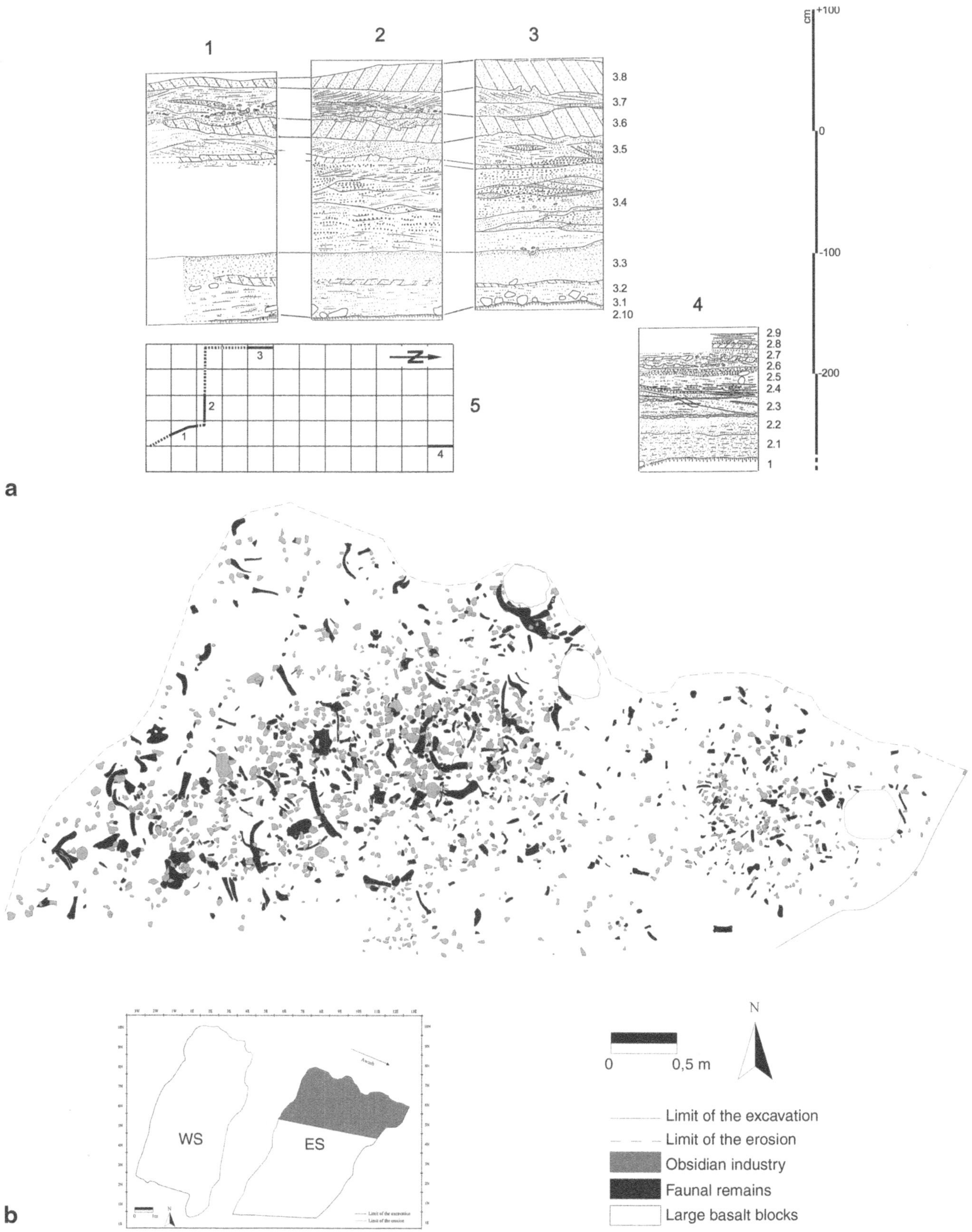

FIGURE 10.4. a: 1-4: stratigraphic sections at Garba IV. 5: localization of the 1-4 sections in the Western Sector of the excavated area at Garba IV; b: a detail of the northern part of the Eastern Sector of Garba IVD archaeological unit, where faunal remains and obsidian industry are concentrated around some large basalt blocks.

10.4.3.2 Archaeological Unit D

Archaeostratigraphic Unit D represents the most important paleo-surface of the entire sequence. The area, which was excavated systematically over about 100 m^2, is divided into two sectors, a western (WS) and an eastern one (ES), separated by an erosion channel formed when a tributary of the Awash River destroyed the central part of the deposit. Of the total number of 19,055 finds, 2,580 are faunal remains and one coprolite, 6,654 are unmodified pebbles and cobbles and 9,821 are lithic artifacts.

The most frequently identified animals (Geraads et al. 2004) are bovids (*Pelorovis* sp., *Connochaetes gentryi, Damaliscus strepsiceras, Gazella* sp.), equids (including *Hipparion* sp.), suids (*Kolpochoerus* and *Metridiochoerus*), giraffe (including *Sivatherium*), hippopotamus, elephant, and a primate related to the modern Gelada Baboon (*Theropithecus*). Some of the bovids differ little from those of Gombore I and suggest that the two sites are not much different in age. The *Connochaetes* is an endemic subspecies of the wildebeest found at Olduvai Bed I to lowermost middle Bed II, later replaced by *C. taurinus*. The striking abundance of its horn-cores, together with the extreme fragmentation of most bone remains, precludes natural deposition and implies anthropic intervention.

Our hypothesis for the genesis of the archaeological unit in the Garba IV D case is that of an anthropic intervention on a lag deposit which contained unmodified cobbles of ignimbrites or other volcanic rocks, including obsidian pieces. The hominins settled on the lag deposit where they used the raw material available on the spot to manufacture lithic tools, after the water had receded. The bone remains belong to animals hunted or scavenged by hominins and consumed at this locality. All elements abandoned on the site have been partly reworked when sands of unit 3.2 deposited (Raynal et al. 2004).

Superficially the spatial distribution of the finds appears totally random. Nevertheless, in contrast to Gombore IB2, when a detailed analysis of the spatial trends is applied to the different categories of finds, it appears possible that some zones of paleo-surface D may harbour a kind of "memory", if not of their original distribution, at least of some associations attributable to anthropogenic action (D'Andrea et al. 2002; Gallotti and Piperno 2004). Although the zones with high and low concentrations of objects remain more or less constant regardless of both raw material and object category, variations in frequencies could provide some interesting information.

The lower half of the WS is the largest high-density zone. Here, all the categories of finds (unmodified materials, lithic artifacts and faunal remains) are present in significant quantities, but without noteworthy differences in their spatial distribution. There is slightly less material in the upper half of the area, but it does not show any significant spatial patterns. The only noteworthy element — with the exception of a semicircular barren area — is the recurrent spatial association of large basalt blocks with large faunal remains. This association is also observable in the ES, but here one also finds a very high number of obsidian cores, modified and unmodified flakes, and especially obsidian tools on flake and small debris (Figure 10.4b). Indeed, the two areas in the ES with the highest densities seem to be connected to, and possibly determined by, the presence of the large basalt blocks more directly than in the WS. In the same zones one also finds many unmodified pebbles strewn all over the surface of the WS, while pebble tools and broken and battered pebbles are extremely rare.

The obsidian industry represents 41% of the total assemblage, which is a higher frequency than seen at Gombore IB2. Excluding the percussion material that is less abundant (27%) than at Gombore IB2, the obsidian artifacts constitute 55% of the total lithic assemblage.

TABLE 10.2. Garba IVD: lithic assemblage components, excluding percussion material.

	Obsidian			Others			Total	
Category	*N*	*% in relation to the total of the category*	*% in relation to the total of the obsidian artifacts*	*N*	*% in relation to the total of the category*	*% in relation to the total of artifacts on other volcanic rocks*	*N*	*%*
Shaping products	50	4.3	1.3	1,110	95.7	34.6	1,160	16.2
Cores	338	55.1	8.5	275	44.9	8.6	613	8.5
Flakes	2,558	64.3	64.3	1,420	35.7	44.1	3,978	55.3
Retouched flakes	212	73.9	5.3	75	26.1	2.3	287	4.0
Tools on flake	405	78.5	10.2	111	21.5	3.5	516	7.2
Debris	414	65.2	10.4	221	34.8	6.9	635	8.8
Total	**3,977**			**3,212**			**7,189**	

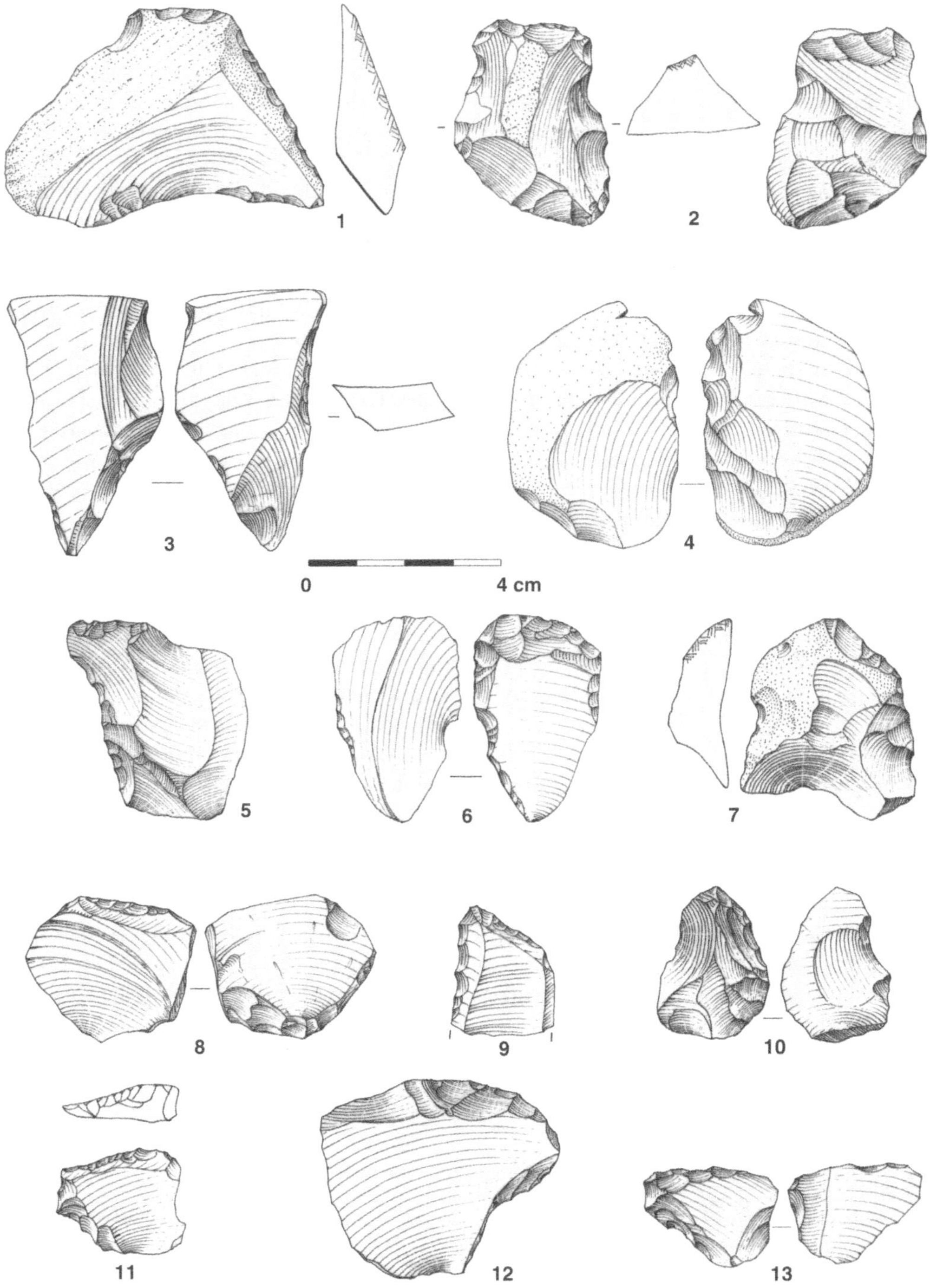

FIGURE 10.5. Garba IVD. Obsidian. 1: simple straight side-scraper; 2, 6, 8, 11, 12: transversal side-scrapers; 3: side-scraper with alternate retouch; 4: inverse side-scraper; 5: simple concave side-scraper; 7, 10: simple convex side-scrapers; 9: convergent side-scraper; 13: déjeté side-scraper.

As shown in Table 10.2, obsidian was used almost exclusively for flake production and only sporadically for pebble tools (rare obsidian choppers). The number of obsidian flakes is almost twice that of other volcanic rocks. Given the small number of shaping products, the flakes in this assemblage seem to be the result of flaking activities, compared to Gombore IB2. Moreover, while at Gombore IB2 the number of the retouched flakes (as defined by Chavaillon et al. 2004) and tools on flakes (e.g. side-scrapers, denticulates and notches) is not very different for obsidian compared to the other raw materials, at Garba IVD 75% of such artifacts are of obsidian. Compared to Gombore IB2, artifacts from Garba IVD are often surprisingly small with intense retouch (Figure 10.5).

TABLE 10.3. Garba IVE: lithic assemblage components.

Category	Obsidian	Others	Total
Flakes	26	11	37
Retouched pieces	2	1	3
Indeterminable fragments	5	12	17
Cores	15	6	21
Bolas	0	1	1
Modified pebbles and cobbles	0	18	18
Modified blocks	0	7	7
Broken pebbles and cobbles	0	11	11
Unmodified pebbles and cobbles	4	105	109
Unmodified blocks	0	1	1
Natural fragments	0	9	9
Total	**52**	**182**	**234**

10.4.3.3 Archaeological Unit E

Archaeostratigraphic Unit E was explored in 1982 on a surface of 4 m^2. At this locality 78 finds were discovered, of which 50 were faunal remains, 27 were lithic artifacts and one was a fragmentary mandible of a 2/3 years *Homo erectus* child (Condemi 2004; Piperno and Bulgarelli 2004). In 2005, a new series of excavations explored E on a surface of about 8 m^2, where 429 pieces were brought to light. Of these 194 were faunal elements.

The palaeontological material is extremely fragmentary and only 30% of the assemblage could be identified to families. Hippopotamus is the best represented genus with 55 finds, mostly dental fragments, followed by bovids, among which *Connochaetes* sp. and *Kobus* sp. were identifiable. Two metatarsal fragments of *Kobus*, discovered in the same square meter, could be refitted. Equids are represented by *Equus* sp.

The composition of the lithic assemblage from the 2005 excavation is shown in the Table 10.3.

Besides an angular block and nine natural fragments of welded ignimbrites, most of the lithic objects are unmodified cobbles: 96 are of welded ignimbrites, 23 of various basalts, four of obsidian, two of trachybasalt, one of trachyte and 10 of a still undetermined lava. The median size of unmodified cobbles is 74.5x 64x 44.5 mm for welded ignimbrite, 107.5x 77x 61 mm for basalts, 97x 87x 60 mm for other lavas and 75x 52x 48 mm for obsidian. Apart from the unmodified material, 11 broken pebbles, 18 modified pebbles and seven modified blocks are present and more than half are of welded ignimbrites (62%). In this case, the word "modified" is used to identify pebbles, cobbles and blocks presenting discrete and dispersed impact marks, but no percussion zone that would enable them to be defined as hammerstones.

Even if the welded ignimbrite is abundantly available in the alluvium at a close range to the site, it has not been used for knapping activities, except for four cortical flakes and the production of a bola stone, a unique testimony of shaping activity in this unit.

A classification of the cores has been carried out with the following criteria: identification of the number of the flaking platforms (unifacial=one flaking surface; bifacial=two flaking surfaces; multifacial=more than two flaking surfaces); the direction of flaking, which allows a distinction between unipolar, bipolar, centripetal (from a continuous peripheral striking platform) and multipolar (from separate striking platforms) exploitation, and the presence or absence of a distinct prepared striking platform. Considering these attributes, the core analysis allowed us to identify exploitation modalities, the management of volumes and the presence or absence of a hierarchical organization of the surfaces. The following flaking strategies have been identified for the two most commonly used raw material groups, namely obsidian and other volcanic rocks.

Six flaking strategies have been distinguished for obsidian (Figures 10.6, 10.7)

- Unifacial bipolar flaking. A single surface of the core has been flaked through two series of bipolar removals. Flaking is carried out from an unprepared striking platform on the cortical surface of a reworked pebble.
- Unifacial centripetal flaking. These cores present a single flaked surface by series of centripetal removals. The original blank consists of a reworked cobble whose cortical surface acts as the striking platform.
- Unifacial centripetal flaking with a prepared striking platform. These cores are characterized by a single flaked surface, through one or two series of centripetal removals which originate from a prepared striking platform generated by unipolar removals.
- Bifacial centripetal flaking. Two options have been recognized:

1) Flakes are alternatively removed from the two flaking surfaces which are convex and share an intersection plane; no hierarchical organization of the surfaces can be seen. This option fits the basic criteria first outlined by Boëda (1993) for the identification of the bifacial discoid method;
2) These cores present two parallel flaking surfaces with centripetal exploitation and several removals intended to rectify a peripheral striking platform rather than to create a hierarchical organization of the surfaces. In one case, in the final phase of core exploitation, a change in the role of the surfaces is obvious: the last removal is localized on the previous striking platform.

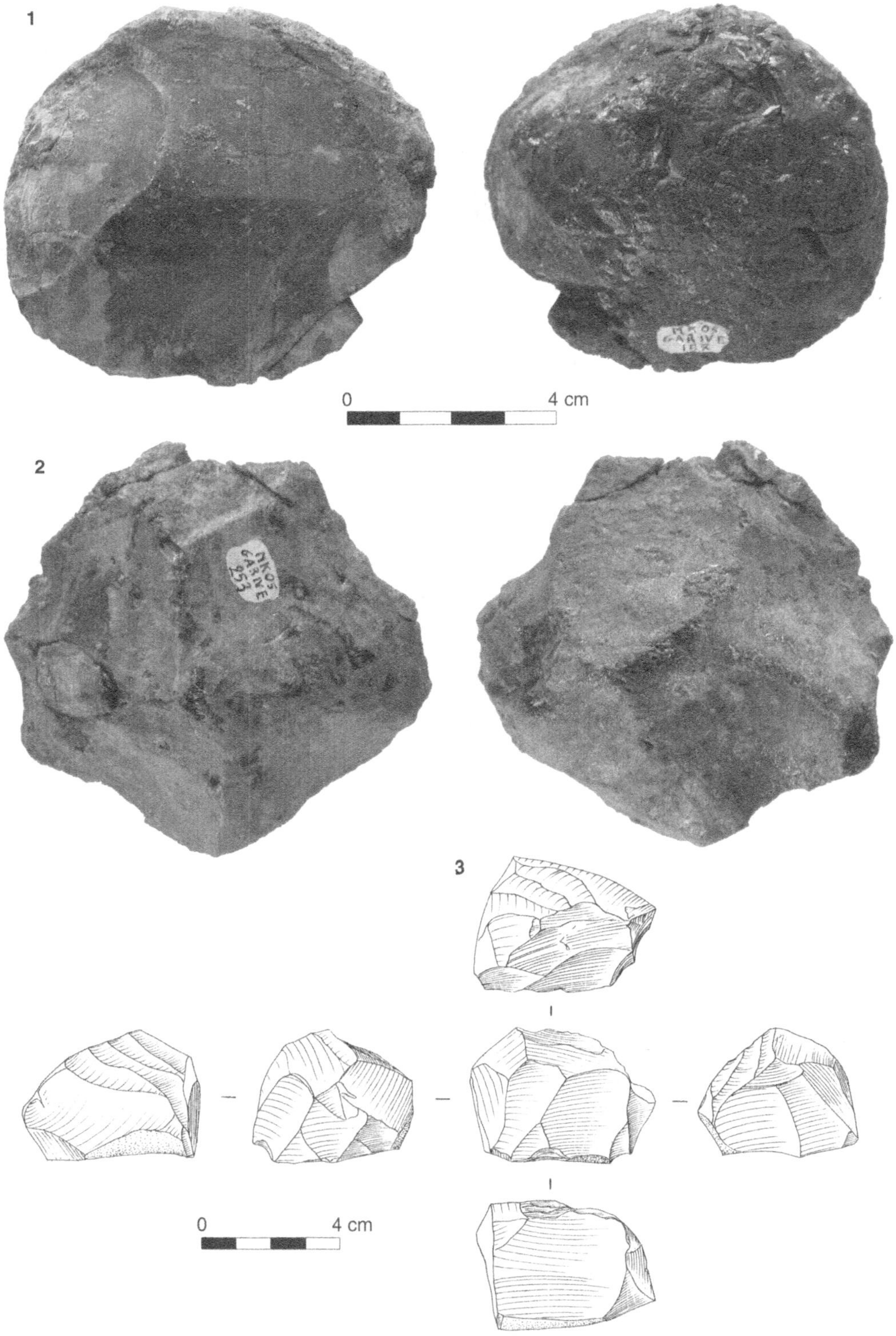

FIGURE 10.6. Cores from Garba IVE. Obsidian. 1: unifacial centripetal flaking; 2: bifacial centripetal flaking with alternate series of removals; 3: multifacial multipolar irregular flaking.

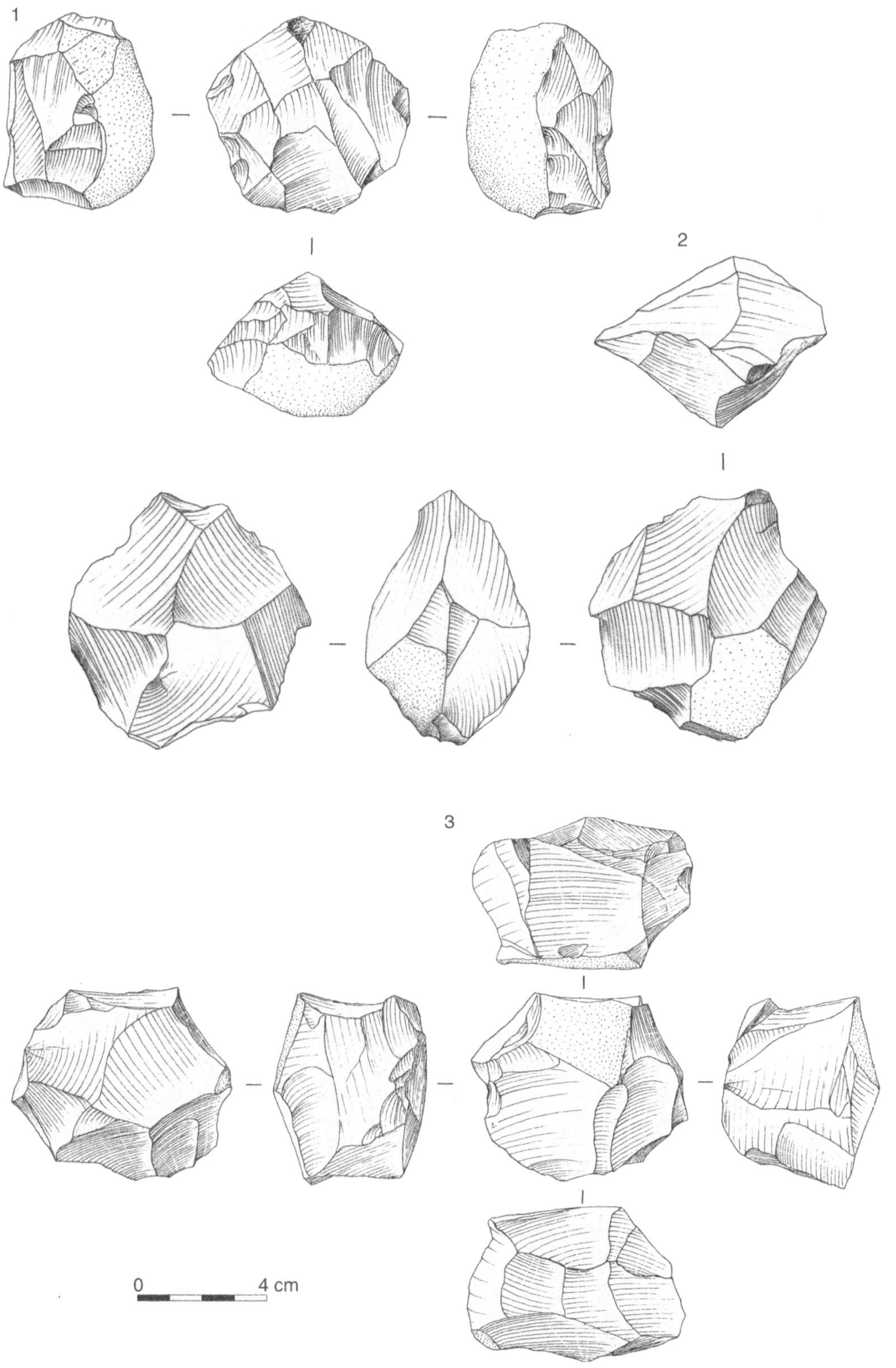

FIGURE 10.7. Cores from Garba IVE. Obsidian. 1: unifacial bipolar flaking; 2: bifacial centripetal flaking. The removals alternate on the two convex flaking surfaces which share an intersection plane; no hierarchical organization of the surfaces can be seen; 3: bifacial centripetal flaking with two parallel surfaces and several removals to rectify a peripheral striking platform; no hierarchical organization of the surfaces. In the final phase of core exploitation, a change in the role of the surfaces is visible: the last removal is localized on the previous striking platform.

- Partial bifacial flaking. Flaking starts from an unprepared striking surface, which is the cortical surface of a cobble, and continues alternatively from the negative scar.
- Multifacial multipolar irregular flaking. These cores present several independent flaking surfaces. In one case the natural blank is an angular block.

The median size of obsidian cores is 63x 51x 41 mm, consistent with the size of the available cobbles.

Three flaking strategies have been identified for other volcanic raw materials:

- Unipolar partial peripheral flaking. One core shows two series of perpendicular removals in relation with a striking platform on a pebble's natural surface.
- Partial bifacial flaking. The process is identical to the one described for obsidian.
- Multifacial multipolar irregular flaking. The cores are of relatively small size and present a great number of removals, suggesting that they were overexploited.

In the case of the multifacial multipolar irregular flaking on different volcanic rocks, the relatively small dimensions of the cores (58x 55.4x 43 mm) could be explained by the intensity of their exploitation.

The number of flakes is not in agreement with the mean number of negative scars observed on the cores (16 for obsidian and eight for other volcanic rocks), especially if we consider that the total number of scars does not realistically reflect the number of flakes detached and that some cores have been heavily exploited, as already pointed out by other authors (Braun et al. 2005, 2006; see also Kimura 2006). Flakes with cortical surface are the more abundant (62%) and indicate the prevalence of the initial reduction stage. In the case of obsidian, the flakes of full reduction stage present almost three negative scars on the dorsal face and the butts, generally cortical ones, are facetted for five items. Only three flakes have been retouched: one of them is a transversal side-scraper of small dimensions.

10.5 Conclusions

Obsidian represents a large, even major, component of the Oldowan lithic assemblages: 31.5% at Gombore IB2 and 55% at Garba IVD, when the percussion material is excluded from the artifact inventories. The data are not statistically significant for level E of Garba IV, due to the insufficient sample size from the small excavated area.

Generally, the patterns of raw material utilization during Oldowan times were determined by local resource availability. In fact, almost all obsidian cores were produced from small to medium sized cobbles and in only a few cases are the original blanks represented by angular blocks, suggesting preferential use of secondary sources. In the case of two obsidian handaxe-like artifact from Garba IVD, the original blanks were large flakes, whose manner of supply is difficult to explain based on the size of obsidian cobbles and blocks currently available in the alluvial deposits.

The utilization of the other volcanic rocks follows the same supply modalities; the only difference being the quantitative availability in the alluvium of unmodified cobbles and pebbles, of which very few specimens are obsidian.

On the basis of the general typological information from Gombore IB2 and Garba IVD, the detached pieces are the most abundant artifact type. In particular, the number of obsidian flakes at Garba IVD is almost twice that of other volcanic rocks. In contrast to Gombore IB2, these flakes seem to be the result of flaking rather than shaping activities. The relative importance of obsidian use at Garba IVD compared to Gombore IB2 is highlighted by the number of retouched pieces, 75% being in obsidian.

We would like to reemphasize that the technological analysis for these sites is forthcoming and that this article is essentially focused on describing the unusual early use of very high-quality raw materials. Nevertheless the data presented concerning the lithic assemblage of Garba IVE permits some tentative conclusion and suggests interesting directions for future research.

Preliminary results of the technological analysis of the lithic assemblage from Garba IVE demonstrate that the raw material composition is the major source of technological variation in contexts where the supply modalities are not dependent on the type of utilized volcanic material.

The differential flaking strategies identified for the two raw material types are characterized by some essential features. First of all, the variability of the obsidian flaking strategies is more significant compared to that on other volcanic rocks. This variability is not linked to the morphological and dimensional variability of the obsidian original blanks. Furthermore, a hierarchical organization of the surfaces is generally lacking, except in the case of unifacial centripetal flaking with prepared striking platforms. The variability of the flaking strategies identified the presence of prepared striking platforms and the change in the role of the surfaces, indicate a good control of the volumes of the obsidian cores.

The completion of the technological analysis of the Gombore IB2 and Garba IVD lithic assemblages will add further information about exploitation strategies in the context of the Oldowan of eastern Africa. In particular, these data allow a quantitative approach to define how the enormous availability of raw material determined a human response to the specificities of each of the lithic resources recognisable within the exploitation strategies.

Finally, the abundance and availability of raw material in the alluvium systems of Melka Kunture, the proximity of the primary supply source of Balchit obsidian and the high quality of this material, could explain the extensive use of obsidian in the lithic assemblages of Gombore IB2 and in particular

Garba IV. The local supply modality is in accordance with the evidence from the East African Oldowan sites of Olduvai (Hay 1976), Koobi Fora (Isaac et al. 1997), Peninj (de la Torre et al. 2003), Nyabusosi (Texier 1995) and Kanjera South (Plummer et al. 1999), which indicate that Oldowan tool-makers systematically used raw materials abundant in local channels. The few documented exceptions are known at Olduvai (Hay 1976; Potts 1988) and at the Late Pliocene KS1 and KS2 Beds at Kanjera South (Plummer et al. 1999; Braun et al. 2008), where the high quality raw material derives from non-local sources. In this framework the availability of a high quality raw material such as the Balchit obsidian in a primary source and in the alluvium system represents a further exception in the Oldowan context.

References

Bardin, G., Raynal, J.-P., Kieffer, G., 2004. Drainage patterns and regional morphostructure at Melka Kunture (Upper Awash, Ethiopia). In: Chavaillon, J., Piperno, M., (Eds.), Studies on the Early Paleolithic site of Melka Kunture, Ethiopia. Origines, Firenze, pp. 83–92.

Barrat, J.A., Blichert-Toft, J., Gillet, Ph., Keller, F., 2000. The differentiation of eucrites: the role of *in situ* crystallization. Meteoritic & Planetary Science 35, 1087–1100.

Bishop, L.C., Plummer, T.W., Ferraro, J.V., Braun, D., Ditchfield, P.W., Hertel, F., Kingston, J.D., Hicks, J., Potts, R., 2006. Recent research into Oldowan Hominin Activities at Kanjera South, Western Kenya. African Archaeological Review 23, 31–40.

Boëda, E., 1993. Le débitage discoïde et le débitage Levallois récurrent centripète. Bulletin de la Societé Préhistorique Française 90, 6, 392–404.

Bonnefille, R., 1976. Végétations et climats des temps oldowayens et acheuléens à Melka-Kunturé (Ethiopie). L'Ethiopie avant l'Histoire 1, CNRS, pp. 55–71.

Braun D.R., Plummer, T., Ditchfield, P., Ferraro, J.V., Maina, D., Bishop, L. C., Potts R., 2008. Oldowan behavior and raw material transport: perspectives from the Kanjera Formation. Journal of Archaeological Science 35, 2329–2345.

Braun, D.R., Tactikos, J.C., Ferraro, J.V., Harris, J.W.K., 2005. Flake recovery rates and inferences of Oldowan hominin behavior: a response to Kimura 1999, 2002. Journal of Human Evolution 48, 525–531.

Braun, D.R., Tactikos, J.C., Ferraro, J.V., Harris, J.W.K., 2006. Archaeological inference and Oldowan behavior. Journal of Human Evolution 51, 106–108.

Chavaillon, J., 1976. Melka-Kunturé : campagnes de fouilles 1971–1973. Annales d'Ethiopie 10, 1–18.

Chavaillon, J., 1979. Essai pour une typologie du matériel de percussion. Bulletin de la Société Préhistorique Française 76, 230–233.

Chavaillon, J., 2004. Discovery, geological introduction and study of percussion material and tools on pebble. In: Chavaillon, J., Piperno, M., (Eds.), Studies on the Early Paleolithic site of Melka Kunture, Ethiopia. Origines, Firenze, pp. 253–369.

Chavaillon, J., Berthelet, A., 2004. The archaeological sites of Melka Kunture. In: Chavaillon, J., Piperno, M., (Eds.), Studies on the Early Paleolithic site of Melka Kunture, Ethiopia. Origines, Firenze, pp. 25–80.

Chavaillon, J., Chavaillon, N., Berthelet, A., 2004. Methodology. In: Chavaillon, J., Piperno, M., (Eds.), Studies on the Early Paleolithic site of Melka Kunture, Ethiopia. Origines, Firenze, pp. 195–209.

Chavaillon, J., Chavaillon, N., Hours, F., Piperno, M., 1979. From the Oldowan to the Middle Stone Age at Melka-Kunture (Ethiopia). Understanding cultural changes. Quaternaria XXI, 87–114.

Chavaillon, J., Coppens, Y., 1986. Nouvelle découverte d'Homo erectus à Melka-Kunturé. Comptes Rendus de l'Académie des Sciences, Paris, 303, 99–104.

Chavaillon, J., Piperno, M., (Eds.) 2004. Studies on the Early Paleolithic site of Melka Kunture, Ethiopia. Origines, Firenze.

Chernet, T., Hart, W.K., Aronson, J.L., Walter, R.C., 1998. New age constraints on the timing of volcanism and tectonism in the northern Main Ethiopian Rift-southern Afar transition zone (Ethiopia). Journal of Volcanology and Geothermal Research 80, 267–280.

Condemi, S., 2004. The Garba IV E mandible. In: Chavaillon, J., Piperno, M., (Eds.), Studies on the Early Paleolithic site of Melka Kunture, Ethiopia. Origines, Firenze, pp. 687–701.

Cressier, P., 1980. Magnétostratigraphie du gisement pléistocène de Melka-Kunturé (Ethiopie). Datation des niveaux oldowayens et acheuléens. Ph. D. Dissertation. University Louis Pasteur. Strasbourg.

D'Andrea, A., Gallotti, R., Piperno, M., 2002. Taphonomic interpretation of the Developed Oldowan site of Garba IV (Melka Kunture, Ethiopia) through a GIS application. Antiquity 76, 991–1001.

de la Torre, I., 2004. Omo revisited. Current Anthropology 45, 439–465.

de la Torre, I., Mora, R., 2005. Technological Strategies in the Lower Pleistocene at Olduvai Beds I & II, Éditions Eraul, Liège.

de la Torre, I., Mora, R., Domínguez-Rodrigo, M., de Luque. L., Alcalá, L., 2003. The Oldowan industry of Peninj and its bearing on the reconstruction of the technological skills of Lower Pleistocene hominids. Journal of Human Evolution 44, 203–224.

Delagnes, A., Roche, H., 2005. Late Pliocene knapping skills: the case of Lokalalei 2C, West Turkana, Kenya. Journal of Human Evolution 48, 435–472.

Gallotti, R., Piperno, M., 2004. The site of Garba IV. Spatial analysis of the lithic material from level D. In: Chavaillon J., Piperno M. (Eds.), Studies on the Early Paleolithic site of Melka Kunture, Ethiopia. Origines, Firenze, pp. 599–635.

Geraads, D., Eisenmann, V., Petter, G., 2004. The large mammal fauna of the Oldowan sites of Melka Kunture. In: Chavaillon J., Piperno M. (Eds.), Studies on the Early Paleolithic site of Melka Kunture, Ethiopia. Origines, Firenze, pp. 169–192.

Gowlett, J.A.J., 1993. Le site Acheuleen de Kilombe: stratigraphie, geochronology, habitat et industrie lithique. L'Anthropologie 97, 69–84.

Gowlett, J.A.J., Crompton, R.H., 1994. Kariandusi: Acheulian morphology and the question of allometry. The African Archaeological Review 12, 3–42.

Hay, R.L., 1976. Geology of the Olduvai Gorge: A Study of Sedimentation in a Semiarid Basin. University of California Press, Berkeley, CA.

Hivernel, F., 1976. Etude préliminaire du gisement de Wofi III (Melka-Kunturé). L'Ethiopie avant l'Histoire 1, CNRS, 45–49.

Hivernel-Guerre, F., 1976. Les industries du Late Stone Age dans la région de Melka-Kunturé. In: Berhanou, J., Chavaillon, J., Sutton,

R. (Eds.), Proceedings of the VII Panafrican Congress of Quaternary Studies, 1971. Addis Ababa, pp. 93–98.

Inizan, M.-L., Reduron-Ballinger, M., Roche, H., Tixier, J., 1999. Technology of Knapped Stone. Préhistoire de la Pierre Taillée 5, CREP, Nanterre.

Isaac, G.L., Harris, J.W.K., Kroll, E., 1997. The stone artefact assemblage: a comparative study. In: Isaac, G., Isaac, B. (Eds.), Koobi Fora Research Project, Vol. 5: Plio-Pleistocene Archaeology. Clarendon Press, Oxford, pp. 262–299.

Kieffer, G., Raynal, J.-P., Bardin, G., 2002. Cadre structural et volcanologiques des sites du Paléolithique ancien de Melka Kunture (Awash, Ethiopie): premiers résultats. In: Raynal, J.-P., Albore-Livadie, C., Piperno, M. (Eds.), Hommes Et Volcans. De l'éruption à l'objet. Les Dossiers de l'Archéologie 2, Proceedings Symposium 15.2 , XIV UISPP Congress, Liège, 2001, pp. 77–92.

Kieffer, G., Raynal, J.-P., Bardin, G., 2004. Volcanic markers in coarse alluvium at Melka Kunture (Upper Awash, Ethiopia). In: Chavaillon, J., Piperno, M. (Eds.), Studies on the Early Paleolithic site of Melka Kunture, Ethiopia. Origines, Firenze, pp. 93–101.

Kimura, Y., 2006. On the critique of flake recovery rate and behavioral inference for early hominids: a reply to Braun et al. (2005). Journal of Human Evolution 51, 102–105.

Leakey, M.D., 1971. Olduvai Gorge, Vol. 3. Excavations in Bed I and II, 1960–1963. Cambridge University Press, Cambridge.

Martínez-Moreno, J., Mora, R., de La Torre, I., 2003. Oldowan, rather more than smashing stones: An introduction to "The Technology of First Humans" workshop. In: Martínez-Moreno, J., Mora, R., de la Torre, I. (Eds.), Oldowan: Rather more than smashing stones. First Hominid Technology Workshop, Bellaterra, December 2001, Barcelona. Treballs d´Arqueologia 9. Universitat Autònoma de Barcelona, Barcelona, pp. 11–35.

Merrick, H.V., Brown, F.H., 1984. obsidian sources and patterns of source utilization in Kenya and northern Tanzania: some initial findings. The African Archaeological Review 2, 129–152.

Merrick, H.V., Brown, F.H., Nash, W. P., 1994. Use and movement of obsidian in the Early and Middle Stone Ages of Kenya and northern Tanzania. In: Childs, S.T. (Ed.), Society, culture, and technology in Africa. MASCA 11, 29–44.

Mohr, P., 1999. Le système des rifts Africains. Environnement géologique et géographique. In: Gallay, A. (Ed.), Comment l'Homme? À la découverte des premiers hominidés d'Afrique de l'Est. Éditions Errance, Paris, pp. 231–288.

Muir, I.D., Hivernel, F., 1976. obsidians from the Melka-Konture prehistoric site, Ethiopia. Journal of Archaeological Science 3, 211–217.

Negash, A., Shackley, M.S., Alene, M., 2006. Source provenance of obsidian artifacts from the Early Stone Age (ESA) site of Melka Kunture, Ethiopia. Journal of Archaeological Science 33, 1647–1650.

Piperno, M., Bulgarelli G.M., 2004. The site of Garba IV. Excavations 1973–1982. In: Chavaillon, J., Piperno, M., (Eds.), Studies on the Early Paleolithic site of Melka Kunture, Ethiopia. Origines, Firenze, pp. 449–458.

Plummer, T.W., Bishop, L.C., Ditchfield, P., Hicks, J., 1999. Research on Late Pliocene Oldowan sites at Kanjera South, Kenya. Journal of Human Evolution 36, 151–170.

Potts, R., 1988. Early Hominid Activities at Olduvai. Aldine de Gruyter. New York.

Poupeau, G., Kieffer, G., Raynal, J.P., Milton, A., Delerue, S., 2004. Trace element geochemistry in Balchit obsidian (Upper Awash, Ethiopia). In: Chavaillon, J., Piperno, M. (Eds.), Studies on the Early Paleolithic site of Melka Kunture, Ethiopia. Origines, Firenze, pp.103–110.

Raynal, J.-P., Kieffer, G., 2004. Lithology, dynamism and volcanic successions at Melka Kunture (Upper Awash, Ethiopia). In: Chavaillon, J., Piperno, M., (Eds.), Studies on the Early Paleolithic site of Melka Kunture, Ethiopia. Origines, Firenze, pp. 111–135.

Raynal, J.-P., Kieffer, G., Bardin, G., 2004. Garba IV and the Melka Kunture Formation. A preliminary lithostratigraphic approach. In: Chavaillon, J., Piperno, M. (Eds.), Studies on the Early Paleolithic site of Melka Kunture, Ethiopia. Origines, Firenze, pp. 137–166.

Raynal, J.-P., Poupeau, G., Kieffer, G., Piperno, M., Milton, A.J., Delerue, S., Le Bourdonnec, F.-X., 2005. Balchit obsidian (Upper Awash, Ethiopia). Colloquium GMPCA "Archéométrie 2005". INSTN-CEA, Saclay, 19–22 April 2005.

Soulier, Ph., 1976. Prospection du site à obsidienne de Balchit. L'Ethiopie avant l'Histoire 1, CNRS, 51–54.

Stout, D., Quade, J., Semaw, S., Rogers, M., Levin, N.E., 2005. Raw material selectivity of the earliest stone toolmakers at Gona, Afar, Ethiopia. Journal of Human Evolution 48, 365–380.

Taieb, M., 1974. Evolution quaternaire du basin de l'Awash (Rift éthiopien et Afar). Ph D. Dissertation, University of Paris VI.

Texier, P.-J., 1995. The Oldowan assemblage from NY18 site at Nyabusosi (Toro-Uganda). Comptes rendus de l'Académie des Sciences, Paris, 320, 647–653.

Texier, P.-J., 2005. L'Oldowayen dans le Grand Rift occidental : le site NY18 à Nyabusosi, Ouganda. In: Sahnouni, M. (Ed.), Le Paléolithique en Afrique. L'histoire la plus longue. Éditions Errance, Paris, pp. 83–98.

Watkins, R.T., 1981. The geochemistry of Rhyolitic and Tholeiitic Rocks from East of Lake Turkana, Northern Kenya. M.Sc. Thesis, University of Leeds.

Westphal, M., Chavaillon, J., Jaeger, J.J., 1979. Magnétostratigraphie des dépots pléistocènes de Melka-Kunturé (Éthiopie) : premières données. Bulletin de la Société géologique de France XXI, 237–241.

WoldeGabriel, G., Aronson, J.A., Walter, R.C., 1992. Geochronology and distribution of silicic volcanic rocks of Plio-Pleistocene age from the central sector of the main Ethiopian Rift. Quaternary International 13/14, 69–76.

11. Are all Oldowan Sites Palimpsests? If so, what can they tell us about Hominid Carnivory?

Manuel Domínguez-Rodrigo*
Department of Prehistory
Complutense University
C/Prof. Aranguren s/n
28040 Madrid
Spain
m.dominguez.rodrigo@gmail.com

Keywords: hunting, scavenging, Oldowan stone tools, Pliocene, Early Pleistocene, site functionality, central-place

Abstract

The present work analyses the main Plio Pleistocene Oldowan sites from a taphonomic point of view. It is concluded that some sites present unrelated depositional histories between stone tools and bones, and that the preservation of faunal remains at other sites is insufficient to establish a functional link between both types of materials. It is argued that only one bone-plus-artifact site (FLK *Zinj*, Olduvai) can be claimed to be almost exclusively anthropogenic during the first million year of archeological record. Through the exceptional preservation of this site, it can further be argued that the carcass acquirement strategies used by hominids did not involve passive scavenging from carnivores but primary access to fleshed carcasses.

11.1 Introduction

Archaeologists have traditionally elaborated hominid behavioral models and interpretations using a Plio-Pleistocene archaeological record characterized by redundancy in the spatial association of stone tools and bones. The most widespread inferential step is the assumption of a functional link between both types of materials as the responsible cause for such a spatial association. Part of this inferential association stems from what could be called "the Zinj effect".

The abundant research carried out at the FLK *Zinj* site at Olduvai Gorge (Tanzania) has prompted some researchers to claim that the site represents the clearest example of "central place" in which it is taphonomically possible to ascertain that most of the carcass remains were transported, processed and exploited by hominids (Isaac 1978, 1983; Potts 1988; Bunn 1981, 1982, 1983, 1991; Bunn, Kroll 1986; Oliver 1994; Blumenschine 1995; Capaldo 1995, 1997; Rose and Marshall 1996; Domínguez-Rodrigo 1997a, b, 2002; Domínguez-Rodrigo and Pickering 2003; Pickering and Domínguez-Rodrigo 2006). The taphonomic sieve that this particular site has undergone for the past 25 years has brought confidence that the other "Type C" sites in Olduvai (and everywhere else) could also represent similar "central places". Most zooarchaeological research on the Olduvai sites tends to support the idea that "Type C" sites are hominid made (Potts 1988; Rose and Marshall 1996). Archaeologists are still arguing whether these sites could be the result of hominids hunting and selectively transporting those carcass parts (Isaac 1978, 1983, 1984; Bunn 1982, 1983, 1991; Bunn and Kroll 1986, 1988; Bunn and Ezzo 1993; Oliver 1994; Rose and Marshall 1996; Domínguez-Rodrigo 1997a, b, 2002; Domínguez-Rodrigo and Pickering 2003; Pickering et al. 2004, 2007), hominids transporting complete skeletons from partially defleshed carcasses (Capaldo 1995, 1997) or hominids scavenging the brain and marrow-bearing bones from defleshed carcasses at felid kills (Blumenschine 1986, 1991).

Type C sites are composed of stone tools and bones belonging to multiple animals (Isaac 1978). More than 20 years ago, Binford (1981, 1988) suggested that those sites could represent palimpsests in which various agents intervened independently, hominids being the most marginal of them. While this has been proven wrong for sites such as the FLK *Zinj* (Bunn 1982, 1991;

* Address for correspondence: m.dominguez.rodrigo@gmail.com

E. Hovers and D.R. Braun (eds.), *Interdisciplinary Approaches to the Oldowan*, DOI: 10.1007/978-1-4020-9059-2,

Bunn and Kroll 1986; Domínguez-Rodrigo and Barba 2006; Domínguez-Rodrigo et al. 2007a), the present work will prove that this is the case for most other Plio-Pleistocene archaeological sites. I believe that untested assumptions still make up an important part of widespread interpretations of early Oldowan sites. I also believe that behavioural reconstructions obtained for specific sites (e.g. FLK *Zinj*) cannot be uncritically applied to all sites in the same category (in this case, "Type C" assemblages). Each site needs to be addressed independently.

However important the archaeological information from the FLK *Zinj* has been to reconstruct hominid behavior, it has, nonetheless, limited archaeologists' views on early human adaptive patterns in more than one way: by neglecting regional variability (Potts 1994) and by deterring the development of explanatory frameworks to account for archaeological site diversity, both intra- and extra-regionally, in the same time periods. If the "Type C" sites have generated a fruitful discussion in the past two decades, researchers have hardly advanced on the interpretation of other sites suggesting alternative and complementary activities related to the use of stone tools that resulted in the formation of a different kind of archaeological record.

Isaac (1978, 1984) defined "Type A" sites as concentrations of artefacts without bones. "Type B" sites are clusters of stone artefacts mixed with bones from a single animal. "Type M" sites are concentrations of bones bearing artefact-inflicted damage, but without the presence of stone tools. It is not known whether "type A" sites represent places where hominids engaged in stone tool-making or sites where bones have not been preserved (Isaac 1984). Sites belonging to "Type B" have also been questioned (Bunn 1982; Potts 1988). Lack of taphonomic analyses also prevent "type M" sites from being clearly understood beyond the assertion that they represent brief episodes of hominids involved in animal carcass manipulation.

The present work will synthesize current information on the taphonomy of most published Oldowan sites from 2.6 Ma (Million years ago) to 1 Ma. It will focus on sites from Omo, Gona, Gadeb, Fejej, Hadar (Ethiopia), Olduvai, Peninj (Tanzania), Koobi Fora (Kenya) and Swartkrans Member 3 (South Africa). It will show that bone assemblages in most "Type C" sites were either naturally formed –involving processes that did not imply hominid participation in the accumulation of bones– or that preservation is too poor to make any functional link between hominid behavior and bone deposition at the site. This leaves archaeologists with a very scanty evidence to reconstruct early hominid behavior during the late Pliocene and early Pleistocene.

11.2 How can the Functional Relationship between Stone Tools and Bones be Reliably Established?

Some authors would argue that the presence of single or few cut-marked bones at any given site is enough to argue that a functional relationship between all lithics and bones at the site can be established. However, middle range theory argues otherwise. Archaeologists can only heuristically (*sensu* Lakatos 1978) support such a claim whenever the archaeological data fit referentially-derived frameworks of bone surface modification frequencies (taking into account differential bone preservation).

There has been a long-lasting debate with regard to the utility of skeletal part profiles for differentiating hominid-made bone assemblages from non-hominid-made bone assemblages, carcass transport strategies by hominids, order of access to carcasses, etc. For reasons that are discussed in detail in Domínguez-Rodrigo (2002) and Domínguez-Rodrigo et al. (2007a), the use of skeletal part profiles for these issues is subjected to equifinality and referential frameworks lacking resolution. Therefore, skeletal part profiles are removed in the present study from the mainstream of discussion.

The only way to link stone tools to all the bone remains found in any given site is to find the traces of butchery on bones in the expected proportion range according to referential experimentally and ethnoarchaeologically derived analogical frameworks. Carcasses that are butchered will result in determined ranges of frequencies of cut marks and percussion marks found on well-preserved bones as well as certain diagnostic criteria of bone breakage (notches, breakage planes).

A number of modern cut-marked long limb bone assemblages are available, derived from both ethnoarchaeological and experimental studies. Importantly, cut mark frequencies on limb bone specimens are remarkably consistent across these varied datasets, ranging between 10% – 30% of all long-limb bone specimens recovered after bone breakage by hammerstone (e.g. Bunn 1982 – in identifiable bones only; Lupo and O'Connell 2002). Although cut mark frequencies in Domínguez-Rodrigo's (1997a) experimental butchery dataset are beyond the range cited above, we note that those experiments were aimed explicitly at "complete" flesh removal. However, substantial scraps of flesh still adhere to defleshed bones after a more "typical" (defined as targetting bulk meat removal and not focusing on the complete removal of every single flesh scrap) episode of human butchery. Most usually (e.g. in ethnoarchaeological situations with actual hunter-gatherer butchers) butchers do not bother to remove these adhering scraps. Indeed, in the one experiment by Domínguez-Rodrigo (1997a, b) in which "typical" butchery was carried-out, cut mark frequencies, at *30%*, agree with the "typical" expected values.

Turning to hammerstone percussion marks, Blumenschine and Selvaggio (1991) calculated frequencies in human-first assemblages, modified subsequently by hyenas. Percussion mark frequencies on limb bone specimens (irrespective of bone portion) varied by carcass body size (following Brain's [1974, 1981] well-known scheme for antelopes): Size Class 1 and 2 limb bone assemblages showed percussion mark frequencies of *~30%*; Size Class 3 had frequencies of 21.5%; Size Class 4 had frequencies of 53.8%. Further, Blumenschine and Selvaggio (1991) established percussion mark to tooth mark ratios in these simulated hammerstone-to-carnivore assemblages that range from 1.5: 1 (in Size Class 3 carcasses) to 2: 1 (in Size Class 2 carcasses). In other words, percussion mark frequency is nearly double in cases in which hyenas act as secondary modifiers of limb bones first demarrowed and abandoned by humans.

Thus, this comprehensive consideration of both cut mark and percussion mark frequencies leads us to suggest that the *relative* contribution of hominids to the formation of prehistoric bone accumulations might be inferred if those frequencies are calculated in relevant fossil assemblages. If cut mark and percussion mark frequencies are lower than in the modern assemblages (discussed above) in which humans were the primary agents of carcass processing, this probably indicates a relatively (and absolutely) minor overall hominid contribution in the studied fossil fauna or the intervention of other processes not contemplated in modern experimental and ethnographic scenarios.

As a further supporting point, the study of tooth marks also sheds light on hominid-carnivore (hyena) interactions. With regard to bone surface modifications, Capaldo (1995) reports that ~70% of the appendicular specimens recovered in experiments in which hyenas were the primary carcass-ravaging agents display tooth marks. This percentage is only slightly lower than the 84% tooth marked value reported by Blumenschine (1988, 1995) in his hyena-first experiments. In addition, both researchers demonstrate that secondary access by hyenas to human-fractured bones results in low percentages of tooth-marked appendicular specimens (20% in Capaldo's experiments and 19% in Blumenschine's). More specifically, the percentages of tooth-marked mid-shaft portions are only 15% Capaldo's hyena-second experiments and 10% in Blumenschine's.

Another way of differentiating a hominid-made bone accumulation from a natural one is the study of bone fracture patterns and measurements. Bunn (1982) was the first to realize that humans and carnivores modify long bone shafts in a different way regarding the amount of section preserved in the fragments surviving marrow extraction. Generally, the articular ends of broken limb bones have a section of shaft attached that is rather complete in circumference (Bunn's type 3), while the isolated shaft fragments are less than complete in circumference –Bunn's types 2 (more than half the circumference) and 1 (less than half the circumference). Experiments have shown that when bones are broken either by humans or carnivores, the percentages of each of the types of circumferences represented in shafts may vary but the proportion of types 3 and 2 together with respect to type 1 range from 0.44 to 0.10; that is, specimens showing type 1 circumference section outnumber the other types (Bunn 1982).

Bone breakage can be studied by differentiating a double process: green/dry fracture and agent of breakage in the "green" broken sub-assemblage. The identification of green and dry (including diagenetic) breakages can be carried out by using Villa and Mahieu's (1991) criteria, namely, dry breaks result in abundant breakage planes that are longitudinal and transverse to the axis of the bone, the angle measured between the cortical and medulary surfaces is close to 90 degrees, and the breakage plane surface is uneven, with micro-step fractures and rough uneven texture in contrast with the more smooth surface of green broken specimens.

This analytical approach is further supported by the measurement of complete notches (Capaldo and Blumenschine 1994), the frequencies of notch types (Domínguez-Rodrigo et al. 2007a), and the measurements of angles in breakage planes (Alcántara et al. 2006; Pickering et al. 2005; Domínguez-Rodrigo et al. 2007a), to estimate if bone breakage was caused by dynamic (hammerstone) or static (carnivore) loading.

11.3 The Pliocene Sites

Sites older than 2.0 Ma play a crucial role in understanding early stone tool emergence and hominid behavior. Among these, the best preserved ones are found in Gona and Hadar (Ethiopia). Archaeological sites (KG 1–4) in Gona were first discovered by Roche and Tiercelin (1980). Research was resumed by Harris (1983) and Harris and Semaw (1989) with the discovery of new sites (WG 1–3) and, currently, by Semaw (1997–present). The archaeological sites (EG 10, EG 12 and WG) reported by Semaw et al. (1997) are in low-energy contexts (clays and silts) and their age is about 2.5 Ma. Therefore, they are among the oldest archaeological sites in the world. They show an excellent preservation of lithic remains. They are also practically devoid of fauna. Since 1997 until now, a series of sites dating from 2.0 to 2.5 Ma. were also found in Gona. Most of them are in undisturbed/primary position. Limited excavations were undertaken in some of these sites, among which OGS-6 and OGS-7 have yielded the oldest spatial association of stone tools and bones (Semaw et al. 2003) (see Table 11.1). At some of these localities, bones with cut marks have been found on the surface (Domínguez-Rodrigo et al. 2005).

However, that sites exist that contain freshly preserved stone tools but are devoid of fauna (with no detectable taphonomic cause for its absence)[1] and that a total of 12 cut-marked bones have been retrieved from an archaeological record spanning more than 500 Ka years, cast a doubt about:

[1]The state of preservation of these two types of finds need not necessarily be related, since they may respond to different preservation factors. It could still be argued stone tools were more immune to the destruction processes that destroyed bones. However, the fact that basalt and lavas are sensitive to the diagenetic processes of high acidity and alkalinity necessary to completely delete bones from the lithologic record argues otherwise since no such processes have been o bserved on the optimally-preserved stone tools at most of the Gona sites. Also the presence of bones in the same strata and in the immediate vicinity of several of these sites (understood as the loci where the tools are clustered) argues against post-burial diagenetic processes affecting exclusively the hypothetical bones accumulated within these stone tool clusters.

TABLE 11.1. Some of the Pliocene archaeological sites found at Gona

Site Name	Location	Age	Observations
ONG 3	Ounda Gona	2 my	Oldowan
OGS 3	Ounda Gona	2.5 my	Oldowan
OGS 6	Ounda Gona	2.6 my	Oldowan (cutmarks)
OGS 7	Ounda Gona	2.6-2.7 my	Oldowan
BSN 6	Busidima	2.5 my	Oldowan
DAN 1	Dana Aoule	2.5 my	Oldowan
DAN 2	Dana Aoule	2.3 my	Oldowan (cut marks)
DAN 5	Dana Aoule	> 2 my	Oldowan
EG 10	Kada Gona	2.5 my	Oldowan
EG 12	Kada Gona	2.5 my	Oldowan
EG 13	Kada Gona	2.5 my	Oldowan (cut marks)
EG 26	Kada Gona	2-1.5 my	Oldowan (cut marks)
WG 9	Kada Gona	2.5 my	Oldowan (cut marks)
WG3	Kada Gona	2,3 my	Oldowan

1. The functional association of stone tools and bones in most of these sites and,
2. That butchery was the main (exclusive) function of early stone tools.

The discovery in Bouri (Ethiopia) of a "Type M" archaeological site, in which bones dated at 2.5 Ma appear cutmarked, adds new data into the discussion (de Heinzelin et al. 1999). The evidence from Bouri shows only 3 cut-marked elements (de Heinzelin et al. 1999), which together with the Gona evidence is not enough to support the claim that stone tools emerged mainly as a result of hominid increased carnivory. There is no doubt that episodes of butchery occurred during the Pliocene, but the preservation of their evidence is so reduced that one could wonder about whether butchery was the only functional purpose of the Oldowan stone-toolkit.

Sites at Hadar also show a spatial relationship between stone artefacts and bone remains (Kimbel et al. 1996; Hovers et al. 2002). However, evidence of hominid modified bones is currently missing.

Archaeological sites in member F in Omo (Ethiopia) (2.3 Ma) seem to be mostly in a derived position, except two of them (FtJi 2 and Omo 123) (Chavaillon 1976; Merrick 1976; Merrick and Merrick 1976). They are found in channels in a context of medium to high energy. Material associations are, therefore, uncertain. Hydraulic transport is shown in several artifacts being polished and abraded in most of these sites. Given that several channels in the member F contain abundant faunal remains, the association of bones and stone tools in some of these sites may be fortuitous. Site FtJi 2 contains lithic artifacts and lacks associated bones. The Omo 123 site is situated adjacent to a channel and also contains stone tools without bones (Chavaillon 1976; Merrick 1976). Both are "Type A" sites. A couple of presumably older sites were found in the Omo research area (Omo 71 and Omo 84) in the member E. Omo 84, containing both stone tools and bones, is in an uncertain stratigraphic position and Omo 71 is composed of surface materials (one lithic artifact and several vertebrate fossils), the stratigraphic position of which is not clear (Howell et al. 1987). Furthermore, the anthropic origins of some of the sites have recently been questioned (de la Torre 2004).

Research in the Lusso Beds (Congo) resulted in the discovery of one site (Senga 5A) where stone tools where found associated with faunal remains. A date slightly over 2.2 Ma was inferred based on biochronology (Harris et al. 1987). However, the association of the materials and even the situation of the site and its age were questioned. It has been argued that it is a palimpsest in secondary position containing both old and more recent remains (Harris et al. 1990).

At Lokalalei (GaJh 5, LA1), West Turkana (Kenya), dated around 2.3–2.1 Ma, bones and stone tools were associated (Kibunjia 1994). The preservation of the fauna at the site makes it difficult to provide clear evidence that the bones and the stone tools are functionally associated. Another site, Lokalalei 2C, contains abundant lithic materials and fewer faunal specimens (Roche et al. 1999). The functional association between both types of materials still remains to be demonstrated. Most of the bone sample consists of reptile remains. Bones belonging to macrovertebrates are scarcer.

Something similar could be said about Fejej (Ethiopia). This Pliocene site, dated around 2.2 Ma, also shows a spatial association of stone tools and bones. No traces of hominid-imparted damage was observed on any bone surface. Claims of the anthropogenic origin the association are exclusively based on the presence of very few long limb bone fragments (most of them from an impala) showing green fractures (Echassoux et al. 2004). No percussion marks were discovered on any bone specimen, but some carnivore-gnawed remains were discovered nearby. For this reason, the green fractures documented could be due to the action of bone-breaking carnivores and not hominids.

The total published evidence of hominid modified bones from sites prior to 2.0 Ma amounts to 15 elements from over 20 archaeological sites spanning more than half a million years. This brings up the issue of meat-eating as the only cause of the emergence of stone tools. Beyries (1993) has analysed the edges of some of the lithic artefacts from Gona and claims they may have been used for plant processing. Were stone tool-making and meat-eating two separate and independent processes in human evolution? The only clear "Type C" site we know prior to 2.0 Ma is located at Kanjera. Artifacts found in association with diverse faunal remains, dated at about 2.1 Ma, were discovered in Excavation 1 (Plummer et al. 1999). However, a taphonomic study of the site is still missing. To sum up, the behavioural meaning of the Pliocene archaeological sites is at present open to varied interpretations.

11.4 The Early Pleistocene Sites

11.4.1 Koobi Fora

The existence of spatially associated bones and stone tools becomes more frequent in sites from 2.0 Ma to 1 Ma. However their preservation is often less than optimal. Faunal remains are often poorly preserved in archaeologically rich areas from this time period. The classical large array of archaeological sites in a prolific region such as Koobi Fora, for instance, is substantially affected by poor preservation of cortical surfaces from faunal remains which are crucial to unravel hominid-carnivore interaction in the formation of sites (Isaac 1997) notwithstanding published evidence of hominid involvement in carcass at FxJj50 with over a dozen cut-marked bones (Domínguez-Rodrigo 2002) and more recent unpublished discoveries involving cut-marked bone (Harris et al. 2002a, b). Of the published sites, no taphonomically-supported claims can be made about the authorship of the bone accumulations other than they often are spatially associated with stone tools[2].

11.4.2 Olduvai

Olduvai is an exceptional case among Oldowan sites. Bone preservation in most Olduvai sites ranges from moderately good to optimal. This explains why Olduvai is at the core of all the modeling of early human behaviour. The models of early hominid behavior that were academically used in the late 1960s and early 1970s were based on information from the archaeology of Olduvai Gorge (Leakey 1971). These models were expanded in Koobi Fora (Isaac 1978), but both the subsequent critique of these models and their restatement were based on the study of sites from Olduvai Gorge (Binford 1981, 1985; Bunn 1981; Potts 1982, 1988; Bunn, Kroll 1986; Blumenschine 1995; Rose, Marshall 1995; Domínguez-Rodrigo 1997a, 2002; Plummer 2004). Most interpretations based on the Olduvai materials have not rigorously established that the stone tool and bone assemblages are functionally related. Leakey (1971) and Isaac (1978) posited that hominids created bone accumulations through the establishment of camps and butchering spots. Binford (1981) claimed that stone tools at Olduvai sites were discarded at scavenging places. Potts (1988) argued that stone tools were crafted *in situ* from previously cached raw material

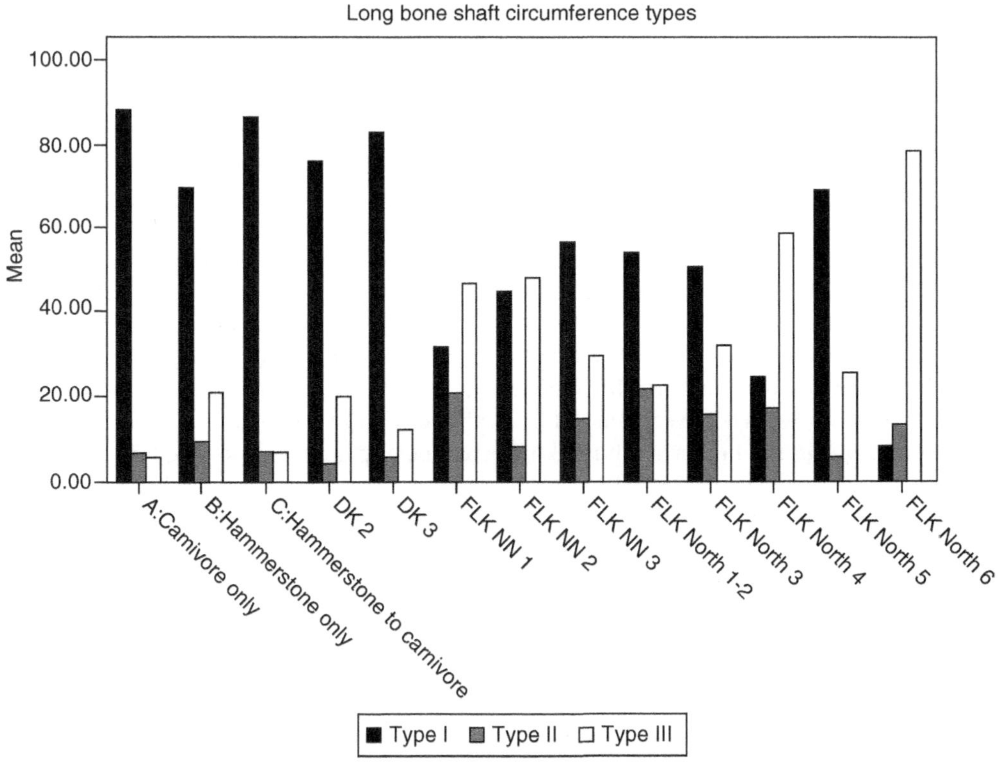

FIGURE 11.1. Distribution (in percentages) of the different types of bovid long bone circumference types (Bunn 1982) in experimental assemblages and at the Olduvai Bed I sites. Data for experimental assemblages are from Marean and Spencer (1991) and Marean et al. (2004). Data for DK, FLK N 5 and FLK NN 2 are from Egeland (in Domínguez-Rodrigo et al. 2007a)

[2]As this work was in press, a new study of two recently found sites with cut-marked bones, FwJj14 and GaJi14, was published (Pobiner et al. 2008).

when carcasses were brought into the site and used for butchery. An extended taphonomic study (the most complete up to date) of all the Bed I sites suggests that most of the previous assertions lack heuristics and do not fit well with current taphonomic evidence (Domínguez-Rodrigo et al. 2007a).

One feature that immediately catches the eye when analysing the Olduvai Bed I sites is that most of them do not fit any experimental pattern of bone breakage by hyenas or humans. Complete bones or elements with more than half of shaft circumference from long limb bones seem to predominate in several sites (Fig. 11.1). This implies that a significant portion of complete bones have survived breakage at these sites. This appears to be unrelated to tooth marking rates on long limb bones (Fig. 11.2). However, if we compare tooth mark percentages with bone-breaking intensity (tallied as the proportion of complete bones versus those that are fragmented, expressed in MNE) in all the FLK N levels, for instance, we can observe a positive correlation (Spearman's rho = 0.55; although p [0.2] is high because of the small sample size). The first question that comes to mind is: why did hominids leave so many bones unbroken if they were supposedly the main responsible agent in their accumulation and exploitation?

A look at tooth mark frequencies is revealing. Basically, long limb bones at all sites are tooth-marked at higher percentages than those obtained in hominid-carnivore (hyena) scenarios for all limb bones and, especially, for mid-shaft sections (Fig. 11.3). This suggests that when hyenas accessed previously discarded remains, they did not break all the limb bones. This excludes them *de facto* from being the primary agents of bone accumulation. Had that been the case, a much higher incidence of tooth marks should be observed both on mid-shafts and all long limb bone portions together. The documented tooth-marking seems to have been the result of a combined process of partial bone breaking, which includes the preservation of complete unmodified bones (Figures 11.2 and 11.3[3]). This requires the participation of other agents besides hyenas.

The Olduvai Bed I sites show a much better match with tooth mark rates documented in unbroken felid-modified bone assemblages (Fig. 11.3). However, with the exception of FLK North 6, they differ from those in the higher tooth-marking on mid-shaft sections. This can only be explained by models in which hyenas must have partially ravaged bones accumulated by another agent (very likely, felids) that abandoned most limb bones unbroken. Given the extremely high number of complete bones surviving breakage at these sites, and the intermediate tooth mark frequencies (between carnivore-only and hominid-carnivore), it can be argued that an initial agent accumulated carcasses over extensive periods of time and hyenas accessed them only intermittently, breaking only a portion of the assemblage. This has been supported by the distribution of notch types, measurements of notches and breakage planes in all Bed I sites but FLK *Zinj* (see below; Domínguez-Rodrigo et al. 2007a). The surprising observation is that the virtual lack of percussion marks excludes hominids from the list of agents that exploited marrow at these sites, whereas the virtual lack of cut marks also exclude them from the list of candidates for carcass defleshing.

[3]Hypothesis testing in modern science is based on Popper's demarcation criteria requiring each hypothesis to be articulated around "refutable" premises which must be subjected to testing prior to its resulting rejection or acceptance

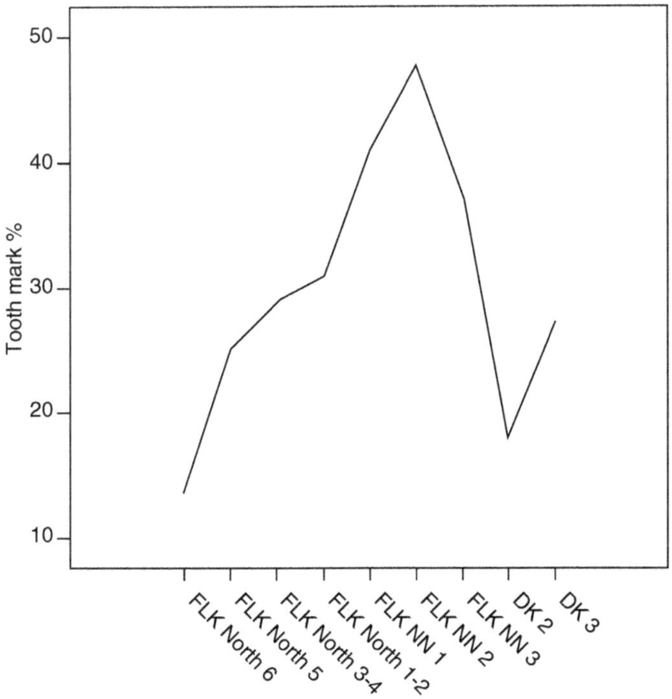

Figure 11.2. Tooth mark frequencies at each Olduvai Bed I site. Data for DK, FLK N 5 and FLK NN 2 are from Egeland (in Domínguez-Rodrigo et al. 2007a).

So, if hominids did not initially accumulate and butcher the carcasses accumulated at these sites, who did?

Domínguez-Rodrigo et al. (2007a) have provided an intriguing answer to this question: felids. The intermediate tooth mark frequency and the high proportion of complete bones are only supportive arguments to a more direct evidence of felid intervention in these assemblages, namely the particular pattern of bone modification on complete bones. It has previously been shown that felids modify carcasses in a distinctively different way from hyenas (Domínguez-Rodrigo et al. 2007b).

It has also been stressed that within each assemblage from the Olduvai sites there are two sub-assemblages. On the one hand, there are carcasses whose bones have undergone intense ravaging by hyenas (right upper corner of Figure 11.4), while others are so complete that it seems that hyenas never were responsible for their accumulation and modification (left side of Fig. 11.4). Bones ravaged by hyenas show typical signatures of deletion of cancellous epiphyses and of mid-shaft breakage in several stages, until the shaft is reduced to a near-epiphyseal section of the element. This behavior is responsible for the tooth mark frequency and distribution observed and for all notches documented. The intriguing part is the modifications observed in complete elements, which are more abundant in small carcasses (one out of three

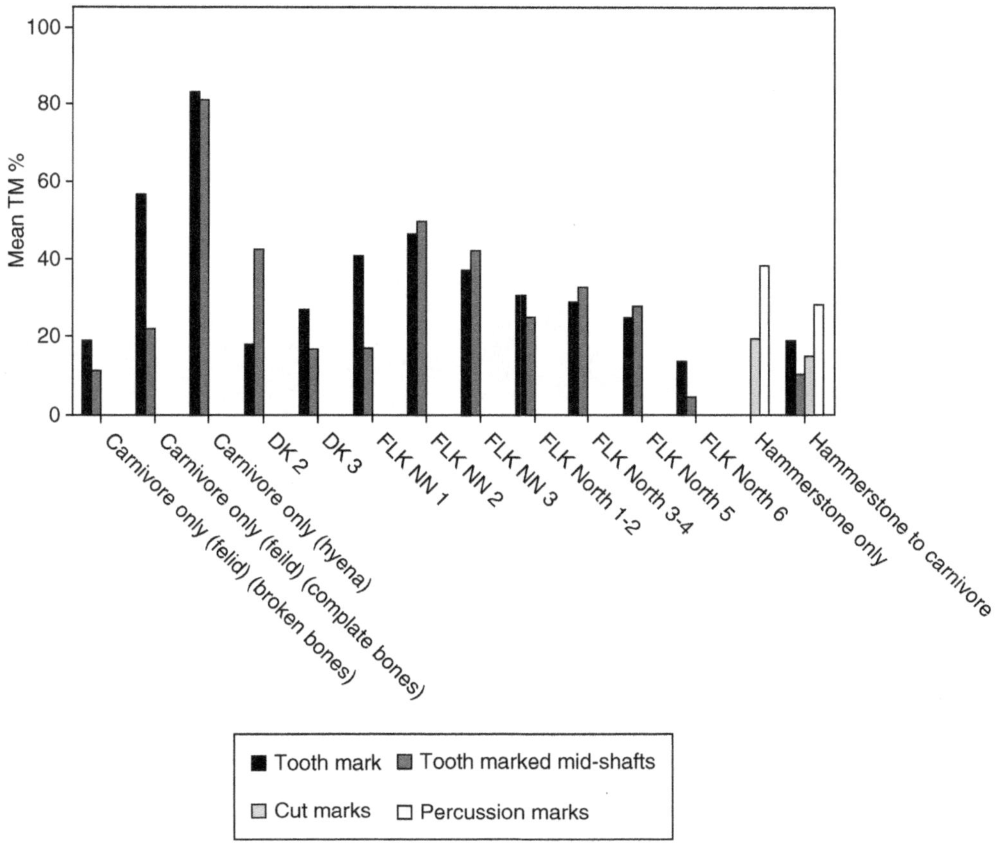

FIGURE 11.3. Frequencies of bone surface modifications on long limb bones at Olduvai Bed I sites and experimental assemblages. CM, cut marks; TM, tooth marks; PM, percussion marks; msh, mid-shafts. Data for experimental assemblages are from Blumenschine (1988, 1995) for carnivore-only, hammerstone only and hammerstone-carnivore and Domínguez-Rodrigo et al. (2007b) for felid-modified assemblages. Data for DK, FLK N 5 and FLK NN 2 are from Egeland (in Domínguez-Rodrigo et al. 2007a).

limb bones was retrieved complete in some sites). Within the complete elements, some show a patterning of furrowing on certain ends (left side of Fig. 11.4). Humeri in small and medium-sized carcasses may appear slightly furrowed on their proximal ends (more heavily so in small animals), affecting either tubercles or part of the articular surfaces. A very common pattern of furrowing is documented on the distal mesio-caudal epicondylar area in both small and large carcasses. This is connected to the furrowing observed in the oleocranon of ulnae, also very common. Complete tibiae also show some minor furrowing on the proximal epiphysis (mainly on the crest) or, depending on carcass size, complete deletion of just the proximal epiphysis. This pattern contrasts with that observed in bones modified by hyenas, especially with bones from small carcasses. For humeri, for instance, hyenas exhibit a pattern of distal epiphyseal deletion which begins with the lateral condyle and trochlea affecting both the cranial and caudal sides equally (Kerbis, personal communication), instead of the medial epicondyle. A pattern like the one described for the Olduvai bones has been reported for felids (Domínguez-Rodrigo et al. 2007a).

Of the extant large-bodied carnivores, felids are the most specialized flesh eaters, with teeth developed almost exclusively for meat-slicing (e.g. Turner and Anton 1997). There are only very weakly expressed mechanical adaptations for bone breaking in the felid dentition. Behavioral observations of carcass consumption by large cats support the tooth morphology-based inference that felids will likely avoid the intentional contact of teeth with bone (Turner and Anton 1997). This, in turn, leads to the prediction that felids will impart fewer tooth marks on bone assemblages than will hyenas (e.g. Brain 1981). Despite this, felids can conspicuously modify long limb bones. The pattern of felid-modified bones is mostly recognized in 3 areas: caudal mesio-distal humerus, proximal ulna and proximal tibia (see lower right corner of Fig. 11.4).

The fact that these modifications are systematically found in several complete elements from both carcass sizes in Olduvai supports the interpretation that those elements belong to carcasses modified by felids as a result of defleshing and, very likely, also transported by them to the site. However, if there is an anatomical section where felid bone modification is observed in detail (given the relative absence of hyena intervention in the surviving elements), that is on the axial skeleton; more specifically, on vertebrae. A large fraction of the vertebrae (between 20% and 30%, depending on the site) that

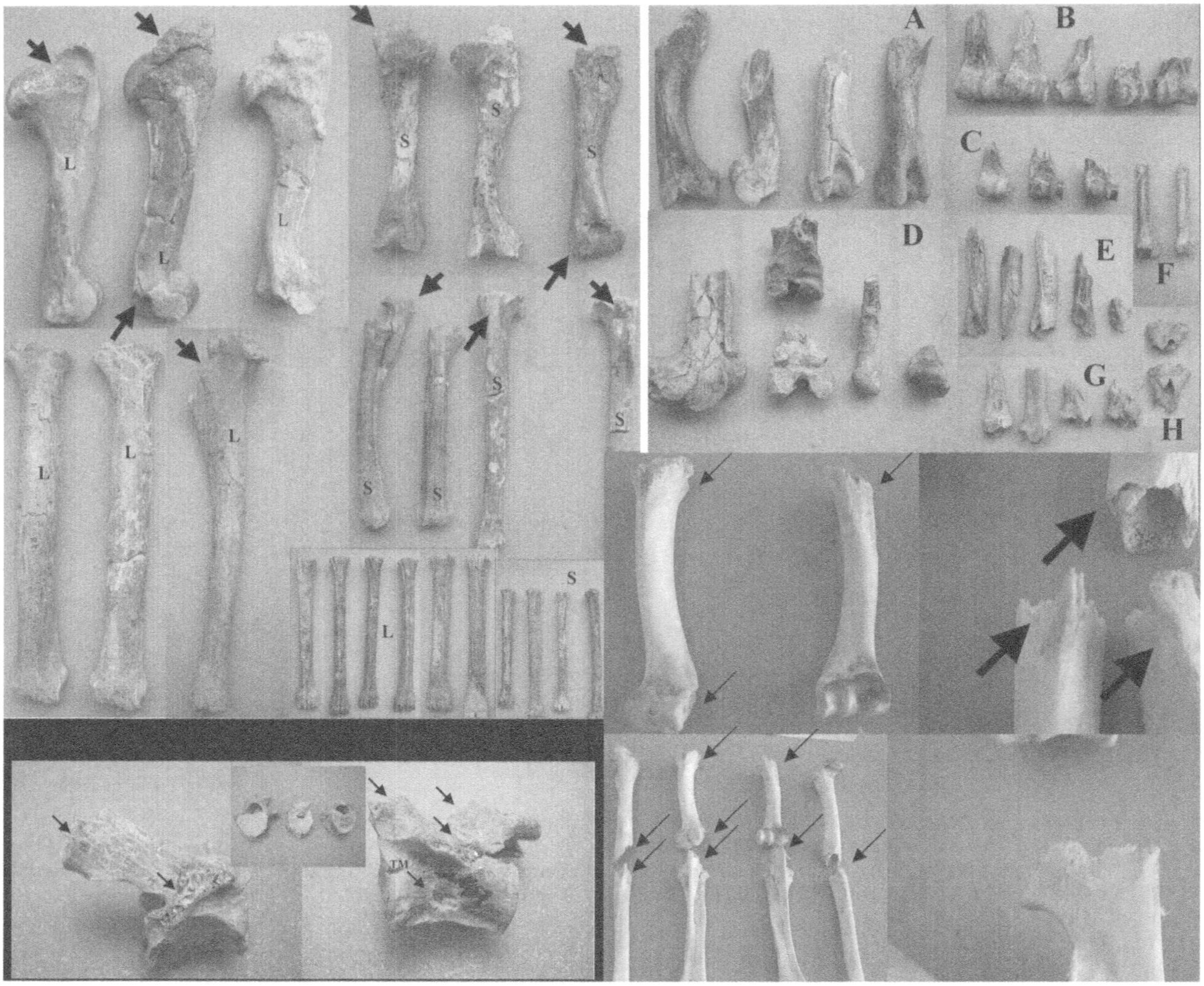

FIGURE 11.4. Upper left, complete long bones from large (L) and small (S) carcasses discovered at FLK N, bearing traces of furrowing of felid bone damage. Upper right, various stages of long bone fragmentation; A and F, proximal ends deleted; B-D, G, only distal ends surviving; E, only shafts surviving; H, only proximal ends surviving. All these patters (A-G) are typical of hyena damage to bones and account for most of the tooth-marked specimens. Lower left, vertebrae from *Antidorcas* with damage on the apophyses and tooth marks on the centra typical of felids. Lower right, Several bones from modern experiments with felids, showing typical damage in small carcasses: deletion of cancellous epiphyses (infrequent) and furrowing on ends (more frequent).

have been preserved bear small tooth marks or furrowing and have their bodies intact. There are even articulated sections of axial skeletons from *Antidorcas*. The only sections that have been deleted are the apophyses (lower left corner of Fig. 11.4). Since these are small carcasses, and none of these elements in small carcass sizes survive hyena ravaging or are present at hyena dens, the only candidate for their modification is a felid.

Most of the stone tools preserved at the sites show two basic features: more than 90% of the raw material weight has been used for artefacts aimed at battering activities. Mora and de la Torre (2005) have shown that these type of activities are predominant in the Bed I and Bed II assemblages. The taphonomic study of these sites shows that such pounding activities were not aimed at bone-crushing and marrow extraction given the virtual lack of percussion marks, percussion notches and breakage planes indicating dynamic loading. An experimental study carried out by the author and colleagues (unpublished) showed that a minimum of 15–20 flakes are necessary to completely butcher small (goat-sized) and large (cow-sized) carcasses. If this ratio (number of flakes:MNI) is applied to the Olduvai sites a picture emerges where the presence of cutting tools (flakes) is obviously not sufficient for efficient cutting and butchering activities of the amount of animals represented (Fig. 11.5). This supports de la Torre's claim (2009) about battering activities being predominant in these assemblages.

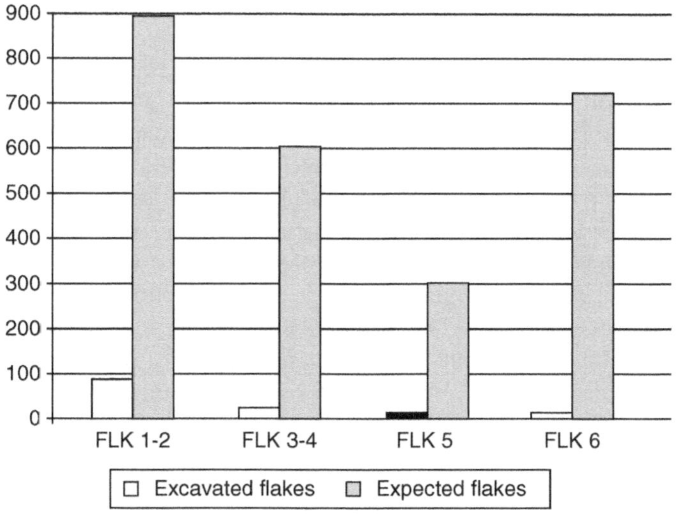

FIGURE 11.5. Frequencies of estimated expected flakes at several Olduvai sites and the actual number of flakes present. Estimates were derived by assuming a maximum of 15 flakes used per carcass documented for complete butchery of all carcasses represented at the site (according to MNI).

This begs the question as to what hominids were smashing, if bones were not the targeted objects. There is no answer to this question at present.

In sum, the spatial association of stone tools and most bones at Olduvai Bed I sites seems to be fortuitous. FLK NN1 and FLK NN3 are clearly carnivore-made accumulations. The tooth mark frequencies, the completeness of most elements, together with a very low degree of fragmentation, seem to support a non-bone cruncher as the main agent responsible for the accumulation.

FLK NN2 has previously been reported as a bone accumulation of carnivore origin (Bunn 1986; Potts 1988). However, the type of carnivore agent was never specified. Bunn (1986) identified the carnivore as a large animal, and probably a scavenger, given the large size (3B) of most of the accumulated carcasses. However, the high frequency of fairly complete limb bones together with a frequency of tooth-marking that is lower than reported for hyena dens, would suggest that a medium to large-sized felid could have been responsible for the accumulation, followed by an important degree of postdepositonal ravaging by hyenas. Given that the under- and overlying levels provide sustained support for this hypothesis, the FLK NN site could, as a matter of fact, show repeated use of the same space by a single type of carnivore. Environmental reconstruction at FLK NN site suggests a surrounding environment of very closed vegetation. This is consistent with a scenario of felids storing carcasses in trees, which could account for bone accumulation on the ground and for moderate modification of discarded bones by hyenas. The few flakes found at FLK NN 1 and FLK NN 3 could be accounted for by brief visits of hominids, during which no activity related to carcass processing took place. Most of the purported artifacts in both levels (Leakey 1971; Potts 1988) are likely natural ecofacts rather than "manuports," as argued by de la Torre and Mora (2005).

More frequently, Bed I sites are palimpsests, with a predominance of carnivore input over that of hominids. Hominids seem to have been more actively involved in the formation of the various levels at FLK N than at FLK NN. There, they performed slightly more activities that required stone knapping and especially battering (de la Torre and Mora 2005). None of these activities were linked to carcass processing, let alone to the bone accumulation at the various levels of FLK N. At FLK N 1–5 most bones from these levels seem to be background scatters or accumulations resulting from predation over time or carcass-storing by felids. The co-occurrence of bones and stone tools in these levels seems to represent several independent, unrelated depositional events. The predominant activity inferred from the study of stone tools (battering) is not related to bone demarrowing, since neither percussion marks nor green fractures caused by dynamic loading have been clearly documented. Most bones appear fairly complete, proving that carnivores also did not demarrow most of them. In summary, the superposition of different archaeological levels, in which faunal remains seem to be natural background scatters and accumulations, suggests that sites were situated in an ecologically favorable place to which herbivores, carnivores and hominids were attracted, creating independent depositional events.

The contribution of hominids seems to be –although still marginal– more significant in the FLK N 1–2 level, though the overall taphonomic signature suggests that carnivores were responsible for most of the accumulations. This is a perfect example of palimpsest. Given the carcass sizes that dominate the bone assemblages at FLK N 1–2–3–4 (*Antidorcas* and *Parmularius*), leopards or *Dinofelis* could account for most of the assemblage. A specialized hunter of size 1–3A carcasses (Bunn 1982) fits well with the type of animals upon which modern leopards prey. Alternatively, leopards are not known to store size 3A carcasses in trees, and a somewhat larger felid could be responsible for the *Parmularius* portion of the assemblage on the ground. In this case *Dinofelis* comes to mind; indeed remains of this felid have been found at the site.

Similar hypothetical claims of the natural nonhominid origin for the bone accumulations in several Olduvai Bed II sites can be made (Egeland and Domínguez-Rodrigo, 2008 [in press]). The only site that is an exception to this is FLK *Zinj*, which is of anthropogenic origin and will be discussed in more detail below.

11.4.3 Gadeb

In 1975 and 1977, archaeological sites dating to 1.5 Ma and younger were found and excavated in the plain of Gadeb in the high plateau of Ethiopia (Clark and Kurashina 1979). The sites showed an alternate sequence of Acheulian and Developed Oldowan dating between 1.5 Ma and 0.7 Ma. At

Site 8F, overlying the Acheulian occupation found in locality 8, dated to 0.7 Ma, nearly 100m^2 were excavated, which yielded a total of 385 artifacts and few faunal remains, among which are several bones from a hippo. The lithic assemblage was assigned to the Developed Oldowan B complex, and was interpreted by Clark and Kurashina (1979:39) as a concentration of bones in primary position, resulting from a butchery episode of a hippo. The authors noted that at this site "the bone waste found on the sites indicates that butchery was more commonly practiced by the makers of Developed Oldowan".

The site 2E contained the largest faunal concentration unearthed at Gadeb. The archaeological materials were found in a "channel fill cut into a pumice mudflow dated to 1.48 Ma" (Clark and Kurashina 1979:34). The lithological matrix was composed of sand and gravel. The excavation of 40 m^2 unearthed 1741 stone artifacts and several bone fragments, including hippopotamus, bovids, equid, suid and proboscidea. Like site 8F, this site is interpreted as being minimally affected by postdepositional sedimentary processes.

Sites 8F and 2E stand out among the Gadeb sites which normally contain very few bones, and also because hippo remains appear spatially associated with the lithics. In 2001, the faunal collections from these two sites were analyzed by the author at the National Museum in Addis Ababa. A total of 36 bone specimens were recorded from 8F and 566 specimens from 2E (work in progress). The higher number of fragments at 2E does not represent a higher number of carcasses represented at this site than at 8F. It is the result of intensive fragmentation due to diagenesis. Out of the 566 bone specimens, 465 bore evidence of dry fragmentation, which in turn skewed bone size representation. There is a virtual lack of green fractures in small specimens, which suggests an underrepresentation due to postdepositional size-sorting by hydraulic processes. This would be in accordance with the information drawn from the stone tools, which in the case of 2E show some abrasion suggestive of moderate "resorting" during sedimentation (Clark and Kurashina 1979:34). Only 15 specimens from 2E and 4 specimens from 8F showed green fractures. Not a single notch was observed; therefore, the agent of breakage remains unknown. A total of 8 specimens from 8F and 49 specimens from 2E showed good cortical preservation. In none of them was any hominid-induced mark (cut or percussion marks) observed. Several tooth marks were documented, in contrast. This is the only tangible clue of the agent of breakage at both sites. Despite Clark and Kurashina's (1979) remark that the polyhedrons, spheroids and some small tools were found in close association with the bones, there is no clear indication that those heavy-duty tools were used for bone breakage at the site. If butchery was carried out, not a single physical evidence thereof has survived. In the light of current evidence, the spatial association of bones and stone tools is not informative of any functional connection between both types of archaeological materials.

11.4.4 Peninj ST4

The area exposed through excavation at the ST4 site from the ST Site Complex shows (together with Olduvai's *Zinj*) some of the earliest likely evidence of carcass transport (complete or partial). Hominid transport and modification (in the form of cut marks and percussion marks) of an assemblage of three clusters of bones belonging to 3 different individuals is documented. The partial semiarticulated distribution of different anatomical units resembles the distribution of bones in Hadza kill sites (O'Connell et al. 1992), with spatially differentiated distributions of taxa, and might be indicative of the place acting as a butchering spot (Domínguez-Rodrigo et al. 2002). The spatial association of bones and stone tools would support this, as well as the low density of artifacts (see Domínguez-Rodrigo et al. 2008). Elements in home bases from modern foragers tend to be more widely mixed and dispersed. Pseudo-articulation of elements in home bases is also marginal or non-existent. Several anatomical units of at least three different animals have been accumulated and manipulated by hominids in a spatially restricted area at the base of the ST4 site (Domínguez-Rodrigo et al. 2008). The evidence from the ST4 site (skeletal part representation, anatomical distribution of cut marks and percussion marks, low frequency of tooth marks) supports that Lower Pleistocene hominids were transporting and processing carcasses and not just dismembered bones. This gives further support to the interpretation of hominid carcass acquisition and carcass processing behaviours as being primary to carnivore intervention. It also discredits interpretations assuming that hominids had only access to marginal carcass resources. Most importantly, it also shows that some sites may have been carcass butchery and preparation places, similarly to modern foragers.

11.4.5 Swatkrans Member 3

Swartkrans has been archaeologically presented as the oldest South African site created by deposition of bones and stone tools by hominids. The Member 3 assemblage preserves an occurrence of bones and stone tools. Bones appear well preserved with an important frequency of natural modifications, including trampling, sediment abrasion and occasional biochemical modifications of cortical surfaces. Bones show a much lower proportion of specimens with the co-occurrences of hominid- and carnivore-derived surface modifications than do modern actualistic assemblages derived by the interdependent actions of both agents (Pickering et al. 2004, 2007, 2008). This suggests that the fossil assemblage can actually be divided into two fairly independently-formed and high integrity sub—assemblages—one created largely by the actions of hominids and the other created largely by the actions of carnivores (Pickering et al. 2004, 2007, 2008). The evidence of the latter is more abundant than the traces left by the former. Cut marks and percussion marks are distributed fairly evenly across all limb elements, suggesting fairly complete processing of whole limb units by hominids.

By extrapolation, this might mean that hominids were acquiring whole carcasses for processing (see below discussion of FLK *Zinj*) (Pickering et al. 2004, 2008).

So, while hominids contributed only minimally compared to carnivores to the overall assemblage that accumulated during the formation of Swartkrans Member 3, the evidence of hominid activity that *is* preserved is informative behaviorally.

11.5 The "*Zinj* effect" and the Best Window to Understand Plio-Pleistocene Hominid Carcass Acquirement Strategies

Archaeologists working on Plio-Pleistocene occurrences in Africa are only left with one site showing spatial association of stone tools and bones at Olduvai (FLK *Zinj*) that can be convincingly argued to be completely anthropogenic. Data on cut mark, percussion mark and tooth mark frequencies and distributions together with analyses of breakage planes and notches show that hominids were the main accumulators of bones at the site, and that bones were collected in the form of carcasses that were butchered and demarrowed by hominids. It has been argued that these carcasses were acquired by primary access to fleshed carcasses (see Bunn 1982, 2006; Bunn and Kroll 1986; Domínguez-Rodrigo 1997a, b; Domínguez-Rodrigo and Pickering 2003; Domínguez-Rodrigo and Barba 2006). Alternatively, it has also been argued that these carcasses might have been obtained by passive scavenging of defleshed remains from felid kills (Blumenschine 1986, 1995) and somewhat more fleshed remains from tree-stored leopard kills (Capaldo 1995; Cavallo 1998). The arguments provided to support the latter claim are:

1. *Tooth mark percentages per long limb bone portion.* Blumenschine's (1988) work showed that carnivore modification of epiphyseal and near-epiphyseal long limb bone specimens is similar in both primary and secondary access contexts. That is because the greasy cancellous bone of these sections make them appealing whether carcasses are fleshed or defleshed. However, a sharp contrast in tooth mark frequencies is observed for mid-shaft portions. When carnivores have access to complete bones and break long bone shafts to obtain the marrow, they leave tooth marks on the shafts in frequencies usually > 70% of all the broken fragments. When carnivores intervene secondarily after humans had access to the marrow they are confronted with only dense mid-shaft fragments, devoid of any edible resource. In this situation, carnivores leave barely any tooth marks on mid-shaft sections. In human-first experimental assemblages carnivores generate on average between 5% (Blumenschine 1988) and 15% (Capaldo 1995) tooth-marked pieces of all mid-shaft fragments when dealing with bones from medium sized carcasses, and a similar range of frequencies for smaller animals. Tooth mark frequency on mid-shafts from the most abundant carcass size at FLK *Zinj* was interpreted as 59% (Blumenschine 1995).
2. *Cut marks are not relevant to the debate* since they are subjected to equifinality (Capaldo 1995) and because we lack proper referential frameworks to understand butchery processes and to differentiate them from marginal flesh scrap removal (Blumenschine 1991, 1995). When referential frameworks are provided (Domínguez-Rodrigo 1997a, b) they are considered insufficient because of small sample sizes and non-replicability of the sample (Blumenschine and Pobiner 2006).

Some researchers see the hunting-scavenging debate with scepticism. They consider that both sides have equally well-supported arguments. However, a growing body of research literature over the past 10 years has provided enough evidence to show that both hypotheses have fairly different heuristics; that is, several arguments have falsified the passive scavenging hypothesis as initially pronounced (showing a large amount of inconsistencies) and have provided very strong arguments in favour of the hypothesis that hominids had primary access to fleshed carcasses. These arguments have not been scientifically addressed by the partisans of passive scavenging hypotheses. At present, they are satisfied by suggesting that the passive scavenging hypothesis still is an alternative to the primary access hypothesis (which has not been falsified at any point of this debate) (Blumenschine and Pobiner 2006). The task that defenders of passive scavenging have ahead of them is to prove that the following arguments are false.

11.5.1 Tooth Marks and Felid Kills

1. Passive scavenging from carnivores, as currently conceived, should be modelled in the following terms: felids were the primary consumers of carcasses exploiting most of the flesh from them; hominids followed them by focusing mostly on marrow extraction and hyenas intervened lastly by deleting grease-bearing bones. From a conceptual point of view, all the Carnivore-only experiments carried out by Blumenschine (1988) that inspired the carnivore-hominid-carnivore model (Selvaggio 1994; Blumenschine 1995) were carried out by using hyenas as carnivores (Blumenschine 1995; Capaldo 1995) or a mix of carnivores involving felids, hyenids and canids and tallying the resulting tooth mark frequencies together when comparing them to FLK *Zinj* (Selvaggio, 1994). Bone crunching carnivores (hyenas) are a bad proxy for flesh-eaters (felids) since they will modify bones differently. If felids and their resultant bone damage are *actually* used as the "first" carnivore in multiple-pattern models (as they ought to be, according to the hypothetical scenario constructed by the proponents of these models), instead of hyenas, then tooth mark percentages drop drastically (especially on mid-shafts) and become non-diagnostic (Domínguez-Rodrigo et al. 2007b).

2. Blumenschine and Pobiner (2006) have argued that the frequency of tooth marks documented at FLK *Zinj* (intermediate between carnivore-only and hammerstone–carnivore models) can be explained by felids initially defleshing carcasses as is experimentally modelled by Selvaggio (1994). It has been argued that about 65% of mid-shafts in Selvaggio's carnivore-hominid experiments are tooth marked (Blumenschine and Pobiner 2006). However, the use of such a percentage is misleading. The bulk of Selvaggio's experiments are based on small carcasses, more easily fragmented by felids than the larger carcasses represented at the *Zinj*. The high frequency cited by Blumenschine and Pobiner (2006) is derived by lumping together broken bone fragments and complete bones. Broken fragments are more highly tooth-marked than complete bones and a hominid would pay no attention to them because they would be resource-free. In addition, the carnivore-hominid-carnivore experimental sample is obtained by lumping together all carnivore types (bone crunching and flesh eaters) and not just felids. The percentage thus obtained does not reflect the basic premises of the carnivore-hominid-carnivore scenario (Selvaggio 1994; Blumenschine 1995; Capaldo 1995), according to which hominids were scavenging *complete* (marrow-bearing) long limb bones (ignoring the fragmented ones with no marrow) from *felid* kills. However, Selvaggio (1994) quantified tooth mark densities (actual number of marks per a bone specimen) in experimental assemblages and showed how different tooth mark frequencies generated by felids on complete bones are from those inflicted by hyenas. For felid-derived assemblages, 75% of the specimens have tooth mark densities <21 (Selvaggio 1994). More specifically, up to 96% of small animal bone assemblages displayed tooth mark densities <20, while 86% of medium-sized animal bone assemblages displayed tooth mark densities <21. These values are significantly lower than carnivore-only experimental assemblages, in which hyenas had been responsible for bone modifications and in which densities >70 were usually reported for both carcass sizes. These results indicate clearly that felids tooth mark limb bones at much lower rates than hyenas. This contrast is even more acute when observing complete bones remaining unbroken by felids after carcass defleshing. For such specimens, 50% displayed not a single tooth mark (Selvaggio 1994).
   In addition, Selvaggio's (1994) data indicate an intra-element difference between felids and hyenas in tooth mark distribution. Of all bone portions in the felid modified sample, mid-shafts exhibited the lowest mean percent of specimens bearing at least one tooth mark. Frequencies are always <50%, much lower than those reported for hyena-created and hyena only-modified assemblages. The contrast is even more marked when one considers that part of the felid tooth mark sample was created by lions on the bones of small gazelles, far outside the larger body size of more "typical" lion prey (e.g. Schaller 1972). This means that if hominids were acquiring the complete bones from this sample, they would be further reducing the frequency of tooth marks by breaking open the bones and generating multiple fragments. This is supported by recent studies by Domínguez-Rodrigo et al. (2007b) who show that tooth mark damage reported for mid-shafts from carcasses consumed by felids (cheetah, leopard and lion) and later broken by humans are <15%, lower than in hammerstone-carnivore scenarios where hyenas were used as secondary scavengers. This is the only study that reproduces defleshing by felids and subsequent removal of marrow by hominids.
   This advises against using tooth mark distribution in complete elements as a proxy for hammerstone broken bones following similar processes of tooth-marking. Complete bones tooth-marked by felids will *always* show higher frequencies of tooth-marked specimens than broken elements.
3. It has been argued that hominids might have scavenged from felid tree-stored kills containing larger amounts of flesh than terrestrial felid kills (Cavallo and Blumenschine 1989). Blumenschine's (1986) carcass consumption sequence showed that the graded nutrient yields of flesh- and marrow-bearing units of a carcass (i.e. parts ranked from highest to lowest in terms of net yield per unit of handling time) appear to "condition" the selection of parts to be consumed and the order of their consumption by carnivores. Importantly, he observed that hindquarters are consumed first by lions and hyenas, followed by forequarters. After consuming head flesh, these carnivores move back to consume first the hindlimb marrow then the forelimb marrow. Finally hyenas only eat the head contents. Cavallo (1998) documented that leopards, like lions, consume the flesh of prey from back to front. An important variation, however, is the leopard's early consumption of facial flesh and nasal marrow and other parts of the head. Although leopard consumption of the lower viscera, pelvic and femur flesh, and femur and humerus marrow corresponds with Blumenschine's sequence, there is a significant deviation in the delayed consumption of forelimb flesh and associated upper marrow-yielding bones by leopards. That is, whereas it is highly unlikely to scavenge flesh from a lion kill (Domínguez-Rodrigo 1999a), it could theoretically be possible to do so from a leopard tree-stored kill. Cavallo (1998) also found that tree-stored kills of small antelopes remained potentially scavengeable resources for 2 to 3 days, as opposed to Blumenschine's (1986) recording of 1 hour for similarly-sized prey that were killed and consumed on the ground by lions and hyenas.
   However, if hominids were accessing carcasses at the late stage of leopard consumption as represented in Cavallo's (1998) collection, bones would be defleshed and no cut marks would appear on their surfaces. In contrast, the analysis of the distribution of cut marks on small-sized carcasses at the FLK *Zinj* site shows a different pattern.

The high incidence of cut marks together with the even anatomical distribution on the appendicular section (Bunn, Kroll 1986) can only be obtained when replicating primary access to fleshed carcasses (Domínguez-Rodrigo 1997a, b; Domínguez-Rodrigo and Barba 2005). This clearly suggests that if hominids were scavenging small-sized carcasses from leopards, they must have done it regularly from the earliest stages of carcass consumption by these felids. This means that if most flesh was still present, leopards would have left even lower tooth mark frequencies than those reported for the last stage of carcass consumption by leopards in Cavallo's collection. If hominids had access to carcasses after the first stage of leopard consumption, the resources that leopards would have deleted from the carcass would have been: all pelvic flesh including upper hindlimbs, viscerae, and facial and mandibular flesh. Therefore, no cut marks would be expected on the internal side of ribs (suggestive of evisceration), nor on the mid-shafts of femora (devoid of the scraps of flesh that may survive carnivores' initial consumption), nor on the facial part of the skull. Likewise, these sections are more likely to bear surface modifications from leopards, namely tooth marks.

Hominid access to tree-stored carcasses in an intermediate stage of carcass consumption by leopards would result in the total absence of cut marks on mid-sections and distal section of ribs (since the ribcage flesh has disappeared), and also lack of hominid-inflicted modifications (cut marks) on the tibia, since it would already be defleshed. Later access of hominids would only allow access to marrow, and therefore, no cut marks would be expected. Some of these anatomical areas are not very likely to be preserved in the archaeological record given their preferential deletion by post-depositional ravaging carnivores (Marean et al. 1992; Capaldo 1995). However, early and intermediate access to carcasses scavenged from leopards should be reflected in very low or nonexistent frequencies of cut marks on the mid-sections of femora and tibiae, respectively (because the front limb is consumed late in the sequence). In contrast, cut marks on these anatomical sections are documented on small carcasses accumulated by hominids at FLK *Zinj*, disproving that they acquired them from leopard kills.

If hominids were exploiting tree-stored kills, they could have used the felid-transported carcasses at such *loci* as inferred from several Olduvai sites (see above) by exploiting the remaining flesh and whole marrow. The fact that a large number of complete bones survived, together with the lack of percussion traces (or cut marks) at most of the Olduvai sites falsifies the hypothesis that hominids were seeking tree-stored kills (Domínguez-Rodrigo et al. 2007a).

4. A recent study of tooth mark frequencies at the FLK *Zinj* shows that the percentage previously identified by Blumenschine and colleagues is extremely inflated and results from misidentifying biochemical marks caused by fungi and bacteria in combination with root etching as carnivore-imparted tooth marks (Domínguez-Rodrigo and Barba 2006). The actual frequency of tooth marks on mid-shafts from all carcasses at FLK *Zinj* is <20%, much lower than any carnivore-first experimental model reproduced by Blumenschine (1988) or Selvaggio (1994). According to Blumenschine (1988, 1995), this would be indicative of primary access to fleshed carcasses by hominids.

11.5.2 Cut Marks

1. *Equifinality I.* From a strictly analytical consideration, Domínguez-Rodrigo (2002) argued that the equifinality reported by Capaldo (1995) in the use of cut marks was merely methodological. Capaldo (1995, 1997, 1998) claims that percentages and distributions of cut marks are indistinguishable in experimental scenarios reproducing early access (defleshing of carcasses) or secondary access (removal of flesh scraps from carnivore kills) to carcasses. The ranges of variation of both experimental samples overlap. Cut marks, thus, would be subjected to equifinality (Capaldo 1995, 1997, 1998). It is clear that Blumenschine's, Capaldo's and Selvaggio's methodology is not suitable for the study of cut marks, because the analyses of bone sections *per se* could never test, for instance, Bunn and Kroll's (1986) hypothesis that cut marks are related to the amount of flesh extracted from bones. Flesh is differentially distributed according to bone element. Similarly the small scraps of flesh that might survive carnivore consumption also show a typical pattern of anatomical distribution according to bone type and bone section (as discussed above). Analyses of cut marks per *bone section* cannot relate the amount of meat removed to the distribution and percentages of the resulting cut marks. Thus two different behaviors may mimic each other when analysed through Blumenschine, Selvaggio and Capaldo's approach. The removal with cutting tools of skin and periostium from lower limb bones of carnivore-defleshed carcasses may leave an abundance of cut marks on all the lower-limb sections (epiphyses, near-epiphyses and mid-shaft fragments). The total percentage and sectional distribution of cut marks per carcass will be similar to cases in which hominids processed fully fleshed upper limb bones and discarded the lower limb elements. The way to differentiate these two scenarios is by quantifying the cut marks by element (upper limb, intermediate limb, lower limb) as well as by bone section. This supports the notion that cut marks should be analyzed as to their anatomical distribution on bones from archaeological sites.
2. *Equifinality II.* Blumenschine, Selvaggio and Capaldo's method of analysis is problematic not only because it does not consider such a differential distribution of flesh, but also due to their definition of bone sections. They consider a bone specimen that has a fragment of epiphysis with a section of the near-epiphysis and mid-

shaft to be an "epiphyseal specimen". Mid-shaft sections of such fragments that bear cut marks are classified as cut-marked "epiphyseal fragments", which gives the incorrect impression that the cut mark is situated on the epiphysis itself. The high survival of epiphyseal fragments attached to shaft sections and the high percentage of marks on these specimens will result in a relatively high number of "epiphyseal fragments" showing marks. Thus, the actual distribution of marks according to bone section is not properly represented. This is of real consequence when evaluating different models of hominid carcass acquisition and processing behavior. For instance, the availability of scraps of flesh from mid-shaft sections of limb bones first consumed by lions has been shown to be rather limited (see above). Implications vis-a-vis a hypothesis of lion followed by hominid consumption would differ if a specimen conserving part of the epiphysis with cut marks on the mid-shaft was classified as a "cut-marked epiphyseal fragment" rather than as a "cut-marked mid-shaft fragment". Domínguez-Rodrigo (1997a, b, 1999b, 2002) argues for a more precise classification of cut mark location that does not lump together mid-shafts and epiphyses.

3. *Equifinality III* . When using a method that incorporates such a distinction (element type and bone section) equifinality is reduced and it is possible to formulate and successfully test alternative hypotheses of primary and secondary access to fleshed or defleshed carcasses (Domínguez-Rodrigo 1997a, 2002). This can be done because access to fleshed carcasses is reflected in a high frequency of cut-marked specimens on mid-shafts, especially on upper limb bones (humerus and femur). When applied to FLK *Zinj* , results match closer the experimental scenario simulating access to fleshed carcasses.
4. *Experimental sample size*. Blumenschine and Pobiner (2006) have qualified Domínguez-Rodrigo's experimental sample as insufficient to understand cut mark frequencies in archaeological assemblages. In Domínguez-Rodrigo's studies, both primary and secondary access to carcasses have been experimentally modelled from analyses of a total of 62 carcasses (33 of them butchered). A sample of almost 30 carcasses from felid kills allowed the study of where and how much meat survive carnivores' initial consumption of carcasses (Domínguez-Rodrigo 1999a), with special emphasis on mid-shafts (especially from upper limb bones). Butchery of 16 carcasses from lions' kills further resulted in a thorough understanding of where cut marks appear when processing felid kills (Domínguez-Rodrigo 1997a, b, c). Other 17 carcasses were used to observe cut mark patterns on fully fleshed carcasses when hominids have primary access to them (Domínguez-Rodrigo 1997a, b; Domínguez-Rodrigo and Barba 2005). Both parametric and non-parametric tests have confirmed the sample is large enough to differentiate primary to secondary access to carcasses using cut marks on shafts (Domínguez-Rodrigo 1999a, 2003; Domínguez-Rodrigo and Pickering 2003). Hence the sample is large enough to constitute a heuristic referential framework to understand the behavioural meaning of cut marks on fossil assemblages. In contrast to the sample size used by Domínguez-Rodrigo, Blumenschine's (1988, 1995) modelling for tooth marks on small carcass bones involved only sample size $n = 2$, and larger carcases a sample size of $n = 7$. Blumenschine and Pobiner (2006) have to justify why their sample size is large enough and Domínguez-Rodrigo's is not.
5. *Replicability of sample*. Blumenschine and Pobiner (2006) mention that they did not replicate previous studies of cut marks in African savannas. Experiments by Domínguez-Rodrigo (1997a, b) were carried out in natural parks with minimal anthropogenic intervention. If Blumenschine and colleagues failed to obtain similar results, they should consider revising if the variables used in Domínguez-Rodrigo's experiments were used in an identical or different way by them (see discussion in Domínguez-Rodrigo 2003). They should also show that the ecological contexts of their experiments afforded enough guarantees to be used as referential frameworks for Plio-Pleistocene savannas. For instance, high anthropogenic modification of trophic dynamics in a given ecosystem (e.g. by removing hyenas from it), enabling lions to remove any competition in their adaptive pattern and avail of much more food than necessary is not a scientifically-supported proxy for a Plio-Pleistocene savanna. The resulting availability of resources in this situation would be higher than in other savannas with minimal anthropogenic impact (which could be referred to as natural savannas to differentiate them from those modified by humans) and the experiments performed on carcasses obtained in this type of altered ecological context would yield different results from those carried out in savannas free from human impact. Experiments carried out in these contexts are of limited validity to understand trophic dynamics in past savannas with a much smaller impact of human behavior. Replicability depends on the similar use of experimental variables. Only when these condictions are met, can differences in resulting cut mark patterns be meaningfully discussed. See a more in-depth description of how different prey consumption behaviors by felids are affected by anthropogenically-modified ecological contexts in Domínguez-Rodrigo (2008a).
6. *Additional evidence of access to meat by hominids*. A new approach analyzing cut mark location on the exact anatomical position and the anatomical distribution of flesh scraps has shown that the location of many cut marks on bones from the FLK *Zinj* are not related to flesh scrap survival on bones from felid kills (Domínguez-Rodrigo et al. 2007a).
7. *Dismembering as an indicator of meat butchery*. There is a substantial number of cut-marked bones from the FLK *Zinj* resulting from long limb bone dismembering (Bunn and

Kroll 1986; Domínguez-Rodrigo et al. 2007a). Dismembering is a frequent behaviour associated with food-sharing dynamics and cooking in modern forager societies. These societies do not usually dismember long limb elements at other type of *loci* (e.g. kill sites) where food is not consumed. If scavenged, dismembering of defleshed bones goes against the logic of optimal foraging theory: why would hominids disarticulate long limb elements if carcasses were already defleshed? Disarticulation remains the most energy-requiring butchering activity. This suggests indirectly that dismembering cut marks on bones from FLK *Zinj* would support access to fleshed carcasses.

The persistent defence of the passive scavenging theory when using cut mark data clings on Selvaggio's (1994) data from experimentally-reproduced carnivore-hominid-carnivore models. Previous work (Domínguez-Rodrigo 1997a) has questioned the value of Selvaggio's analysis as a proper referential framework for Oldowan behavior at FLK Zing for the reasons summarized below:

1. Following the analytical procedures of Blumenschine (1988), Selvaggio's analyses are based on the distribution of marks according to bone section. For the reasons discussed above (e.g. disregard to differential flesh distribution which leads to equifinality), this type of approach is not useful for testing Bunn and Kroll's (1986) hypothesis that cut marks are the result of flesh extraction..
2. Several of Selvaggio's experiments are made up of single bones and many of them are sets consisting of one to three elements. Capaldo's (1995) experiments (and this author's personal observation) suggest that carnivore post-ravaging of bone assemblages is influenced by the availability of remains. Thus, two or three bones do not realistically simulate the large amount of bones that hyenas found at sites abandoned by hominids (especially at *Zinj*), nor the results of their intervention
3. Selvaggio removed the scraps of flesh and the marrow from bones herself, under the assumption that "differences in the frequencies of butchery marks are not related to different butchers" (Selvaggio 1994), implying that there should be no differences in technique based on butchery experience. The result is the creation of cut-mark patterns that are not necessarily realistic. Domínguez-Rodrigo (1997c) –like many other researchers- documented the variation of cut-mark frequencies relative to the experience of the butcher. Selvaggio's percentages, thus, may be inflated.
4. With the exception of metapodials, periostium does not need to be removed from long bones before breaking them, Selvaggio does not make it clear whether she removed the periostium from all bones in her experiments. If that was the case (following the protocol of Blumenschine (1988) and Capaldo (1995), the percentages of cut marks may have been inflated.
5. To construct her overall percentages of mark distribution in the three-patterned experiments, Selvaggio combines the data obtained from all carcasses, irrespective of their size and the predator(s) responsible for their capture and processing. This creates a double bias. First, the result of carcass consumption by solitary predators (more likely to abandon animals with higher remaining yields of flesh) is blended with that of social predators (more likely to deflesh carcasses more efficiently and extensively). Second, the availability of flesh on smaller animals that are preferentially hunted by solitary predators (Cavallo and Blumenschine 1989) is not differentiated from that of larger carcasses,usually hunted and processed by gregarious predators. This creates an artificial frequency of marks, *in which the resulting cut-marking from processing scavenged felid kills alone is not reproduced* .
6. Finally, the experimental sample used by Selvaggio (1994) is mainly composed of small carcasses. Only 4 out of 47 carcasses are medium-sized animals (two zebras, one wildebeest and one buffalo). Of these, only two were used in the three-patterned experiments under controlled circumstances. One of them consisted of six bones and the other experiment of only one bone. Clearly this sample is statistically flawed.

In addition to these circumstances, Selvaggio acknowledges that

"Limbs were abandoned by lions with little or no flesh on humeri and femora. Occasionally the skin was not completely consumed from the tibiae or the radius–ulnae and small scraps of flesh remained near the distal epiphyses" (Selvaggio 1994:54–55).

She further claimed that

"Long bones abandoned by large groups of carnivores were usually disarticulated from the carcass and while the marrow cavity remained intact, the bones were usually encountered completely defleshed" (Selvaggio 1994:124).

Therefore, a high percentage of cut marks on upper limb bones from early sites would suggest defleshing of large muscle masses rather than the removal of scraps of flesh (Domínguez-Rodrigo 1997a). Selvaggio's and Domínguez-Rodrigo's studies both lead to the same conclusion with respect to carcasses processed by lions. Upper limb bones are utterly defleshed on most occasions. According to her, this observation could also be extended to other predators and smaller carcasses: "Rarely were scraps of flesh available on proximal long bones abandoned by carnivores in the carnivore-hominid sample" (Selvaggio 1994:122). Had she used a different methodological approach, she would have documented the contrasting cut mark frequencies of these elements compared to the other limb bones.

The low tooth mark frequency at FLK *Zinj* is coupled with a widespread distribution of cut marks on all anatomical parts, including mid-shafts from upper limb bones showing a diversity of butchery behaviours. Those include dismembering, filleting and even evisceration (Domínguez-Rodrigo et al. 2007a). The evidence is clearly supportive of a scenario of primary access to fully-fleshed carcasses undisturbed by previous carnivore consumption.

Both Kuhn (1962) and Lakatos (1978) claimed that supporters of paradigms rarely give them up without resilience. Recent debates about the *Zinj* assemblage are a good example of this. Blumenschine and Pobiner (2006:184) claim: "A more peripheral and subordinate role for hominins as passive scavengers of abandoned felid kills is the most conservative hypothesis that has yet to be rejected". This claim can only be sustained if an important part of scientific litterature produced over the past ten years is not taken into consideration. In the case of FLKN1-2, Blumenschine and Pobiner (2006:182) argue that a low frequency of tooth-marked mid-shafts would be indicative of a hominid-carnivore model, in which hominids would have presumably obtained fleshed carcasses. It is their own conclusion that they fail to apply to interpret the data from the FLK *Zinj* assemblage. Rather than accepting the implications from their own results, their commitment to the passive scavenging hypothesis leads them to conclude that in this case the low frequency of tooth-marked mid-shafts "*underscores the likelihood that carcass foods were acquired from multiple sources in a variety of ways*". This is an untested ambiguity that goes unsupported by their experimental scenarios.

As it stands, the available evidence has already falsified the passive scavenging hypothesis, following Popperian demarcation criteria[4]. It is only waiting for a new switch of paradigm.

11.6 Conclusions

The abundance of Pliocene sites with pristinely-preserved stone tools but no faunal remains, combined with the *meagre* evidence of butchering traces and the existence of an archaeological record in derived positions, do not warrant claims of functional links between lithics and whole bone assemblages of any currently available site during the first 600 Ka of existence of archaeological record (regardless of cases of spatial association of the two types of finds). If the full implications of these assertions are acknowledged, it cannot be claimed that butchery is the sole reason for the emergence of the earliest stone tools, if only because there is currently no solid support for this position.

The "Olduvai effect", caused by the taphonomic revision of the Olduvai sites, not only demonstrates that incidental associations of stone tools and bones were possible during the Plio-Pleistocene, but that they were fairly common. The paleosurfaces where stone tools were discarded were exposed for prolonged periods of time during which successive unrelated depositional events by non-hominid biotic agents might have generated the spatial association of bones and stone artifacts that archaeologists unearthed. This was previously stressed by Binford (1981). This is also true in a large portion of the archaeological record from Bed I and Bed II in Olduvai, where finds are often vertically dispersed, spanning large depositional periods. Therefore, no functional assumptions can be made based solely on spatial associations of bones and stone tools.

For this reason, and because of poor bone preservation, the nature of the association of both types of materials in most sites from Koobi Fora remains unknown (see though recent discoveries in Harris et al. 2002a, b; Pobiner et al. 2008), despite the fact that at some sites (e.g. FxJj 82) taphonomic analyses suggest also a natural depositional history of bones unrelated to the deposition of stone tools (personal observation; work carried out with C. P. Egeland)

I agree with Bunn's (2006) assertion that meat made us human, but good evidence of hominid butchery during the Plio-Pleistocene is found only in a limited sample from ST4 at Peninj and Swartkrans member 3, and more extensively, at FLK *Zinj*. There are also cut-marked bones found without the association of stone tools (Bouri and Koobi Fora) that suggest butchery was performed at various *loci*. All this evidence is suggestive of meat and marrow eating by hominids. However, this marginal sample embodies over one million years of archaeological record. Hominids might have been butchering carcasses all over the landscape but the traces preserved do not allow us to scientifically support that claim. While various chapters in this volume have shown a wide range of analytical studies of Oldowan stone tools, I suggest that we bear in mind that we still do not know what Oldowan stone tools were used for in most cases. Butchery comes to mind because bones were preserved in the archaeological record, but an alternative array of possibilities (most of them unlikely to leave any preserved traces) is also possible. There is no doubt that carcass butchery was part of hominid behavioural repertoire as far back as 2.6 Ma (Domínguez-Rodrigo et al. 2005). But if Keeley and Toth (1981) were right in arguing that there were more traces of plant-processing than meat-processing on the flakes in their sample of flakes with use traces, is it not time to consider in a more serious manner the role that stone tools may have had in plant-processing, wood-working or other activities in addition to butchery during the Plio-Pleistocene?

Acknowledgements. I thank D. Braun and E. Hovers for inviting me to take part in the Oldowan symposium and this book. Suggestions made by T. R. Pickering, D. Braun, E. Hovers and one anonymous reviewer are much appreciated. I am indebted to the institutions (Antiquities, COSTECH, National Museums of Kenya, National Museum of Ethiopia) that granted me permit to conduct research and to the institutions that granted it: the Spanish Ministries of Culture and Science and Technology (for their financial support through annual grants for archaeological projects abroad and the Development and Technology projects (I+D): BHA 2000-0405, BHA2002-11667-E and

[4]Hypothesis testing in modern science is based on Popper's demarcation criteria requiring each hypothesis to be articulated around "refutable" premises which must be subjected to testing prior to its resulting rejection or acceptance.

BHA2003-00839), the Complutense University and the Paleontological Network Foundation of Teruel.

References

Alcántara, V., Barba, R., Barral, J., Crespo, A. Eiriz, A., Falquina, A., Herrero, S., Ibarra, A., Megías, M., Pérez, M., Pérez, V., Rolland, J., Yravedra, J., Vidal, A., Domínguez-Rodrigo, M., 2005. Determinación de procesos de fractura sobre huesos frescos: un sistema de análisis de los ángulos de los planos de fracturación como discriminador de agentes bióticos. Trabajos de Prehistoria 63, 25–38.

Beyries, S., 1993. Are we able to determine the function of the earliest Paleolithic tools? In: Berthelet, A., Chavaillon, J. (Eds.), The Use of Tools by Non-human Primates. Clarendon Press, Oxford, pp. 225–236.

Binford, L.R., 1981. Bones: Ancient Men, Modern Myths. Academic Press, New York.

Binford, L.R. 1985. Human ancestors: changing views of their behavior. Journal of Anthropological Archaeology 4, 292–327.

Binford, L.R. 1988. Fact and fiction about the Zinjanthropus floor: data, arguments and interpretations. Current Anthropology 29, 123–135.

Blumenschine, R.J., 1986. Early Hominid Scavenging Opportunities. Implications of Carcass Availability in the Serengeti and Ngorongoro Ecosystems. BAR International Series 283. Oxford.

Blumenschine, R.J., 1988. An experimental model of the timing of hominid and carnivore influence on archaeological bone assemblages. Journal of Archaeological Science 15, 483–502.

Blumenschine, R.J., 1989. A landscape taphonomic model of the scale of prehistoric scavenging opportunities. Journal of Human Evolution 18, 345–371.

Blumenschine, R.J., 1991. Hominid carnivory and foraging strategies, and the socio-economic function of early archaeological sites. Philosophical Transactions of the Royal Society (London) 334, 211–221.

Blumenschine, R.J., 1995. Percussion marks, tooth marks and the experimental determinations of the timing of hominid and carnivore access to long bones at FLK Zinjanthropus, Olduvai Gorge, Tanzania. Journal of Human Evolution 29, 21–51.

Blumenschine, R.J., Pobiner, B., 2006. Zooarchaeology and the ecology of oldowan hominid carnivory. In: Ungar, P. (Ed.), Evolution of the Human Diet: the Known, the Unknown and the Unknowable. Oxford University Press, Oxford, pp. 167–90.

Blumenschine, R.J., Selvaggio, M.M., 1991. On the marks of marrow bone processing by hammerstones and hyenas: their anatomical patterning and archaeological implications. In: Clark, J.D. (Ed.), Cultural Beginnings: Approaches to Understanding Early Hominid Life Ways in the African Savanna. R. Habelt. Bonn, pp. 17–32.

Brain, C.K., 1981. The Hunters or the Hunted? Chicago University Press, Chicago.

Bunn, H.T., 1981. Archaeological evidence for meat-eating by Plio-Pleistocene hominids from Koobi Fora, Kenya. Nature 291, 574–577.

Bunn, H.T., 1982. Meat-eating and Human Evolution: Studies on the Diet and Subsistence Patterns of Plio-Pleistocene Hominids in East Africa. Ph. Dissertation. University of California, Berkeley.

Bunn, H.T., 1983. Evidence on the diet and subsistence patterns of Plio-Pleistocene hominids at Koobi Fora, Kenya, and at Olduvai Gorge, Tanzania. In: Clutton-Brock, J. (Ed.), Animals and Archaeology: Hunters and their Prey. BAR International Series 163. Oxford, pp. 21–30.

Bunn, H.T., 1986. Patterns on skeletal representation and hominid subsistence activities at Olduvai Gorge, Tanzania, and Koobi Fora, Kenya. Journal of Human Evolution 15, 673–690.

Bunn, H.T., 1991. A taphonomic perspective on the archaeology of human origins. Annual Review of Anthopology 20, 433–467.

Bunn, H.T., 2006. Meat made us human. In: Ungar, P. (Ed.), Evolution of the Human Diet: The Known, the Unknown and the Unknowable. Oxford University Press, Oxford, pp. 191–211.

Bunn, H.T., Ezzo, J.A., 1993. Hunting and scavenging by Plio-Pleistocene hominids: nutritional constraints, archaeological patterns, and behavioural implications. Journal of Archaeological Science 20, 365–398.

Bunn, H.T., Kroll, E.M., 1986. Systematic butchery by Plio-Pleistocene hominids at Olduvai Gorge, Tanzania. Current Anthropology 27, 431–452.

Bunn, H.T., Kroll, E.M., 1988. A reply to Binford. Current Anthropology 29, 123–149.

Capaldo, S.D., 1995. Inferring Hominid and Carnivore Behavior from Dual-Patterned Archaeological Assemblages. Ph. D. Thesis. Rutgers University, New Brunswick.

Capaldo, S.D., 1997. Experimental determinations of carcass processing by Plio-Pleistocene hominids and carnivores at FLK 22 (Zinjanthropus), Olduvai Gorge, Tanzania. Journal of Human Evolution 33, 555–597.

Capaldo, S.D., 1998. Simulating the formation of dual-patterned archaeofaunal assemblages with experimental control simples. Journal of Archaeological Science 35, 311–330.

Capaldo, S.D., Blumenschine, R.J., 1994. A quantitative diagnosis of notches made by hammerstone percussion and carnivore gnawing in bovid long bones. American Antiquity 59, 724–748.

Cavallo, J.A., 1998. A Re-Examination of Isaac's Central-Place Foraging Hypothesis. Unpublished Ph. D. thesis, Rutgers University, New Brunswick.

Cavallo, J.A., Blumenschine, R.J., 1989. Tree-stored leopard kills: expanding the hominid scavenging niche. Journal of Human Evolution 18, 393–399.

Chavaillon, J., 1976. Evidence for the technological practices of early Pleistocene hominids, Shungura Formation, Lower Omo Valley, Ethiopia. In: Coppens, Y. , Howell, F.C., Isaac, G.L., Leakey, R.E.F. (Eds), Early Man and Environments in the Lake Rudolf Basin. University of Chicago Press, Chicago, pp 565–573.

Clark, J.D., Kurashina, H., 1979. Hominid occupation of the East-Central highlands of Ethiopia in the Plio-Pleistocene. Nature 282, 33–39.

de Heinzelin, J., Clark, J.D., White, T., Hart, W., Renne, P., WoldeGabriel, G., Beyene, Y., Vrba, E., 1999. Environment and behavior of 2.5-million-year-old Bouri hominids. Science 284, 625–629.

de la Torre, I., 2004. Omo revisited. Evalating the technological skills of Pliocene hominids. Current Anthropology 45, 439–468.

de la Torre, I., Mora, R., 2005. Unmodified Lithic Material at Olduvai Bed I: Manuports or Ecofacts? Journal of Archaeological Science 32, 273–285.

de la Torre, I., Mora, R., 2009. Remarks on the current theoreticla and methodologicla approaches to the study of technologicla strategies of early humans in eastern Africa. In: Hovers, E. and Braun, D.R. (Eds.), Interdisciplinary Approaches to the Oldowan. Springer, Dordrecht, pp. 15–24.

Domínguez-Rodrigo, M., 1996. A landscape study of bone conservation in the Galana and Kulalu (Kenya) ecosystem. Origini 20, 17–38.

Domínguez-Rodrigo, M., 1997a. Meat-eating by early hominids at the FLK 22 Zinjanthropus site, Olduvai Gorge, Tanzania: an experimental approach using cut mark data. Journal of Human Evolution 33, 669–690.

Domínguez-Rodrigo, M., 1997b. A reassessment of the study of cut mark patterns to infer hominid manipulation of fleshed carcasses at the FLK *Zinj* 22 site, Olduvai Gorge, Tanzania. Trabajos de Prehistoria 54, 29–42.

Domínguez-Rodrigo, M., 1997c. Testing meat-eating in early hominids: analysis of cut-marking processes on defleshed carcasses. Human Evolution 12, 169–182.

Domínguez-Rodrigo, M., 1999a. Flesh availability and bone modification in carcasses consumed by lions. Palaeogeography, Palaeoclimatology and Palaeoecology 149, 373–388.

Domínguez-Rodrigo, M., 1999b. Meat eating and carcass procurement by hominids at the FLK *Zinj* 22 site, Olduvai Gorge, Tanzania: a new experimental approach to the old hunting-versus-scavenging debate. In: Ullrich, H. (Ed.), Lifestyles and Survival Strategies in Pliocene and Pleistocene Hominids. Edition Archaea, Schwelm, Germany, pp. 89–111.

Domínguez-Rodrigo, M., 2002. Hunting and scavenging by early humans: the state of the debate. Journal of World Prehistory 16, 1–54.

Domínguez-Rodrigo, M., 2003. Bone surface modifications, power scavenging and the "display" model at early archaeological sites: a critical review. Journal of Human Evolution 45, 411–416.

Domínguez-Rodrigo, M., 2008. Conceptual premises in experimental design and their bearing on the use of analogy: an example from experiments on cut marks. World Archaeology 40, 67–82.

Domínguez-Rodrigo, M., Alcalá, L., Luque, L., 2008 (in press). Peninj: an archaeological project on the origins of human behavior (1995–2005).

Domínguez-Rodrigo, M., Barba, R., 2005. A study of cut marks on small-sized carcasses and its application to the study of cut-marked bones from small mammals at the FLK Zinj site. Journal of Taphonomy 2-3, 111–124.

Domínguez-Rodrigo, M., Barba, R., 2006. New estimates of tooth marks and percussion marks from FLK *Zinj*, Olduvai Gorge (Tanzania): the carnivore-hominid-carnivore hypothesis falsified. Journal of Human Evolution 50, 170–194.

Domínguez-Rodrigo, M., Egeland, C.P., Barba, R., 2007a. Deconstructing Olduvai. Springer, New York.

Domínguez-Rodrigo, M., Egeland, C.P., Pickering, T.R., 2007b. Models of passive scavenging by early hominids: problems arising from equifinality in carnivore tooth mark frequencies and the extended concept of archaeological palimpsests. In: Pickering, T.R. , Toth, N., Shick, K. (Eds.), Breathing Life into Fossils: Taphonomic Studies in Honor of C.K. ("Bob") Brain. Stone Age Institute Press, Gosport, Indiana, pp. 255–268.

Domínguez-Rodrigo, M., Luque, L., Alcalá, L., de la Torre Sáinz, I., Mora, R., Serrallonga, J., Medina, V., 2002. The ST Site Complex at Peninj, West Lake Natron, Tanzania: implications for early hominids behavioural models. Journal of Archaeological Science 29, 639–665.

Domínguez-Rodrigo, M., Pickering, T.R., 2003. Early hominids, hunting and scavenging: a summary of the discussion. Evolutionary Anthropology 12, 275–282.

Domínguez-Rodrigo, M., Pickering, T.R., Semaw, S., Rogers, M. 2005. Cutmarked Bones from Pliocene Archaeological Sites at Gona, Afar, Ethiopia: implications for the Function of the World's Oldest Stone Tools. Journal of Human Evolution 48, 109–121.

Echassoux, A., Moullé, P. E., Desclaux, E., Alemseged, Z. 2004. The Plio-Pleistocene fauna from the Fejej site FJ-1. In: de Lumley, H., Beyene, Y. (Eds.), Les sites prehistoriques de la regio de Fejej, sud-Omo, Ethiopie, dans leur contexte stratigraphique et paleontologique. ADPF, Paris.

Egeland, C.P., Domínguez-Rodrigo, M. 2008 (in press). Taphonomic perspectives on hominid site use foraging strategies during Bed Ii times at olduvai Gorge, Tanzania. Journal of Human Evolution.

Harris, J.W.K., 1983. Cultural beginnings: Plio-Pleistocene archaeological occurrences from the Afar, Ethiopia. African Archaeological Review 1, 3–31.

Harris, J.W.K., Braun, D., McCoy, J.T., Pobiner, B.L., and Rogers, M. 2002a. New Directions and Preliminary Results from a Landscape Approach to the Study of Archaeological Traces for the Behavior of Plio-Pleistocene Hominids at Koobi Fora. Journal of Human Evolution 42, A15.

Harris, J.W.K., Braun, D.R., McCoy, J.T., Pobiner, B.L., Rogers, M.J., Cachel, S. M., 2002b. Oldowan hominid behavior: recent landscape aArchaeology research at Koobi Fora, Kenya. Nyame Akume 57, 54. [Abs.]

Harris, J.W.K., Semaw, S. 1989. Pliocene archaeology in the Gona River, Hadar. Nyame Akuma 31, 19–21.

Harris, J.W.K., Williamson, P.G., Morris, P.J., de Heinzelin, J., Veniers, J., Helgren, D., Bellomo, R.V., Laden, G., Spang, T.W., Stewart, K., Tappen, M.J., 1990. Archaeology of the Lusso beds. In: Boaz, N.T. (Ed), Evolution of the Environments and Hominidae in the African Western Rift Valley. Museum of Natural History, Martinsville (VA), Virginia, pp. 237–272.

Harris, J.W.K., Williamson, P.G., Verniers, J., Tappen, M.J., Stewart, K., Helgren, D., de Heinzelin, J., Boaz, N.T., Bellomo, R.V., 1987. Late Pliocene hominid occupation of the Senga 5A site, Zaire. Journal of Human Evolution 16, 701–728.

Hovers, E., Schollmeyer, K., Goldman, T., Eck, G., Reed, K., Johanson, D., Kimbel, W., 2002. Late Pliocene archaeological sites in Hadar, Ethiopia. Journal of Human Evolution 42, A17.

Howell, F.C., Haesaerts, P., de Heinzelin, J., 1987. Depositional environments, archaeological occurrences and hominids from Members E and F of the Shungura Formation (Omo Basin, Ethiopia). Journal of Human Evolution 16, 665–700.

Isaac, G.L., 1978. The food-sharing behavior of protohuman hominids. Scientific American 238, 90–108.

Isaac, G.L., 1983. Bones in contention: competing explanations for the juxtaposition of Early Pleistocene artifacts and faunal remains. In: Clutton-Brock, J. , Grigson, C. (Eds.), Animals and Archaeology 1. Hunters and Their Prey. BAR International Series 163. Oxford, pp. 3–19.

Isaac, G.L. 1984. The archaeology of human origins: studies of Lower Pleistocene in East Africa, 1971–1981. Advances in World Archaeology 3, 1–87.

Isaac, G. L., 1997. Plio-Pleistocene archaeology. Koobi Fora Research Project, Vol. 5. Cambridge University Press, Cambridge.

Keeley, L.H. ,Toth, N.P., 1981. Microwear polishes on early stone tools from Koobi Fora, Kenya. Nature 293, 464–465.

Kibunjia, M., 1994. Pliocene archaeological occurrences in the Lake Turkana basin. Journal of Human Evolution 27, 159–172.

Kimbel, W.H., Walter, R.C., Johanson, D.C., Reed, K.E., Aronson, J.L., Assefa, Z., Marean, C.W., Eck, G.G., Bobe, R., Hovers, E., Rak, Y., Vondra, C., Yemane, T., York, D., Chen, Y., Evensen, N.M., Smith, P.E. 1996. Late Pliocene *Homo* and Oldowan tools from the Hadar Formation (Kada Hadar Member), Ethiopia. Journal of Human Evolution 31, 549–561.

Kuhn, T.S., 1962. The Structure of Scientific Revolutions. Princeton, Princeton University Press.

Lakatos, I., 1978. The Methodology of Scientific Research Programmes. Cambridge, Cambridge University Press.

Leakey, M., 1971. Olduvai Gorge, vol. 3. Excavations in Bed I and II, 1960–63. Cambridge University Press, Cambridge.

Lupo, K.D., O'Connell, J.F., 2002. Cut and tooth mark distributions on large animal bones: ethnoarchaeological data from the Hadza and their implications for current ideas about early human carnivory. Journal of Archaeological Science 29, 85–109.

Marean, C.W., Domínguez-Rodrigo, M., Pickering, T.R., 2004. Skeletal element equifinality in zooarchaeology begins with method: The evolution and status of the "shaft critique". Journal of Taphonomy 2, 69–98.

Marean, C.W., Spencer, L.M., 1991. Impact of carnivore ravaging of bone in archaeological assemblages. Journal of Archaeological Science 18, 677–694.

Marean, C.W., Spencer, L.M., Blumenschine, R.J. , Capaldo, S., 1992. Captive hyaena bone choice and destruction, the Schlepp effect and Olduvai archaeofaunas. Journal of Archaeological Science 19, 101–121.

Merrick, H.V., 1976. Recent archaeological research in the Plio-Pleistocene deposits of the Lower Omo, southwestern Ethiopia. In: Isaac G.L., McCown, I. (Eds), Human Origins: Louis Leakey and the East African Evidence. W.A. Benjamín, Menlo Park, pp. 461–482.

Merrick, H.V., Merrick, J.P.S 1976. Archaeological ocurrences of earlier Pleistocene ages from the Shungura Formation. In (Y. Coppens, F.C. Howell, G. L. Isaac & R.E.F. Leakey, eds.) Earliest man and environments in the Lake Rudolf Basin. Chicago, Chicago University Press, pp. 574–584.

Mora, R., de la Torre, I., 2005. Percussion tools in Olduvai Beds I and II (Tanzania): Implications for early human activities. Journal of Anthropological Archaeology 24, 179–192.

O'Connell, J.F., Hawkes, K., Blurton-Jones, N., 1992. Patterns in the distribution, site structure and assemblage composition of Hadza kill-butchering sites. Journal of Archaeological Science 19, 319–345.

Oliver, J.S., 1994. Estimates of hominid and carnivore involvement in the FLK Zinjanthropus fossil assemblage: some socioecological implications. Journal of Human Evolution 27, 267–294.

Pickering, T.R., Domínguez-Rodrigo, M., 2006. The acquisition and use of large mammal carcasses by Oldowan hominins in eastern and southern Africa: a selected review and assessment. In: Toth, N., Schick K. (Eds.), The Oldowan: Case Studies into the Earlierst Stone Age, Stone Age Institute Press, Bloomington, pp. 113–128.

Pickering, T.R., Domínguez-Rodrigo, M., Egeland, C., Brain, C.K., 2004. New data and ideas on the foraging behavior of Early Stone Age hominids at Swartkrans cave. South African Journal of Science 100, 215–220.

Pickering, T.R., Domínguez-Rodrigo, M., Egeland, C.P., Brain, C.K., 2005. The contribution of limb bone fracture patterns to reconstructing early hominid behavior at Swartkrans Cave (South Africa): Archaeological application of a new analytical method. International Journal of Osteoarchaeology 15, 247–260.

Pickering, T.R., Domínguez-Rodrigo, M., Egeland, C., Brain, C.K., Kuman, K., 2007. Carcass foraging by early hominids at Swartkrans Cave (South Africa): A new investigation of the zooarchaeology and taphonomy of member 3. In: Pickering, T.R., Toth, N., Shick, K. (Eds.), Breathing Life into Fossils: Taphonomic Studies in Honor of C.K. ("Bob") Brain. Stone Age Institute Press, Gosport, Indiana, pp. 233–254.

Pickering, T.R., Egeland, C.P., Domínguez-Rodrigo, M., Brain, C.K., Schnell, A. 2008. Testing the "shift in the balance of power" hypothesis at Swartkrans, South Africa: hominid cave use and subsistence behavior in the early Pleistocene. Journal of Anthropological Archaeology 27, 30–47.

Plummer, T.W., 2004. Flaked stones and old bones: biological and cultural evolution at the dawn of technology. Yearbook of Physical Anthropology 47, 118–164.

Plummer, T.W., Bishop, L.C., Ditchfield, P., Hicks, J., 1999. Research on Late Pliocene Oldowan sites at Kanjera South, Kenya. Journal of Human Evolution 36, 151–170.

Pobiner, B.L., Rogers, M., Monahan, C., Harris, J.W.K. 2008. New evidence for hominin carcass processing strategies at 1.5 Ma, Koobi Fora, Kenya. Journal of Human Evolution 55, 103–130.

Potts, R., 1988. Early Hominid Activities at Olduvai. New York, Aldine.

Potts, R., 1994. Variables versus models of Early Pleistocene hominid land use. Journal of Human Evolution 27, 7–24.

Roche, H., Delagnes, A., Brugal, J.-P., Feibel, C., Kibunjia, M., Mourre, V., Texier, P.-J., 1999. Early hominid stone tool production and technical skill 2.34 Myr ago in West Turkana, Kenya. Nature 399, 57–60.

Roche, H., Tiercelin, J.J., 1980. Indiustries lithiques de la formation Plio-Pleistocene d'Hadar: Champagne 1976. In: Leakey, R.E.F., Ogot , B.A. (Eds), Proceedings of the VIIIth Pan-African Congress of Prehistory and Quaternary Studies.International Louis Leakey memorial Institute for African Prehistory, Nairobi, pp. 194–199.

Rose, L., Marshall, F., 1996. Meat eating, hominid sociality, and home bases revisited. Current Anthropology 37, 307–338.

Schaller, G.B., 1972. The Serengeti Lion. University of Chicago Press, Chicago.

Schick, K.D., Toth, N., 1993. Making Silent Stones Speak. Simon and Schuster, New York

Selvaggio, M.M. 1994. Identifying the Timing and Sequence of Hominid and Carnivore Involvement with Plio-Pleistocene Bone Assemblages from Carnivore Tooth Marks and Stone-Tool Butchery Marks on Bone Surfaces. Unpublished Ph. D. Dissertation, Rutgers University.

Semaw, S., Renne, P.R., Harris, J.W.K., Feibel, C.S., Bernor, R.L., Fesseha N. , Mowbray, K., 1997. 2.5-million-year-old stone tools from Gona, Ethiopia. Nature 385, 333–336.

Semaw, S., Rogers., M., Quade, J., Renne, P.R., Butler, R.F., Domínguez-Rodrigo, M., Stout, D., Hart, W.S., Pickering, T.R., Simpson, S.W., 2003. 2.6 million-year-old stone tools and associated bones from OGS-6 and OGS-7, Gona, Afar, Ethiopia. Journal of Human Evolution 45, 169–177.

Turner, A., Anton, M., 1997. The Big Cats and Their Fossil Relatives. Columbia University Press, New York.

Villa, P., Mahieu, E., 1991. Breakage patterns of human long bones. Journal of Human Evolution 21, 27–48.

12. The Environmental Context of Oldowan Hominin Activities at Kanjera South, Kenya

Thomas W. Plummer*
Department of Anthropology
Queens College, CUNY and NYCEP
65-30 Kissena Blvd
Flushing, NY 11367
USA
thomas.plummer@qc.cuny.edu

Laura C. Bishop
Research Centre in Evoltuionary Anthropology and Palaeoecology
School of Natural Sciences and Psychology
Liverpool John Moores University
Byrom Street, Liverpool, L3 3AF
United Kingdom
L.C.Bishop@ljmu.ac.uk

Peter W. Ditchfield
Research Laboratory for Archaeology, 6 Keble Road
Oxford, OX1 3QT
United Kingdom
peter.ditchfield@archaeology-research.oxford.ac.uk

Joseph V. Ferraro
Department of Anthropology
Baylor University
One Bear Place
Waco, TX, 76798
USA
joseph_ferraro@baylor.edu

John D. Kingston
Department of Anthropology, Emory University
1557 Pierce Drive
Atlanta, GA 30322
USA
jkingst@emory.edu

Fritz Hertel
Department of Biology
California State University Northridge
18111 Nordhoff Street

* Address for correspondence: thomas.plummer@qc.cuny.edu

E. Hovers and D.R. Braun (eds.), *Interdisciplinary Approaches to the Oldowan,* DOI: 10.1007/978-1-4020-9059-2,

Northridge, CA 91330-8303
USA
fritz.hertel@csun.edu

David R. Braun
NSF-IRFP, Department of Archaeology
University of Cape Town,
Rondebosch 7701
South Africa
dbraun@uct.ac.za

Keywords: Oldowan, Plio-Pleistocene, Africa, early *Homo*, Paleoenvironmental reconstruction

Abstract

The earliest archaeological traces and two new hominin genera (*Homo* and *Paranthropus*) appear in the late Pliocene of Africa. These first appearances may reflect novel hominin adaptive responses to shifting resource bases over geological time and/or an increasingly seasonal distribution of food over the annual cycle. Whereas regional environmental change has been documented during the Plio-Pleistocene of East Africa, it is difficult to resolve relative proportions of specific habitats at a given place and time, how these may have changed over time, and the explicit nature of particular habitats. Detailed reconstructions of paleohabitats based on paleontological, geological and geochemical evidence are necessary in order to better understand the interplay between environmental change and hominin biological and behavioral evolution.

Reconstruction of the habitats in which archaeological sites were formed provides a window on habitat utilization by early *Homo*, independent of inferred hominin adaptations to specific environmental settings based on hominin morphology or the postmortem distribution of hominin fossils. Here we report on the paleoenvironmental setting of the ca. 2.0 Ma archaeological occurrences at Kanjera, southwestern Kenya. Sedimentological analysis indicates that the site was formed in an alluvial fan, probably near the margin of a lake. Isotopic analysis of pedogenic carbonates indicates that the site complex was formed in an open habitat. Bovid dietary category and taxonomic representation demonstrates that a preponderance of animals grazed and preferred open habitats. Site formation occurred in a grassland-dominated ecosystem, rather than an isolated patch of grassy vegetation within a more wooded setting.

12.1 Introduction

The late Pliocene (ca. 1.95–3.0 Ma) of Africa saw important developments in the evolutionary history of African environments. Environmental change driven by global cooling and drying (Vrba 1985, 1995; Prentice, Denton 1988; deMenocal 1995, 2004), increased global climatic variability (Potts 1998), as well as regional tectonic uplift (Partridge et al. 1995; Sepulchre et al. 2006) appear to have transformed the relatively wooded ecosystems found at many Late Miocene and Early Pliocene sites into complex mosaics incorporating larger amounts of C_4 grasses adapted to warm, dry conditions (WoldeGabriel et al. 1994; Leakey et al. 1995; Wesselman 1995; Reed 1997; Bobe and Behrensmeyer 2004; Bonnefille et al. 2004; Kingston and Harrison 2007). This increase in xerophytic vegetation is reflected in morphological changes across many African large mammal lineages related to the consumption of tougher and possibly more abrasive foods, as well as the dispersal of the Eurasian grazer *Equus* across Africa around 2.3 Ma (Coppens and Howell 1976; Turner and Wood 1993; Wood 1995; Bobe and Behrensmeyer 2004).

Evolutionary change within the Hominini during the Late Pliocene is coincident with, and may be causally related to, these environmental changes (Vrba 1985). The last gracile australopithecines (*A. africanus* and *A. garhi*) disappeared between 2.4–2.5 Ma, and *Paranthropus* and *Homo* made their first appearances between 2.3 and 2.7 Ma (Walker et al. 1986; Hill et al. 1992; Wood 1992; Kimbel 1995; Kimbel et al. 1996; White 2002). Profound changes in hominin behavior, including the first manufacture of stone-tools, large mammal butchery, transport of toolstone and carcass parts, and the discard of lithic and faunal debris at locales now recognized as archaeological sites attributed to the Oldowan Industrial Complex began ca. 2.6 Ma (Isaac 1978; Rogers et al. 1994; Semaw et al. 1997; de Heinzelin et al. 1999; Roche et al. 1999; Plummer 2004; Dominguez-Rodrigo et al. 2005). The dental and gnathic specializations seen in *Paranthropus* and the appearance of Oldowan sites may reflect novel hominin adaptive responses to both a changing resource base over geologic time as well as an increasingly seasonal distribution of food over the annual cycle (Foley 1987; Potts 1998; Plummer 2004).

The postcranial skeletons of the oldest hominin(s) attributable to the genus *Homo* (e.g. *H. habilis*) are poorly known (Wood 1992; Wood and Collard 1999; Dunsworth and Walker 2002). In contrast, the African hypodigm of *H. erectus* sensu lato includes several partial skeletons, and documents a tall, narrow body form proportioned similarly to modern humans living in hot, open, dry environments (Ruff 1991, 2000). Combined with an essentially modern nose configuration (Franciscus and Trinkaus 1988) and the ability to endurance run (Carrier 1984; Bramble and Lieberman 2004; Hilton and Meldrum 2004), it is likely that this taxon thermoregulated like humans, and was capable of prodigious sweating, insuring efficient cooling during bouts of high activity during the heat of the day (Foley 1987). The possibility that forms of *Homo* preceding *H. erectus* exhibited femoral elongation (Hausler and McHenry 2004), the ability to endurance run (Bramble and Leiberman 2004) and a modern nose configuration (Franciscus and Trinkaus 1988) may indicate that earlier members of the genus were also increasingly utilizing open, arid settings. Indeed, a number of researchers have suggested that the evolution of the genus *Homo* was climatically forced by African aridification (Vrba 1985; Stanley 1992; deMenocal 1995). Definitive evidence of Plio-Pleistocene hominin activities within such open settings has thus far eluded paleoanthropologists. The location of hominin fossils themselves may not provide a clear indication of habitat preference during life, as taphonomic factors unrelated to hominin behavior can affect the preservation and distribution of fossils (Behrensmeyer, Hill 1978; Brain 1981; White 1988; Lyman 1994). Moreover, paleoenvironmental reconstructions at many localities frequently lack the resolution necessary to ascertain the habitat structure present during the time periods of interest. Undisturbed archaeological sites provide a record of hominin activity at a defined locale. By reconstructing the environmental context of an archaeological occurrence, a clear association between hominin activity and environmental context can be made. Here we present our reconstruction of the paleoenvironmental context of hominin activities at the Late Pliocene Oldowan site of Kanjera South, Kenya. Both *Homo* and *Paranthropus* are known from East Africa during this time. As discussed in detail in Plummer (2004), we feel it is unlikely that *Paranthropus* produced Oldowan tools and so ascribe the activities at Kanjera to an undetermined species of early *Homo*.

12.2 Physical Setting

The Late Pliocene Oldowan occurrences at Kanjera South (0°20'24" S, 34°32'16" E) are found on the northern margins of the Homa Mountain Carbonatite Complex, Homa Peninsula, southwestern Kenya (Figure 12.1). The history of paleoanthropological research on the Peninsula is summarized in Pickford 1984; Behrensmeyer et al. 1995; Plummer and Potts 1995; Ditchfield et al. 1999. Volcanic activity began with doming of the central portion of the edifice in the Late Miocene and shifted from the center to peripheral vents during the Pliocene and Pleistocene (Saggerson 1952; Le Bas 1977). Today, the heavily eroded edifice of Homa Mountain is 1754m high, approximately 600m above the level of Lake Victoria. The mountain's lower slopes are incised by a radial drainage system exposing Late Miocene through recent sediments (Kent 1942; Pickford 1984; Ditchfield et al. 1999). Evergreen woodland and bushes cover portions of the upper slopes undisturbed by human activity.

Fossiliferous deposits outcrop at Kanjera in three areas, termed the Northern, Middle and Southern Exposures (Behrensmeyer et al. 1995) (Figure 12.1). The deposits in the North and South are members of the Kanjera Formation (Fm); the Middle Exposures deposits postdate the others and have not been formally described. The Oldowan archaeological occurrences are known from the Southern Member of the Kanjera Fm, discussed in more detail below.

12.3 The Chrono- and Litho-Stratigraphy of Kanjera South

The Southern Member of the Kanjera Formation has six Beds, from oldest to youngest KS-1 to KS-6 (Behrensmeyer et al. 1995; Ditchfield et al. ms). The Oldowan archaeological occurrences are largely restricted to Beds KS-1 to KS-3. A combination of biostratigraphy (co-occurrence of *Equus* sp., the suid *Metridiochoerus andrewsi* and the proboscidean *Deinotherium* sp.) as well as magnetostratigraphy (the presence of the Olduvai subchron [1.77–1.95 Ma] in Beds KS-5 and KS-6) indicate that the archaeological occurrences must predate the base of the Olduvai Subchron at 1.95 Ma (Plummer et al. 1999; Ditchfield et al. ms). Here we only discuss Beds relevant to the Oldowan occurrences, KS-1 to KS-4 (Figure 12.1).

The base of KS-1 has not been seen, but the unit is at least 4m thick. Lower KS-1 is a thick, poorly bedded, poorly sorted agglomerate. Upper KS-1 exhibits evidence of alluvial reworking, beds up to 50 cm thick and weak to moderate pedogenic alteration of the pyroclastic parent material.

KS-2 has a poorly defined base that is often gradational from the upper part of KS-1. KS-2 is a moderately pedogenically altered, micaceous, fine sand to silt, with abundant granule to pebble grade clasts. At several horizons in Upper KS-2 there are thicker, laterally discontinuous pebble conglomerates (KS-2 CP) lacking channelization and showing very weakly erosive bases.

KS-3 is a clay-rich medium to fine grained sand with an often strongly bioturbated base, which, along with other soft sediment deformation, points to a wetter depositional environment than the underlying beds. It shows moderate to well-developed paleosols. Towards the northern part of

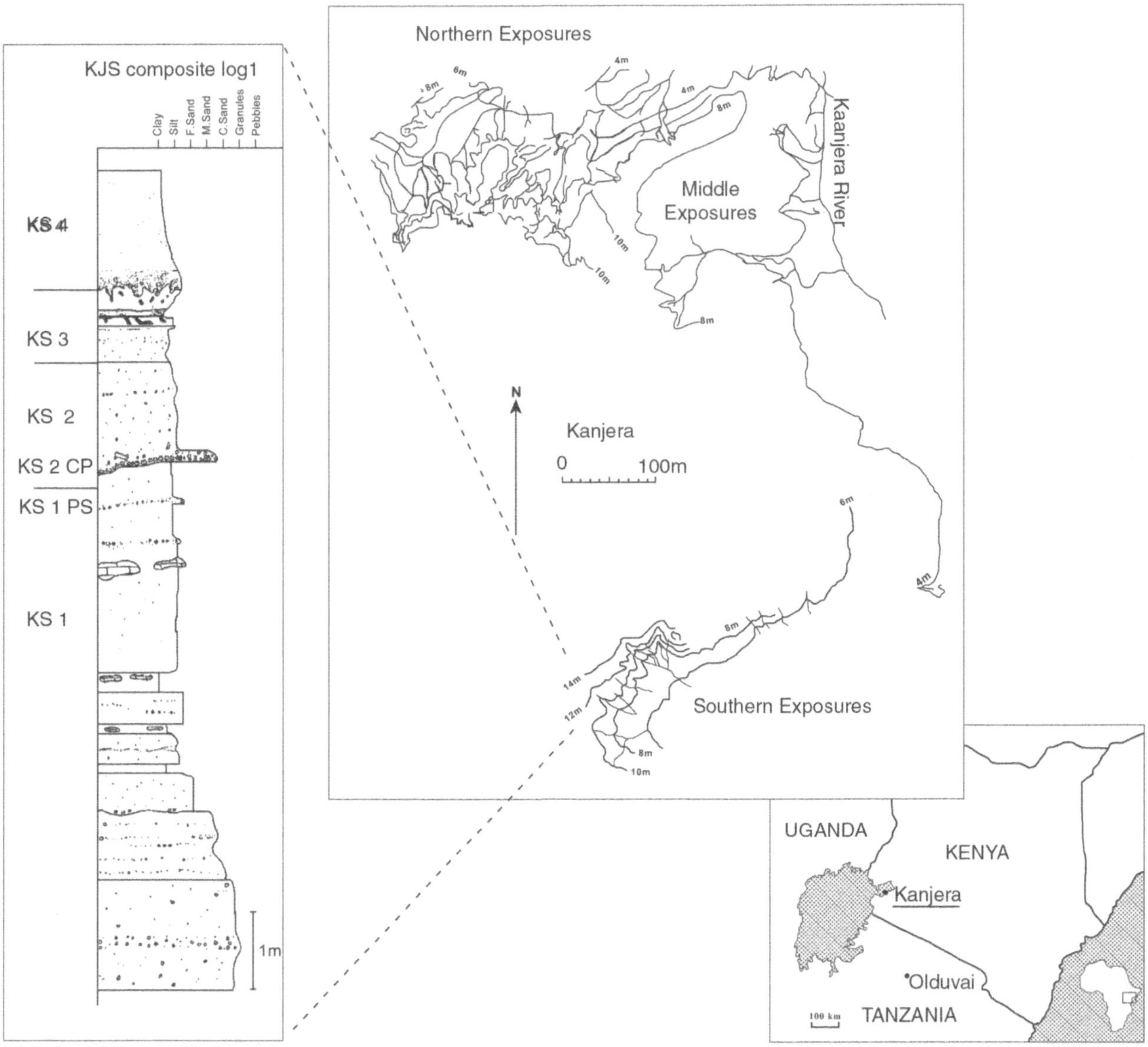

FIGURE 12.1. Placement map and stratigraphic diagram showing the location of Kanjera in southwestern Kenya and of the Southern Exposures at Kanjera. Oldowan artifacts and zooarchaeological fauna are found in beds KS-1 to KS-3.

the Southern Exposures a moderately sized (approximately 1m deep and 2m wide) KS-3 channel feature displaying clear crossbedding with flow directions to the North is present.

KS-4 is a grey to brown, plastic, fine-grained, silty clay with occasional pedogenically altered horizons. It contains few terrestrial fossils and artifacts; however otoliths, fish teeth and fresh water gastropods are relatively common. The pedogenically altered layers become more common in the upper portion of the bed. The entire KS-1 to KS-4 sequence has yielded pedogenic carbonate nodules suitable for isotopic analysis.

12.4 Interpretation of the KS-1 to KS-4 Depositional Environment

Basal KS-1 was probably deposited in one or more relatively large flows of remobilized pyroclastic material possibly as lahars (volcanic debris flows) from the Homa Mountain complex in the South towards a depositional center in the North. These show little internal stratification and no pedogenic development, thus it is likely that they represent rapid, possibly catastrophic deposition across a preexisting landscape. By contrast the well bedded, better sorted and pedogenically modified upper parts of KS-1 yielding artifacts and fossils represent slow reworking of the pyroclastic flow deposits by ephemeral streams running across the fan of the original pyroclastic flows. KS-2 represents a further development of this mode of deposition with more widespread and better developed pedogenesis. Stream flow through the area was restricted to poorly channelized shallow streams that were only weakly erosive into the interchannel deposits. This indicates the area probably had a low depositional slope and represents deposition in a low angle distal fan/lake flat setting. KS-3 documents a wetter depositional environment, reflected both in the style of pedogenic alteration and the increased hippopotamid representation in the fossil assemblage (Table 12.1). The channel feature seen in KS-3 is the first noted in the stratigraphic sequence. KS-4 represents a continuation of this wetter trend as lake deposits transgressed from North to South over the area, sealing the archaeological occurrences. This lake system was periodically dry enough for minor palaeosol development to take place at multiple horizons. These horizons become more common in the upper part of the KS-4 sequence, likely documenting a shallowing/drying trend.

12.5 The Oldowan Excavations

Since 1995 the Homa Peninsula Paleoanthropological Project has undertaken a number of excavations across the gently sloping outcrops of Kanjera South (Excavations 1, 2 and 5; 169m^2, 15m^2 and 4m^2 in size, respectively). Excavation 1 samples KS-1 through KS-3, whereas the other excavations only sample KS-3. Fossil and artifact-bearing horizons were dug with awls and dental picks in 5 cm spits. A Topcon laser theodolite was used for the precise determination of specimen 3D coordinates and in contour mapping. Item dip and orientation was measured with Brunton compasses, and sediments were dry sieved through a 1 mm mesh. To date, thousands of lithic and fossil specimens have been recovered (Table 12.1). Taphonomic and zooarchaeological analyses indicate that the site assemblages formed predominantly through hominin activity (Plummer et al. 1999; Ferraro 2007; Ditchfield et al. ms.).

12.6 Paleoenvironmental Indicators

The stable isotopic analyses of paleosol carbonates and analysis of the archaeological fauna can be used to investigate the paleoenvironmental context of hominin activities at Kanjera. Stable isotopic analysis of pedogenic carbonates at archaeological sites provides an indication of vegetation structure at points on the landscape where archaeological accumulations formed (Sikes 1994; Sikes et al. 1999). Results of stable isotopic analysis of pedogenic carbonates from the archaeological horizons are indicative of habitats with a significant (>75%) proportion of C_4 plant biomass (Plummer et al. 1999; Plummer et al. ms). These values are consistent with dwarf shrub grassland, wooded grassland or open grassland habitats (Sikes 1994). The only other Oldowan site with published stable isotopic data of pedogenic carbonates is FLK I Level 22 (FLK Zinj), from Bed I Olduvai Gorge, Tanzania (Cerling and Hay 1986; Sikes 1994). The stable isotopic composition of carbonates from the Zinj archaeological horizon is indicative of a wooded (riparian woodland or grassy woodland) setting. Thus, isotopic data from these two sites suggest that Oldowan hominins were discarding archaeological debris in both relatively open as well as wooded contexts.

Analysis of archaeological fauna provides an additional method of assessing the paleoenvironmental context of hominin activities (Tables 12.1 and Tables 12.2). In contrast to the locus specific signal from stable isotopic analysis of pedogenic carbonates, macromammalian taxa range broadly during life (Estes 1991; Kingdon 1997) and so provide information on the variety of habitats within a region. The relatively small faunal sample from KS-1 is composed primarily of grazing or mixed feeding ungulates probably (based on modern analogues) favoring relatively open settings, particularly the alcelaphine

TABLE 12.1. Summary of excavated materials from Kanjera South

Bed	Total NISP*	Macromammal NISP	Macromammal MNI	Principal fauna (%NISP, %MNI)	Lithics
KS-1	982	975 (525)	25	Bovid (92.4, 72.0), Equid (4.4, 8.0), Suid (1.5, 8.0), Hippo (0.2, 4.0)	179
KS-2	2190	2153 (886)	40	Bovid (82.6, 67.5), Equid (11.6, 10.0), Suid (0.9, 5.0), Hippo (1.0, 2.5)	2533
KS-3	491	470 (172)	16	Bovid (77.9, 68.8), Equid (4.7, 6.3), Suid (0.6, 6.3), Hippo (14.0, 12.5)	171

*NISP refers to the number of identifiable specimens (Lyman 1994). Here "Total NISP" reflects the sum of specimens recovered with coordinate data. Tens of thousands of non-identifiable bone fragments <2cm are not included in the present study. MNI refers to the minimum number of individual animals necessary to account for the bones in the assemblage. Fossil from conglomeratic facies (CP levels of Plummer et al. 1999) are poorly preserved and are likewise excluded from these counts. Macromammals weigh > 5 kg. Macromammal NISP values are total sums of specimens identified to Linnean class and, in parentheses, the sum of specimens identified beyond class. %NISP and %MNI include macromammals only. Faunal data from Ferraro (2007). Lithic data from Braun (2006), also excluding artifacts from the CP levels.

TABLE 12.2. Macromammalian fauna from Excavation 1, by bed. Several important surface collected taxa noted as well

Taxonomic group	KS-1	KS-2	KS-3	KS-1 and KS-2 Surface
Cercopithecidae				
Cercopithecine indet.		X		
Cercopithecus sp.		X		
Theropithecus sp.		X		
Carnivora				
Felidae, size 3				X
Felidae, size 2		X		
Hyaenidae		X		
Crocuta cf.dietrichi	X			
Suidae				
Suidae, size 2	X		X	
Suidae, size 3a	X			
Metridiochoerus andrewsi				X
Metridiochoerus modestus		X		
Hippopotamidae	X	X	X	
Giraffidae		X		
Bovidae				
Tragelaphini, size 3a		X		
Reduncini, size 1		X		
Reduncini, size 2		X	X	
Kobus sp.		X	X	
Hippotragus sp. size 3		X		
Alcelaphini cf. *Parmularius altidens*	X	X	X	
Alcelaphini, size 3b	X	X	X	
Antidorcas recki	X	X	X	
Equidae			X	
Equus sp.	X	X		
Eurygnathohippus sp.	X	X		
Rhinocerotidae		X		
Proboscidea				
Elephantidae			X	
Deinotherium sp.				X

Parmularius cf. *altidens*, the antilopine *Antidorcas recki* and the equid genera *Equus* and *Eurygnathohippus*. The recovered remains of *Struthio* (ostrich) and a lagomorph (cf. *Lepus*) also suggest grassland habitats.

The largest and most diverse faunal sample is known from KS-2. Monkeys attributable to the genera *Theropithecus* and *Cercopithecus* reflect the presence of C_4 vegetation as well as C_3 food and shelter, respectively. The giraffid and tragelaphine bovid fossils indicate the proximity of woodland, as perhaps does the suid *Metridiochoerus modestus* (Bishop 1999). The bovid *Hippotragus* may reflect the ecotone between C_3-rich and C_4-rich habitats, whereas the reduncine bovids are suggestive of seasonally waterlogged (edaphic) grassland. The presence of standing water is consistent with the recovery of fish, crocodile and hippopotamus remains.

The limited KS-3 sample also exhibits a preponderance of grassland indicators as well as evidence of rising lake levels noted in the discussion of the depositional environment. In addition to antilopines, alcelaphines, and equids presumably signaling secondary grasslands, two different reduncine bovids suggest the presence of edaphic grass. Hippopotamid fossils are more common in KS-3 than underlying Beds, which corresponds with the increasing dominance of the lake in the landscape during KS-3 times.

Although there are macromammal indicators of multiple habitats in each faunal assemblage, the representation of taxa preferring grassland settings is most salient. In the rest of our paleoecological discussion we will focus on bovids, which provide the bulk of the macromammal sample by Bed (78–92%; Table 12.1). Zooarchaeological analysis indicates that the bovids were predominantly collected by hominins (Ferraro 2007). The extent to which hominin foraging practices would have biased paleoenvironmental reconstructions from these samples, however, is unclear. Certainly, issues of processing decisions, transport dynamics, etc., may result in discordance between initially-acquired and subsequently-deposited faunal assemblages. Subsequent biotic and abiotic taphonomic biases may further modify assemblage compositions. Lastly, and perhaps most importantly, the extent to which Early Stone Age hominin prey preferences track local faunal abundances remain wholly unknown. On the other hand, analysis of death assemblages from a range of modern African predators demonstrates that relatively accurate habitat reconstructions can be made from them — specifically when dealing with the presence/absence of taxa (Reed 1997). The wide array of taxa from Oldowan sites (Potts 1988; Plummer 2004) suggests that Oldowan hominins had a broad search image, and that their carcass collecting behavior may provide an accurate paleoenvironmental signal.

The frequency of the tribes Alcelaphini and Antilopini within a bovid sample is often used as a grassland indicator termed the alcelaphine plus antilopine criterion, or AAC (Vrba 1980). Comparison with modern analogues suggests that sites with AAC values of greater than 60% are open (frequently grassland-dominated) settings, whereas those with AAC values less than 30% have considerable bush or tree cover (Vrba 1980; Kappelman 1984; Potts 1988). Whether calculated using NISP or MNI, alcelaphines and antilopines at Kanjera South greatly outnumber members of the other bovid tribes, suggesting that the local ecosystem was grassland-dominated.

Shipman and Harris (1988) elaborated on the AAC, proposing a method of habitat reconstruction utilizing the frequency of alcelaphines and antilopines (AA) as well as the summed frequencies of reduncines and bovines (RB) and tragelaphines and aepycerotines (TA). They demonstrated a good correspondence between bovid tribal representation and regional habitat representation in modern wildlife areas and extended this relationship to the fossil record. Wildlife areas with high AA frequencies tend to be arid and often have a relatively large proportion of open habitats (Figure 12.2). Parks with high TA frequencies generally also have low rainfall but were dominated by a variety of woodland/bushland habitats. Parks with high RB frequencies tend to have high rainfall and generally include riparian grassland and woodland. Figure 12.2a presents the ternary diagram for the modern wildlife areas discussed by Shipman and Harris (1988). Figure 12.2b presents a plot of the Kanjera tribal values, by Bed, compared with relevant modern and fossil samples. The Kanjera samples sort with several modern open or grassland-dominated ecosystems (e.g., Serengeti, Tazania), consistent with the large proportion of grazers. Also grouped with them are the Olduvai site samples from the driest portion of the Bed I sequence (FLK NI L/4–6) (Potts 1988; Fernandez-Jalvo et al. 1998). A number of analyses have suggested that the FLK NN I levels from Bed I Olduvai were deposited under moist conditions (Potts 1988; Plummer and Bishop 1994) and here they plot out with well wooded areas having high frequencies of reduncines (e.g. Lake Nakuru, Kenya; Arli, Burkina Faso). FLK I Zinj (L/22), the only Bed I Olduvai archaeological site assemblage universally viewed as having been predominantly formed through hominin activity, is plotting with Fina, Mali and reasonably close to Kainji, Nigeria, both wet, well-wooded areas with reasonably high AA and RB frequencies. As noted with the analysis of pedogenic carbonates, the depositional context at Kanjera seems to have been in a more open setting than that documented for FLK Zinj. The Lake Turkana Basin provides additional paleontological and archaeological samples from multiple localities. Some of the largest paleontological samples as well as several Oldowan occurrences have been found in the Late Pliocene deposits of the Shungura Fm, Ethiopia (Howell et al. 1987; Bobe and Eck 2001; de la Torre 2004). Members E through G span the interval from roughly 1.9 to 2.4 Ma and *Homo* as well as *Paranthropus* appear to be present through much of this sequence (Suwa et al. 1996). Oldowan archaeological sites in Member F are approximately 2.3 Ma, and seem to be near ecotones between riparian woodland and forest and more open savanna (Howell et al. 1987; de la Torre 2004). Although the surface collected Shungura Fm faunas, deposited in sediments from a large, meandering river system, may not be directly comparable to the archaeological faunal assemblages from Kanjera, they do suggest paleoenvironmental settings considerably different than those present during KS-1 through KS-3 deposition (Figure 12.2b). The low frequencies of alcelaphines and antilopines are in stark contrast to what has been recovered at Kanjera South. The Member E and F faunas have very high TA frequencies, suggesting a considerable amount of dry bushland or woodland. Lower Member G appears a bit wetter with a higher frequency of reduncines. Upper Member G, roughly coeval with the Kanjera occurrences, is dominated by reduncines, and still has a reasonably high TA frequency. This is suggestive of a near- water setting with extensive riparian grass-

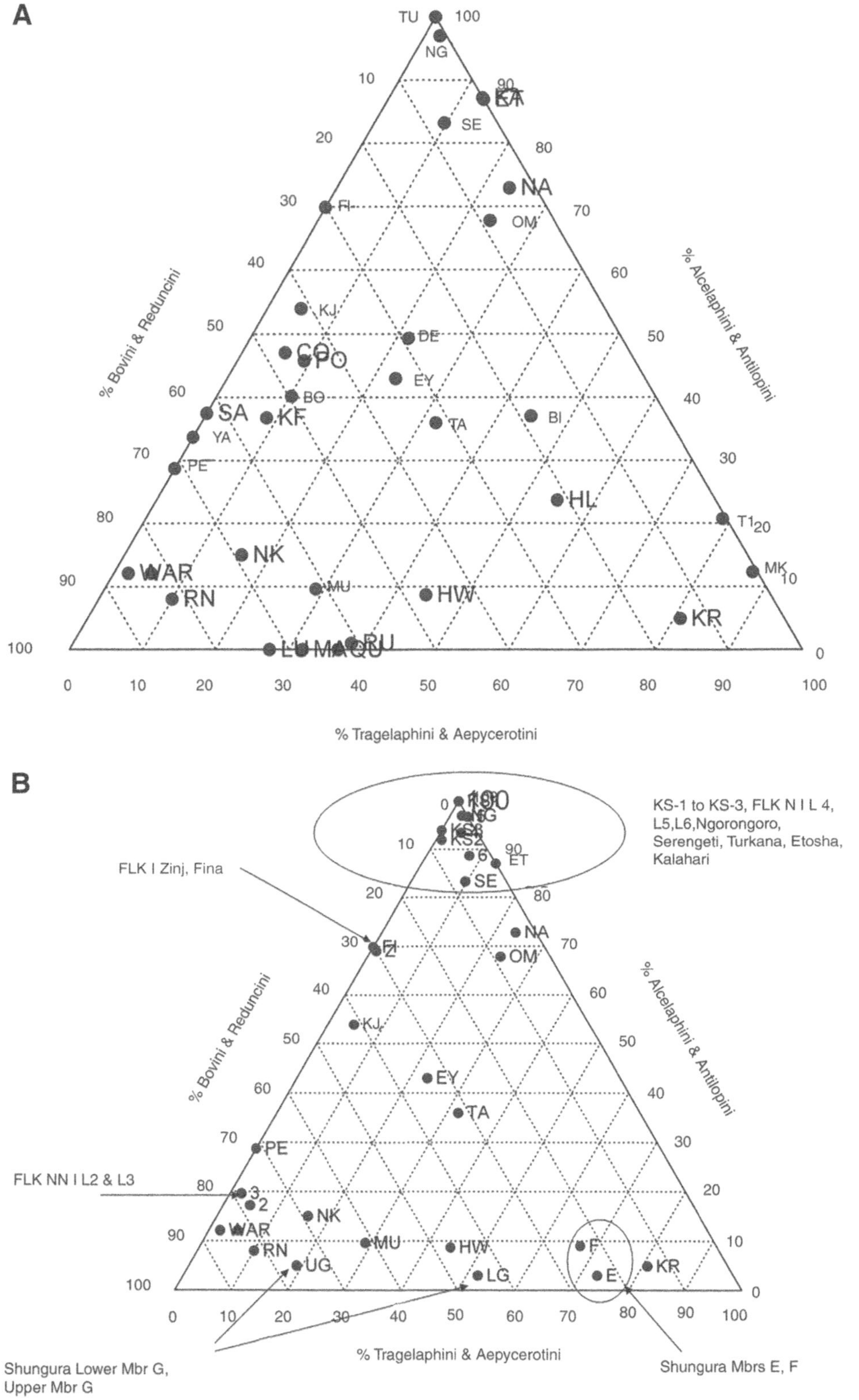

FIGURE 12.2. Ternary diagrams showing the relative frequencies of three sets of bovid tribes, after Shipman and Harris (1988). High Alcelaphini and Antilopini frequencies signal high proportions of open, arid environments, high Bovini and Reduncini frequencies signal closed, frequently moist environments and high Tragelaphini and Aepycerotini frequencies signal closed, frequently dry environments. A). Ternary plot of bovid frequencies from modern wildlife areas described by Shipman and Harris (1988). B). Ternary plot of fossil bovid frequencies from KS-1, KS-2 and KS-3 at Kanjera, select sites from Bed I Olduvai Gorge, Tanzania, Members E, F and lower and upper Member G from the Omo Shungura Formation, Ethiopia and representative modern wildlife areas. Fossil site tribal frequencies generated from NISP values. Olduvai data from Kappelman (1984) and Bunn and Kroll (1986), the Omo Shungura data is from Bobe and Eck (2001) and the Kanjera NISP data is from Ferraro (2007). The three Kanjera samples cluster with modern dry, relatively open wildlife areas as well as the fossil samples from the driest period of deposition at Bed I Olduvai (FLK NI L4 – L6).

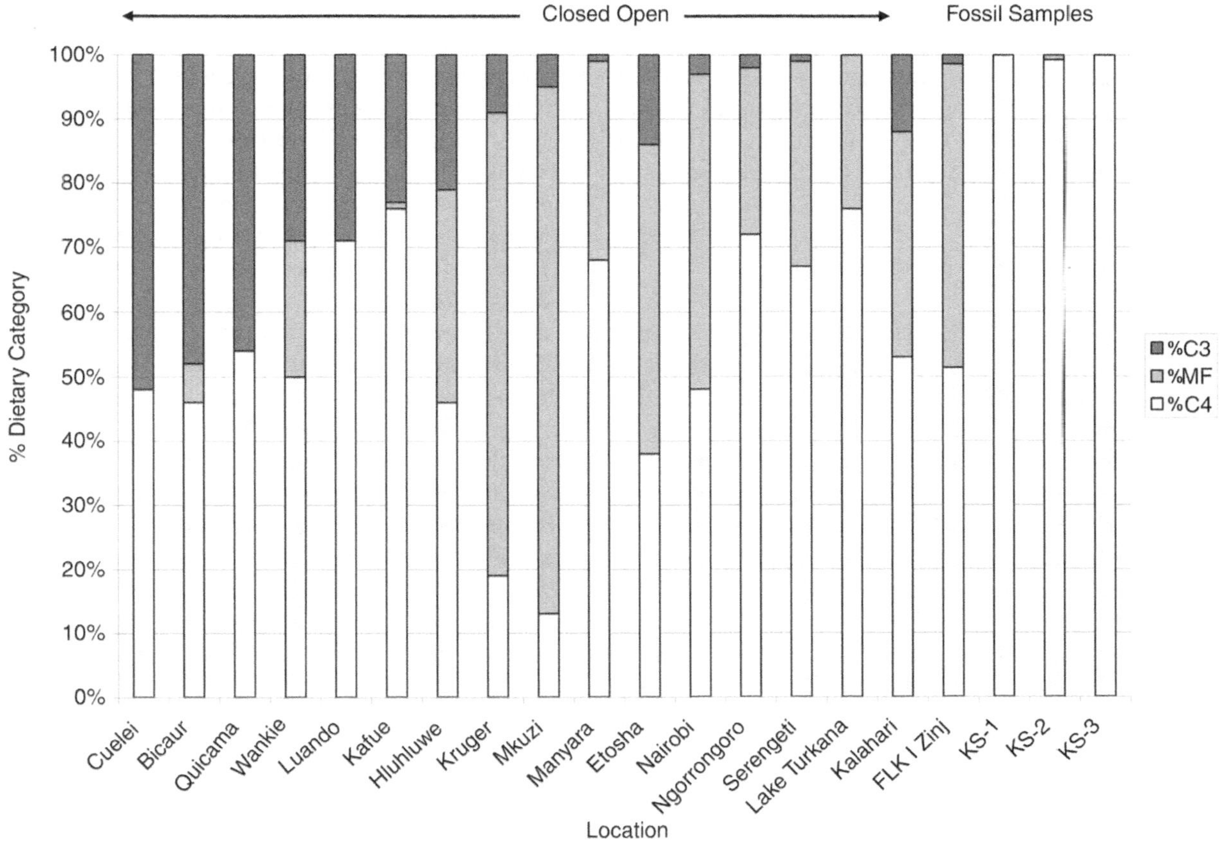

FIGURE 12.3. Stacked bar charts showing the frequency of different bovid dietary categories for parks with low frequencies of Alcelaphini and Antilopini and a high proportion of C_3 vegetation (closed) and parks with high frequencies of Alcelaphini and Antilopini and a low proportion of C_3 vegetation (open). Dietary category frequencies for four fossil assemblages (FLK I Zinj and Kanjera bovid assemblages by bed) are shown on the right side of the plot. The Kanjera assemblages are dominated by grazing bovids, whereas the FLK Zinj assemblage exhibits a higher frequency of bovids with a mixed browse and graze diet. Modern data from Sponheimer and Lee-Thorpe (2003), FLK I Zinj NISP data from Bunn and Kroll (1986) and Kanjera NISP data are from Ferraro (2007).

lands and some woodland, with drier bushland/woodland found away from the main axis of the river. Comparisons between Kanjera South, Bed I Olduvai and the Omo Shungura Fm illustrate the diversity of environments inhabited by early *Homo* in the Late Pliocene.

Carbon stable isotope data from tooth enamel provides an additional method of paleoenvironmental reconstruction complementing analyses discussed above. Sponheimer and Lee-Thorp (2003) defined three dietary classes (C_3 consumers, C_4 consumers, mixed feeders) using isotopic data and a literature review of bovid diets and habitat preferences. In their scheme, C_3 consumers eat 80% or more C_3 vegetation, C_4 consumers eat 80% or more C_4 vegetation, and bovids with diets intermediate between these extremes are mixed feeders (Figure 12.3). These isotopically-defined dietary boundaries were then used to categorize the diets of modern African bovid tribes (e.g. Alcelaphini are C_4 consumers, Aepycerotini are mixed feeders and Tragelaphini are C_3 consumers). The relative frequency of bovid tribes from a particular wildlife refuge are transformed into frequencies of different consumer types, with percentages of C_3 consumers being used to distinguish parks with a high percentage of open versus closed habitat. They found that relatively "open" wildlife areas had a bovid fauna with less than 15% C_3 consumers, whereas parks with a substantial amount of woodland or bushland habitat always had more than 20% C_3 consumming bovids.

Carbon stable isotopic data have been obtained from the teeth of the full spectrum of taxa from Kanjera (Plummer et al. ms.) allowing us to place our data into Sponheimer and Lee-Thorp's (2003) framework. The Kanjera bovid sample is composed almost entirely of C_4 consumers, consistent with other indications of open habitat for the locality. This strong "grass" dietary signal is not simply the result of hominin selectivity for several taxa of grazing bovids. Taxa that frequently are mixed feeding at other localities (e.g. *A. recki*) are grazing at Kanjera, and taxa that tend to browse elsewhere (e.g. the tragelaphines) are mixed feeding with a strong grass component. Using the relationship between tribe and diet to convert the FLK I Zinj bovid assemblage into dietary categories suggests that the Zinj bovid assemblage was also not drawn from a densely wooded area, but that the relatively high frequency of mixed feeders signals a greater availability of C_3 vegetation than was found in the region around Kanjera during KS-1 to KS-3 deposition.

12.7 Summary and Discussion

Multiple lines of evidence indicate that Late Pliocene Kanjera was a lightly-wooded to open grassland habitat, with a lake to the North and presumably bushes and woods lining nearby hills and perhaps some drainages. Wash from the foothills of Homa Mountain drained toward the lake, burying faunal and lithic materials on a generally low-energy alluvial plain during KS-1 through KS-3 deposition. The lake transgressed from North to South through time, completely covering the locality during KS-4 deposition (Behrensmeyer et al. 1995; Ditchfield et al. ms).

Paleoenvironmental information is playing an increasingly important role in interpreting the pattern of hominin biological and behavioral evolution. Faunal, isotopic and pollen data suggest that Plio-Pleistocene ecosystems were more heterogeneous and frequently contained more C_4 vegetation than those of the early Pliocene ecosystems (Reed 1997; Bobe and Behrensmeyer 2004; Kingston and Harrison 2007). Prominent hypotheses for how global climatic change impacted biotic evolution in Africa emphasize shifts in the relative proportion of C_4 plants (particularly grasses) versus C_3 woody vegetation in floral communities (deMenocal 1995; Behrensmeyer 2006). The morphology of fossils of the genus *Homo* from at least the beginning of the early Pleistocene shows morphological adaptations to open, arid environments. Yet definitive evidence of grassland dominated settings, and of hominin activities within open habitats has remained elusive.

Our limited review of several Plio-Pleistocene faunal assemblages and more extensive reviews elsewhere (e.g. Reed 1997; Bobe and Behrensmeyer 2004) illustrate a high degree of environmental variability across East Africa during the Plio-Pleistocene. The paleoenvironmental significance of research at Kanjera South is that it provides the first clear documentation of recurrent hominin activities in an open habitat within a grassland-dominated ecosystem, confirmed by the isotopic geochemistry of paleosols and stable isotope analysis reconstruction of faunal diet. This complements work at Olduvai, where sites such as FLK *Zinj* appear to have formed in a woodland, and the Plio-Pleistocene settings during Shungura Fm deposition, where archaeological occurrences have been found near woodland/savanna ecotones and where the presence of edaphic grassland and/or woodland was more salient than secondary grassland. Several important things are suggested by these data. First, there seems to be a great deal of unevenness in the distribution of secondary grassland, which is likely to have been a regionally variable phenomenon. Variability in habitat representation over geologic time, rather than a unidirectional increase in C_4 vegetation, may have been an important factor shaping hominin evolution (Potts 1998). Second, if these Late Pliocene archaeological sites reflect the range of settings occupied by one species of tool-using hominin, it would suggest that tool use buffered the effects of environmental variability by allowing hominins to expand the range of foods being acquired or acquire certain types of food more efficiently.

Finally, reconstructing the overall environmental context of Plio-Pleistocene ecosystems and the specific settings of archaeological site formation are important, because our understanding of Oldowan hominin habitat utilization impacts our consideration of hominin group size and cohesion, diet, degree of competition with carnivores on site, the place of hominins within the carnivore competitive hierarchy and the potential selective pressures important in transforming an early form of *Homo* into *Homo erectus* by 1.8 Ma (Foley 1987; Plummer 2004). The evidence from Kanjera South suggests that at least one species of the genus *Homo* was repeatedly utilizing an open setting, prior to the earliest evidence for the emergence of *H. erectus*.

Acknowledgments. We are grateful to the Office of the President of Kenya, and the National Museums of Kenya for permission to study the Kanjera fossils and artifacts. The Homa Peninsula field research is conducted through the cooperative agreement between the National Museums of Kenya and the Smithsonian Institution. Logistical support and funding was also provided by the Smithsonian's Human Origins Program. Funding from the L. S. B. Leakey Foundation, the National Geographic Society, the National Science Foundation, the Professional Staff Congress-City University of New York Research Award Program and the Wenner-Gren Foundation for Kanjera field and laboratory work is gratefully acknowledged. We thank Rick Potts and the Human Origins Program for support during all phases of the Kanjera research.

References

Behrensmeyer, A.K., 2006. Climate change and human evolution. Science 311, 476–478.

Behrensmeyer, A.K., Hill, A.P., 1980 (Eds.). Fossils in the Making. University of Chicago Press, Chicago.

Behrensmeyer, A.K., Potts, R., Plummer, T., Tauxe, L., Opdyke, N., Jorstad, T., 1995. The Pleistocene locality of Kanjera, Western Kenya: stratigraphy, chronology and paleoenvironments. Journal of Human Evolution 29, 247–274.

Binford, L.R., 1988. Fact and fiction about the Zinjanthropus floor: data, arguments and interpretation. Current Anthropology 29, 123–135.

Bishop, L., 1999. Suid paleoecology and habitat preferences at African Pliocene and Pleistocene hominid localities. In: Bromage, T.G., Schrenk, F. (Eds.), African Biogeography, Climate Change, and Human Evolution. Oxford University Press, Oxford, pp. 216–225.

Blumenschine, R.J., 1995. Percussion marks, tooth marks, and experimental determinations of the timing of hominid and carnivore access to long bones at FLK Zinjanthropus, Olduvai Gorge, Tanzania. Journal of Human Evolution 29, 21–51.

Bobe, R., Behrensmeyer, A.K., 2004. The expansion of grassland ecosystems in Africa in relation to mammalian evolution and the origin of the genus Homo. Palaeogeography, Palaeoclimatology and Palaeoecology 207, 399–420.

Bonnefille, R., Potts, R., Chalie, F., Jolly, D., Peyron, O., 2004. High-resolution vegetation and climate change associated with Pliocene Australopithecus afarenisis. Proceedings of the National Academy of Sciences 101, 12125–12129.

Bramble, D.M., Leiberman, D.E. 2004. Endurance running and the evolution of Homo. Nature 432, 345–352.

Brain, C.K., 1981. The Hunters or the Hunted. University of Chicago Press, Chicago.

Braun, D.R., 2006. Ecology of Oldowan Technology: Perspectives from Koobi Fora and Kanjera South. Unpublished Ph.D. thesis, Rutgers University, New Brunswick, NJ.

Bunn, H.T., Kroll, E.M., 1986. Systematic butchery by Plio/Pleistocene hominids at Olduvai Gorge, Tanzania. Current Anthropology 27, 431–452.

Carrier, D.R., 1984. The energetic paradox of human running and hominid evolution. Current Anthropology 25, 483–495.

Cerling, T.E., Hay, R.L., 1986. An Isotopic Study of Paleosol Carbonates from Olduvai Gorge. Quaternary Research 25, 63–78.

Coppens, Y., Howell, F.C., 1976. Mammalian faunas of the Omo Group: distributional and biostratigraphical aspects. In: Coppens, Y., Howell, F.C., Isaac, G.L., Leakey, R. E.F. (Eds.), Earliest Man and Environments in the Lake Rudolf Basin. University of Chicago Press, Chicago, pp. 177–192.

de Heinzelin, J., Clark, D., White, T., Hart, W., Renne, P., WoldeGabriel, G., Beyene, Y., Vrba, E., 1999. Environment and Behavior of 2.5-Million-Year-Old Bouri Hominids. Science 284, 625–629.

deMenocal, P.B., 1995. Plio-Pleistocene African climate. Science 270, 53–59.

deMenocal, P.B., 2004. African climate change and faunal evolution during the Pliocene-Pleistocene. Earth and Planetary Science Letters 220, 3–24.

de la Torre, I., 2004. Omo revisited. Current Anthropology 45, 439–465.

Ditchfield, P., Hicks, J., Plummer, T.W., Bishop, L., Potts, R., 1999. Current research on the Plio-Pleistocene deposits north of Homa Mountain, Southwestern Kenya. Journal of Human Evolution 36, 123–150.

Ditchfield, P., Plummer, T., Bishop, L., Ferraro, J., Braun, D.R., ms. Physical context of site formation at Kanjera South, Kenya.

Dominguez-Rodrigo, M., Pickering, T. R., Semaw, S., Rogers, M.J., 2005. Cutmarked bones from Pliocene archaeological sites at Gona, Afar, Ethiopia: implications for the function of the world's oldest stone tools. Journal of Human Evolution 48, 109–121.

Dunsworth, H., Walker, A., 2002. Early genus *Homo*. In: Hartwig, W. (Ed.), The Primate Fossil Record. Cambridge University Press, Cambridge, pp. 419–436.

Estes, R.D., 1991. The Behavior Guide to African Mammals. University of California Press, Berkeley.

Fernandez-Jalvo, Y., Denys, C., Andrews, P., Williams, T., Dauphin, Y., Humphrey, L., 1998. Taphonomy and palaeoecology of Olduvai Bed-I (Pleistocene, Tanzania). Journal of Human Evolution 34, 137–172.

Ferraro, J.V., 2007. Broken Bones and Shattered Stones: On the Foraging Ecology of Oldowan Hominins. Unpublished Ph.D. dissertation. University of California, Los Angeles, Los Angeles, California.

Foley, R., 1987. Another Unique Species: Patterns in Human Evolutionary Ecology. John Wiley and Sons, New York.

Franciscus, R.G., Trinkaus, E., 1988. Nasal morphology and the emergence of Homo erectus. American Journal of Physical Anthropolgy 75, 517–527.

Gentry, A.W., Gentry, A., 1978. Fossil bovidae (mammalia) of Olduvai Gorge, Tanzania, Part 2. Bulletin of the British Museum of Natural History (Geology Series) 30, 1–83.

Haeusler, M., McHenry, H., 2004. Body proportions of *Homo habilis* reviewed. Journal of Human Evolution 46, 433–465.

Hill, A., Ward, S., Deino, A., Curtis, G., Drake, R., 1992. Earliest Homo. Nature 355, 719–722.

Hilton, C.E., Meldrum, D.J., 2004. Striders, runners and transporters. In: Meldrum, D.J., Hilton, C.E. (Eds.), From Biped to Strider: The Emergence of Modern Human Walking, Running and Resource Transport. Kluwer Academic, New York, pp. 1–8.

Howell, F.C., Haesaerts, P., De Heinzelin, J., 1987. Depositional environments, archeological occurrences and hominds from Members E and F of the Shungura Formation (Omo Basin, Ehtiopia). Journal of Human Evolution 16, 665–700.

Isaac, G.L., 1978. The food-sharing behavior of protohuman hominids. Scientific American 238, 90–108.

Kappelman, J., 1984. Plio-Pleistocene environments of Bed I and Lower Bed II, Olduvai Gorge, Tanzania. Paleogeography, Palaeoclimatology, Palaeoecology 38, 171–196.

Kimbel, W.H., 1995. Hominid speciation and Pliocene climatic change. In: Vrba, E.S., Denton, G.H., Partridge, T.C., Burckle, L.H. (Eds.), Paleoclimate and Evolution, with Emphasis on Human Origins. Yale University Press, New Haven, pp. 425–437.

Kimbel, W.H., Walter, R.C., Johanson, D.C., Reed, K.E., Aronson, J.L., Assefa, Z., Marean, C.W., Eck, G.G., Bobe, R., Hovers, E., Rak, Y., Vondra, C., Yemane, T., York, D., Chen, Y., Evensen, N.M., Smith, P.E., 1996. Late Pliocene Homo and Oldowan tools from the Hadar Formation (Kada Hadar Member), Ethiopia. Journal of Human Evolution 31, 549–561.

Kingdon, J., 1997. The Kingdon Field Guide to African Mammals. Academic Press, London.

Kent, P.E., 1942. The Pleistocene beds of Kanam and Kanjera, Kavirondo, Kenya. Geology Magazine 79, 117–132.

Kingston, J.D., Harrison, T., 2007. Isotopic dietary reconstructions of Pliocene herbivores at Laetoli: Implications for early hominin paleoecology. Palaeogeography, Palaeoclimatology, Palaeoecology 243, 272–306.

Leakey, M.G., Feibel, C., McDougall, I., Walker, A., 1995. New four-million-year-old hominid species from Kanapoi and Allia Bay, Kenya. Nature 376, 565–571.

Lyman, R.L., 1994. Vertebrate Taphonomy. Cambridge University Press, Cambridge.

Le Bas, M.J., 1977. Carbonatite-Nephelinite Volcanism. John Wiley and Sons, New York.

Oliver, J.S., 1994. Estimates of hominid and carnivore involvement in the FLK Zinjanthropus fossil assemblage: some socioecological implications. Journal of Human Evolution 27, 267–294.

Partridge, T.C., Bond, G.C., Hartnady, C.J.H., deMenocal, P.B., Ruddiman, W.F., 1995. Climatic effects of late Neogene tectonism and volcanism. In: Vrba, E.S., Denton, G.H., Partridge, T.C., Burckle, L.H. (Eds.), Paleoclimate and Evolution, with Emphasis on Human Origins. Yale University Press, New Haven, pp. 8–23.

Pickford, M., 1984. Kenya Palaeontology Gazetteer, Vol. 1, Western Kenya. National Museums of Kenya, Nairobi.

Plummer, T.W., 2004. Flaked stones and old bones: biological and cultural evolution at the dawn of technology. Yearbook of Physical Anthropology 47, 118–164.

Plummer, T.W., Bishop, L.C., 1994. Hominid paleoecology at Olduvai Gorge, Tanzania as indicated by antelope remains. Journal of Human Evolution 27, 47–75.

Plummer, T.W., Bishop, L., Ditchfield, P., Hicks, J., 1999. Research on late Pliocene Oldowan sites at Kanjera South, Kenya. Journal of Human Evolution 36, 151–170.

Plummer, T., Ditchfield, P., Bishop, L., Kingston, J., Ferraro, J., Braun, D., Hertel, F., Potts, R.B., ms. Oldest evidence of toolmaking hominins in a grassland-dominated ecosystem.

Plummer, T.W., Potts, R., 1995. The hominid fossil sample from Kanjera, Kenya: description, provenance and implications of new and earlier discoveries. American Journal of Physical Anthropology 96, 7–23.

Potts, R., 1988. Early Hominid Activities at Olduvai. Aldine De Gruyter, New York.

Potts, R., 1998. Environmental hypotheses of hominin evolution. Yearbook of Physical Anthropology 41, 93–136.

Prentice, M.L., Denton, G.H., 1988. The deep-sea oxygen isotope record, the global ice sheet system and hominid evolution. In: Grine, F.E., (Ed.), Evolutionary History of the "Robust" Australopithecines. Aldine de Gruyter, New York, pp. 383–403.

Reed, K.E., 1997. Early hominid evolution and ecological change through the African Plio-Pleistocene. Journal of Human Evolution 32, 289–322.

Roche, H., Delagnes, A., Brugal, J.-P., Feibel, C., Kibunjia, M., Mourre, V., Texier, P-J., 1999. Early hominid stone tool production and technical skill 2.34 Myr ago in West Turkana, Kenya. Nature 399, 57–60.

Rogers, M.J., Harris, J.W.K., Feibel, C.S., 1994. Changing patterns of land use by Plio-Pleistocene hominids in the Lake Turkana Basin. Journal of Human Evolution 27, 139–158.

Ruff, C., 1991. Climate and body shape in hominid evolution. Journal of Human Evolution 21, 81–105.

Ruff, C., 2000. Body size, body shape, and long bone strength in modern humans. Journal of Human Evolution 38, 269–290.

Saggerson, E.P., 1952. Geology of the Kisumu District. Geological Survey of Kenya, Report 21, 86 pp.

Semaw, S., Renne, P., Harris, J.W.K., Feibel, C.S., Bernor, R.L., Fesseha, N., Mowbray, K., 1997. 2.5 million-year-old stone tools from Gona, Ethiopia. Nature 385, 333–336.

Sepulchre, P., Ramstein, G., Fluteau, F., Schuster, M., Tiercelin, J., Brunet, M., 2006. Tectonic uplift and eastern African aridification. Science 313, 1419–1423.

Shipman, P., Harris, J.M., 1988. Habitat preference and paleoecology of *Australopithecus boisei* in Eastern Africa. In: Grine, F.E. (Ed.), The evolutionary History of the "Robust" Australopithecines. Aldine de Gruyter, New York, pp. 343–380.

Sikes, N., 1994. Early hominid habitat preferences in East Africa: paleosol carbon isotopic evidence. Journal of Human Evolution 27, 25–45.

Sikes, N., Potts, R., Behrensmeyer, A.K., 1999. Early Pleistocene habitat in Member 1 Olorgesailie based on paleosol stable isotopes. Journal of Human Evolution 37, 721–746.

Sinclar, A.R.E., Norton-Griffith, M. (Eds.) 1979. Serengeti: Dynamics of an Ecosystem. University of Chicago Press: Chicago.

Sponheimer, M., Lee-Thorp, J.A., 2003. Using carbon isotope data of fossil bovid communities for paleoenvironmental reconstruction. South African Journal of Science 99, 273–275.

Stanley, S.M., 1992. An ecological theory for the origin of *Homo*. Paleobiology 18, 237–257.

Suwa, G., White, T.D., Howell, F.C., 1996. Mandibular postcanine dentition from the Shungura Formation, Ethiopia: Crown morphology, taxonomic allocation, and Plio-Pleistocene hominid evolution. American Journal of Physical Anthropology 101, 247–282.

Turner, A., Wood, B., 1993. Comparative palaeontological context for the evolution of the early hominid masticatory system. Journal of Human Evolution 24, 301–318.

Vrba, E.S., 1980. The significance of bovid remains as indicators of environment and predation patterns. In: Behrensmeyer, A. K., Hill, A.P., (Eds.), Fossils in the Making. University of Chicago Press, Chicago, pp. 247–272.

Vrba, E.S., 1985. Ecological and adaptive changes associated with early hominid evolution. In: Delson, E. (Ed.), Ancestors: the hard evidence. Alan R. Liss, New York, pp. 63–71.

Vrba, E., 1995. On the Connections between Paleoclimate and Evolution. In: Vrba, E.S., Denton, G.H., Partridge, T.C., Burckle, L.H. (Eds.), Paleoclimate and Evolution, with Emphasis on Human Origins. Yale University Press, New Haven, pp. 24–45.

Walker, A., Leakey, R.E.F., Harris, J.M., Brown, F.H., 1986. 2.5 million year old *Australopithecus boisei* from west of Lake Turkana, Kenya. Nature 322, 517–522.

Wesselman, H., 1995. Of mice and almost-men: regional paleoecology and human evolution in the Turkana Basin. In: Vrba, E.S., Denton, G.H., Partridge, T.C., Burckle, L.H. (Eds.), Paleoclimate and Evolution, with Emphasis on Human Origins. Yale University Press, New Haven, pp. 356–368.

White, T.D., 1988. The comparative biology of "robust" *Australopithecus:* clues from context. In: Grine, F.E., (Ed.), Evolutionary History of the "Robust" Australopithecines. Aldine de Gruyter, New York, pp. 449–484.

White, T.D., 2002. Earliest hominids. In: Hartwig, W. (Ed.), The Primate Fossil Record. Cambridge University Press, Cambridge, pp. 407–417.

WoldeGabriel, G., White, T., Suwa, G., Renne, P., de Heinzelin, J., Hart, W.K., Helken, G., 2004. Ecological and temporal placement of early Pliocene hominids at Aramis, Ethiopia. Nature 371, 330–333.

Wood, B., 1992. Origin and evolution of the genus *Homo*. Nature 355, 783–790.

Wood, B., 1995. Evolution of the early hominin masticatory system: mechanisms, events and triggers. In: Vrba, E.S., Denton, G.H., Partridge, T.C., Burckle, L.H. (Eds.), Paleoclimate and Evolution, with Emphasis on Human Origins. Yale University Press, New Haven, pp. 438–450.

Wood, B., Collard, M., 1999. The human genus. Science 284, 65–71.

Index

GPSR Compliance

The European Union's (EU) General Product Safety Regulation (GPSR) is a set of rules that requires consumer products to be safe and our obligations to ensure this.

If you have any concerns about our products, you can contact us on ProductSafety@springernature.com

In case Publisher is established outside the EU, the EU authorized representative is:

Springer Nature Customer Service Center GmbH
Europaplatz 3
69115 Heidelberg, Germany

Batch number: 10824451

Printed by Printforce, the Netherlands